AF334404

Bioremediation of Nitroaromatic and Haloaromatic Compounds

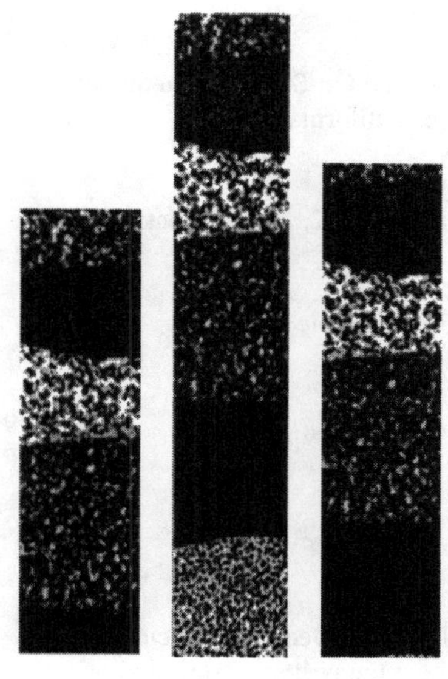

Editors

Bruce C. Alleman
and Andrea Leeson
Battelle

The Fifth International In Situ and
On-Site Bioremediation Symposium

San Diego, California, April 19–22, 1999

BATTELLE PRESS

Columbus • Richland

Library of Congress Cataloging-in-Publication Data

Bioremediation of nitroaromatic and haloaromatic compounds / editors, Bruce C.
    Alleman and Andrea Leeson
            p.   cm.
        Proceedings from the Fifth International In Situ and On-Site Bioremediation
    Symposium, held April 19–22, 1999, in San Diego, California.
        Includes bibliographical references and index.
        ISBN 1-57477-080-2 (hardcover : alk. paper)
        1. Aromatic compounds--Biodegradation Congresses. 2. Bioremediation
    Congresses.
    I. Alleman, Bruce C., 1957–   . II. Leeson, Andrea, 1962–   .
    III. International Symposium on In Situ and On-Site Bioremediation
    (5th : 1999 : San Diego, Calif.)
    TD196.A75B56   1999
    628.5'2--dc21   99-23399
                                                                                        CIP

Printed in the United States of America

Battelle Press
505 King Avenue
Columbus, Ohio 43201, USA
614-424-6393 or 1-800-451-3543
Fax: 1-614-424-3819
Internet: press@battelle.org
Website: www.battelle.org/bookstore

For information on future environmental conferences, write to:
    Battelle
    Environmental Restoration Department, Room 10-123B
    505 King Avenue
    Columbus, Ohio 43201-2693
    Phone: 614-424-7604
    Fax: 614-424-3667
    Website: www.battelle.org/conferences

CONTENTS

FOREWORD

The Fifth International In Situ and On-Site Bioremediation Symposium was held in San Diego, California, April 19–22, 1999. The program included approximately 600 platform and poster presentations, encompassing laboratory, bench-scale, and full-scale field studies being conducted worldwide on a variety of bioremediation and supporting technologies used for a wide range of contaminants.

The author of each presentation accepted for the program was invited to prepare a six-page paper, formatted according to specifications provided by the Symposium Organizing Committee. Approximately 400 such technical notes were received. The editors conducted a review of all papers. Ultimately, 389 papers were accepted for publication and assembled into the following eight volumes:

Natural Attenuation of Chlorinated Solvents, Petroleum Hydrocarbons, and Other Organic Compounds – Volume 5(1)

Engineered Approaches for In Situ Bioremediation of Chlorinated Solvent Contamination – Volume 5(2)

In Situ Bioremediation of Petroleum Hydrocarbon and Other Organic Compounds – Volume 5(3)

Bioremediation of Metals and Inorganic Compounds – Volume 5(4)

Bioreactor and Ex Situ Biological Treatment Technologies – Volume 5(5)

Phytoremediation and Innovative Strategies for Specialized Remedial Applications – Volume 5(6)

Bioremediation of Nitroaromatic and Haloaromatic Compounds – Volume 5(7)

Bioremediation Technologies for Polycyclic Aromatic Hydrocarbon Compounds – Volume 5(8)

Each volume contains comprehensive keyword and author indices to the entire set.

This volume deals with the application of bioremediation technologies at sites contaminated with explosives, pesticides, herbicides, PCBs, and other aromatic compounds. Such sites present formidable technical, regulatory, and financial challenges. Bioremediation offers the promise of cost-effective site remediation that can serve as a key component of a well-formulated strategy for achieving site closure. This volume presents the results of bench-, pilot-, and field-scale projects focused on the use of biological approaches to remediate many problem compounds, such as RDX, HMX, TNT, DDT, 2,4-D, nitro- and chlorobenzenes, nitroaniline, chloroaniline, hexachlorobenzene, PCPs, PCBs, and dichlorophenol in soils and groundwater.

We would like to thank the Battelle staff who assembled the eight volumes and prepared them for printing. Carol Young, Lori Helsel, Loretta Bahn, Gina Melaragno, Timothy Lundgren, Tom Wilk, and Lynn Copley-Graves spent many hours on production tasks—developing the detailed format specifications sent to each author; examining each technical note to ensure that it met basic page layout requirements and making adjustments when necessary; assembling the volumes; applying headers and page numbers; compiling the tables of contents and author and keyword indices, and performing a final check of the pages before submitting them to the publisher. Joseph Sheldrick, manager of Battelle Press, provided valuable production-planning advice and coordinated with the printer; he and Gar Dingess designed the covers.

The Bioremediation Symposium is sponsored and organized by Battelle Memorial Institute, with the assistance of a number of environmental remediation organizations. In 1999, the following co-sponsors made financial contributions toward the Symposium:

Celtic Technologies	U.S. Microbics, Inc.
Gas Research Institute (GRI)	U.S. Naval Facilities Engineering
IT Group, Inc.	Command
Parsons Engineering Science, Inc.	Waste Management, Inc.

Additional participating organizations assisted with distribution of information about the Symposium:

Ajou University, College of Engineering	U.S. Air Force Center for Environmental Excellence
American Petroleum Institute	U.S. Air Force Research Laboratory
Asian Institute of Technology	Air Base and Environmental
Conor Pacific Environmental Technologies, Inc.	Technology Division
Mitsubishi Corporation	U.S. Environmental Protection Agency
National Center for Integrated Bioremediation Research & Development (University of Michigan)	Western Region Hazardous Substance Research Center (Stanford University and Oregon State University)

The materials in these volumes represent the authors' results and interpretations. The support of the Symposium provided by Battelle, the co-sponsors, and the participating organizations should not be construed as their endorsement of the content of these volumes.

Bruce Alleman and Andrea Leeson, Battelle
1999 Bioremediation Symposium Co-Chairs

BIODEGRADATION OF
RDX AND HMX BY A METHANOGENIC ENRICHMENT CULTURE

Neal R. Adrian and *Anna Lowder* (US Army CERL, Champaign, IL)

ABSTRACT: The biotransformation of RDX and HMX by a methanogenic enrichment culture was studied. The enrichment culture only degraded RDX when ethanol was included as an electron donor. Methane production, however, was only observed after RDX had been depleted. The addition of BESA inhibited methane production, but not ethanol fermentation nor RDX degradation. The addition of H_2 gas supported RDX degradation. HMX was also degraded, but the degradation rate (0.12 μM day^{-1}) was nearly 10-fold less than observed for RDX (1.6 μM day^{-1}). HMX degradation slowed when ethanol was depleted, but resumed after reamending the bottles with ethanol. The total number of anaerobes in the enrichment culture ranged from 1.5 x 10^5 to 1.6 x 10^6 ml^{-1}. Ethanol-fermenting syntrophs ranged from 1.5 x 10^4 to 1.6 x 10^5 ml-1; H_2-utilizing methanogens ranged from 7.0 x 10^1 to 7.6 x 10^2 ml^{-1}; acetoclastic methanogens were < 1.0 x 10^1 ml^{-1}; and H_2-utilizing acetogens ranged from 3.8 x 10^3 to 4.2 x 10^4 ml^{-1}. These findings indicate ethanol may serve as a source of H_2 for the RDX and HMX degrading bacteria. The presence of acetogens and a lack of CH_4 production during ethanol degradation may indicate their involvement in the biodegradation of nitramine explosives.

INTRODUCTION

Hexahydro-1,3,5-trinitro-1,3,5-triazine (RDX) and octahydro-1,3,5,7-tetrazocine (HMX) are nitramine explosives widely used by the military (Gorontzy et al., 1994). Improper disposal of RDX and HMX contaminated wastewater in the past has led to environmental contamination near ordnance sites and Army ammunition manufacturing plants (Funk et al., 1993). Despite the Army's apparent need for information on the biodegradation of nitramine explosives, relatively little is known (Gorontzy et al., 1994). RDX is reported to be more easily biodegraded under anaerobic, rather than aerobic conditions (Funk et al., 1993; Kitts et al., 1994; McCormick et al., 1981). The few exceptions include RDX biodegradation by a white rot fungus (Fernando and Aust, 1991), by *Stenotrophomonas maltophilia* PB1 when using RDX as the sole source of nitrogen (Binks et al., 1995), and during composting of explosives contaminated soil (Williams et al., 1992). Studies demonstrating RDX biodegradation under anaerobic conditions, however, were carried out where the electron donors and electron acceptors were not firmly established. The lack of adequate information makes it difficult to develop a biological approach for wastewater treatment or cleanup of contaminated groundwater. Our primary objectives for this study were to study the microbiology and biodegradation of RDX and HMX in a methanogenic enrichment culture. In this presentation, we report on the biodegradation of RDX and HMX and the potential involvement of acetogens in their degradation.

MATERIALS AND METHODS

A five-tube most-probable-number procedure was used to enumerate total anaerobes, H_2-utilizing and acetoclastic methanogens, ethanol-metabolizing syntrophs, and H_2-utilizing acetogens. The MPN tubes were incubated at ambient room temperature and scored after 21 days incubation. Biodegradation studies were carried out in serum bottles (160 ml) containing 80 ml of a basal salts medium and 20 mls of the RDX-degrading enrichment culture. Filter-sterilized resazurin (0.0002%) was added as a redox indicator. The basal salts medium and enrichment culture were dispensed to the serum bottles using strict anoxic techniques, sealed with butyl rubber stoppers and then crimped with aluminum seals. The headspace of the bottles was evacuated and filled three times with $N_2:CO_2$ (80:20), and then pressurized to 1.3 ATM. The study was conducted in triplicate and employed sterile and RDX unamended controls.

Liquid samples were taken periodically and analyzed by reverse phase high-pressure liquid chromatography (HPLC) and gas chromatography (GC) for RDX and acetate, and ethanol, respectively. The headspace of the serum bottles was monitored for the formation of CH_4 by GC. Methane produced from unamended controls was subtracted from that produced in substrate-amended bottles. This amount was compared to the theoretically expected amount of CH_4 (Gottschalk, 1986).

RESULTS

The number of total anaerobes and those comprising four major metabolic groups of bacteria were enumerated in an RDX-degrading culture (Table 1). Total anaerobes ranged from 7.0×10^5 to 7.6×10^6 ml^{-1}. Interestingly, we did not observe any colony forming units on the plate count agar plates incubated aerobically, indicating there were no aerobic bacteria or facultative anaerobes capable of growing under the test conditions. Acetoclastic methanogens were present at relatively low numbers (<10 ml^{-1}) compared to the H_2-utilizing methanogens (3.3×10^2 ml^{-1}). H_2-utilizing acetogens were present in numbers similar to the hydrogenotrophic methanogens, ranging from 3.9×10^2 to 4.2×10^3 ml^{-1}. The ethanol fermenting syntrophs ranged from 3.9×10^2 to 4.2×10^3 ml^{-1}.

A major band was observed in denaturing gradient gel electrophoresis (DGGE) of the 16S rDNA gene in a sample taken from the culture. This band was sequenced and closely resembled *Acetobacterium malicum*. A second major band was observed and sequenced, closely resembling the sequence for *Geobacter akaganeitreducens*.

The enrichment culture only degraded RDX when ethanol was included as an electron donor (Figure 1). Approximately 32 μM RDX was biodegraded in 17 days. Methane production was only observed after RDX had been depleted, while RDX unamended controls experienced no lag in methane production.

Table 1. The number of total anaerobes and those bacteria comprising four major metabolic groups in the RDX-degrading culture. Numbers reported are bacteria ml^{-1}. ND = not detected.

Metabolic Group	Culture A
Aerobes	ND
Total anaerobes	2.3×10^6
Proton-reducing syntrophs	1.3×10^3
H_2-using methanogens	3.3×10^2
Acetoclastic methanogens	$<1 \times 10^1$
Acetogens	2.3×10^2

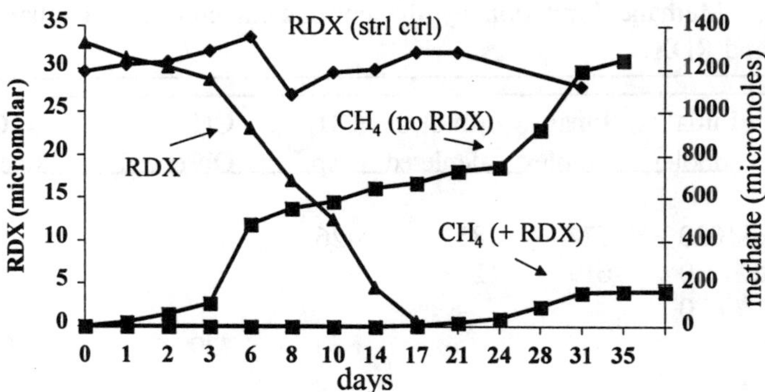

Figure 1. RDX degradation and methane formation by culture. Initial ethanol concentration was 10 mM.

Acetate formation was only observed in bottles amended with ethanol (Figure 2). In bottles amended with ethanol and RDX, on day 21 106% of the expected amount of acetate was produced from the transformation of ethanol to acetate and H_2 (Figure 2). Acetate continued to accumulate to approximately 11.5 mM, accounting for 114% of the expected amount of acetate (Figure 2). After day 35, methane production started to increase and acetate concentrations decreased, indicating acetoclastic methanogens were responsible for the observed activity.

Approximately 110% of the expected amount of methane was observed from ethanol in RDX unamended bottles, while only 14% in bottles containing RDX (Table 2).

The addition of BESA inhibited methane production, but not ethanol fermentation nor RDX degradation. RDX degradation was more rapid when H_2 gas was the electron donor (4.4 μM day^{-1}) compared to ethanol (1.9 μM day^{-1}). There was no loss of RDX in sterile or electron donor unamended controls.

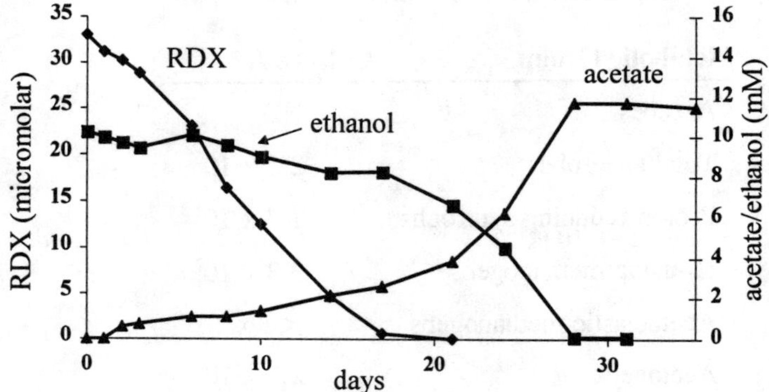

Figure 2. RDX degradation and acetate formation in methanogenic enrichment culture fed ethanol.

Table 2. Methane formation by the enrichment culture fed ethanol only or ethanol and RDX.

	Initial μmoles	Final μmoles[a]	μmoles depleted	CH_4 Exp.[b]	CH_4 Observed	% CH_4 Recovered[c]
Ethanol						
Ethanol	1030	177	853	426		
Acetate	0	619	233	233		
RDX	0	-	-	-		
Total				660	729	110
Ethanol + RDX						
Ethanol	1030	775	255	127		
Acetate	0	272	0	0		
RDX	3	0	3			
Total				127	18	14

(a) μmoles substrate remaining after 21 days incubation
(b) CH_4 expected based on the following equations: 2 ethanol + HCO_3^- $\leftrightarrow$ 2 acetate$^-$ + CH_4 + H_2O + H^+; acetate $\leftrightarrow$ CH_4 + CO_2
(c) % CH_4 recovered = (expected / observed)*100

HMX was more resistant to degradation (Figure 3). Approximately 8 μM HMX was degraded in 55 days (0.15 μM day^{-1}). HMX degradation slowed when ethanol was depleted, but resumed after reamending the bottles with ethanol. There was no loss of HMX in sterile or ethanol unamended controls.

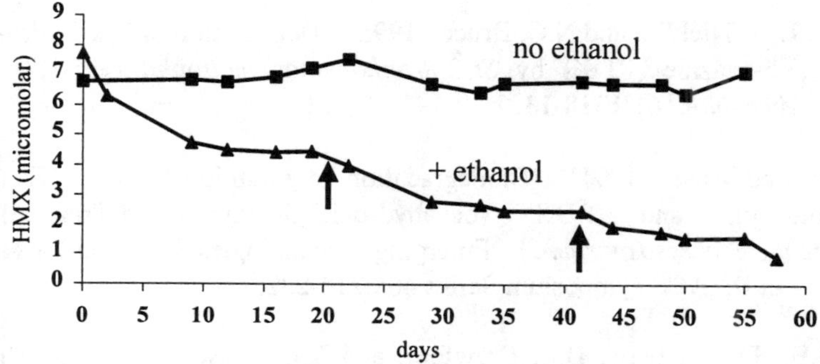

Figure 3. HMX degradation by methanogenic enrichment culture. The initial ethanol concentration was 7 mM. Arrows indicate the reamendment of ethanol to the culture.

DISCUSSION

Our results demonstrate ethanol is required by the enrichment culture for RDX and HMX degradation. In a methanogenic culture, such as the RDX-degrading enrichment one, H_2 must be removed for ethanol degradation to occur, and when CO_2 is the only available electron acceptor, CH_4 production will be observed. However, there was no methane production by the culture from ethanol in the presence of RDX. Ethanol was degraded (Figure 2) and acetate concentrations increased (Figure 3), therefore H_2 must have been removed for ethanol to be degraded. We believe the acetogens may be responsible for this, as well as degrading the explosives. In support of this, a major band in DGGE gels closely matched the sequence from *Acetobacterium malicum*, an acetogen common to fresh water sediments (Tanaka and Pfennig, 1988). Interestingly, RDX degradation rates for culture A are almost 2-fold greater than culture B, a culture similar to A in many respects, but lacks a major band in DGGE gels corresponding to an acetogen. Future studies will investigate in more detail the role of acetogens in degrading nitramine explosives.

Acknowledgements

We are thankful to Greg Davis from Microbial Insights for identifying bacteria by DGGE analysis of polymerase chain reaction-amplified genes coding for 16S rDNA.

This research was supported in part by an appointment at the Research Participation Program administered by the Oak Ridge Institute for Science and Education through an interagency agreement between the U.S. Department of Energy and USACERL.

REFERENCES

Binks, P.R., S. Nicklin, and N.C. Bruce. 1995. "Degradation of hexahydro-1,3,5-trinitro-1,3,5-triazine (RDX) by *Stenotrophomonas maltophilia* PB1." *Appl. Environ. Microbiol.* 61:1318-1322.

Fernando and Aust. 1991. "Biodegradation of munition waste, TNT (2,4,6-trinitrotoluene), and RDX (hexahydro-1,3,5-trinitro-1,3,5-Triazine) by *Phanerochaete chrysosporium*." Emerging Technologies in Hazardous Waste Management II. ACS symposium Series 468:214-232.

Funk, S.B., D.J. Roberts, D.L. Crawford, and R.L. Crawford. 1993. "Initial-phase optimization for bioremediation of munition compound-contaminated soils." 1993. *Appl. Environ. Microbiol.* 59:2171-2177.

Gorontzy, T., O. Drzyzga, M.W. Kahl, D. Bruns-Nagel, J. Breitung, E. von Loew, and K.-H. Blotevogel. 1994. "Microbial degradation of explosives and related compounds." *Crit. Reviews Microbiol.* 20:265-284.

Gottschalk, Gerhard. 1986. *Bacterial Metabolism.* 2nd edition. Springer-Verlag, New York.

Kitts, C.L., D.P. Cunningham, and P.J. Unkefer. 1994. "Isolation of three hexahydro-1,3,5-trinitro-1,3,5-triazine-degrading species of the family *Enterobacteriaceae* from nitramine explosive-contaminated soil." *Appl. Environ. Microbiol.* 60:4608-4711.

McCormick, N.G., J.H. Cornell, and A.M. Kaplan. 1981. "Biodegradation of hexahydro-1,3,5-trinitro-1,3,5-triazine." *Appl. Environ. Microbiol.* 42:817-823.

Williams, R.T., P.S. Ziegenfuss, and W.E. Sisk. 1992. "Composting of explosives and propellant contaminated soils under thermophilic and mesophilic conditions." *J. Indus. Microbiol.* 9:137-144.

Tanaka, K., and N. Pfennig. 1988. "Fermentation of 2-methoxyethanol by *Acetobacterium malicum* sp. Nov. and *Pelobacter venetianus*." *Arch. Microbiol.* 149:181-187.

FIELD DEMONSTRATION OF FBR FOR TREATMENT OF NITROTOLUENES IN GROUNDWATER

Jim C. Spain (USAF Armstrong Laboratory, Tyndall AFB, Florida)
Shirley F. Nishino (USAF Armstrong Laboratory, Tyndall AFB, Florida)
Mark R. Greene , Jon E. Forbort, Nick A. Nogalski, and Ronald Unterman (Envirogen, Inc, Lawrenceville, New Jersey)
Warren M. Riznychok and Scott E. Thompson (Malcolm Pirnie, Inc, White Plains, New York)
Pamela M. Sleeper and Mimi A. Boxwell (USACE Baltimore District, Baltimore, Maryland)

ABSTRACT: A pilot-scale field demonstration was conducted to collect reliable cost and performance data for an aerobic, biological fluidized bed reactor (FBR) system that treats groundwater contaminated with nitrotoluenes. This project is an extension of a previous bench-scale FBR study on the simultaneous degradation of 2,4- and 2,6-dinitrotoluene. The FBR was inoculated with a mixed culture of bacteria that had been acclimated to a mixture of mono- and dinitrotoluenes. The contaminated groundwater was drawn from wells at the Volunteer Army Ammunition Plant (VAAP) in Chattanooga, TN. The nitrotoluene (NT) contaminants identified in the groundwater and their typical concentrations were: 2,4,6-trinitrotoluene (2,4,6-TNT) at 1.1 mg/L, 2,4-dinitrotoluene (2,4-DNT) at 9.2 mg/L, 2,6-dinitrotoluene (2,6-DNT) at 8.6 mg/L, 2-nitrotoluene (2-NT) at 2.9 mg/L, 3-nitrotoluene (3-NT) at 0.4 mg/L, and 4-nitrotoluene (4-NT) at 3.0 mg/L.

A range of loading rates and a variety of operating conditions were used to evaluate the FBR performance. The COD and NT loading rates, based on the volume of the fluidized bed, ranged from 0.6-11 kg COD/m^3/d and 0.45-4.5 kg NT/m^3/d. The best FBR performance was measured at the lower loading rates; a 96% reduction in NT and 93% reduction in COD were found at a loading rate of 1.0 kg NT/m^3/d (2.2 kg COD/m^3/d). The effluent at this loading rate contained an average of 0.5 mg 2,4,6-TNT and 0.8 mg 2,6-DNT per liter. The other NT compounds were below detection levels (0.05 mg/L). At the highest steady loading rate, 3.5 kg NT/m^3/d (7.6 kg COD/m^3/d), a 62% reduction in NT and 49% reduction in COD were measured. The effluent at this loading contained an average of 0.8 mg/L 2,4,6-TNT, 7.0 mg/L 2,6-DNT and 1.3 mg/L 2,4-DNT. The mononitrotoluenes were still below detection levels at the highest loading rate. Therefore, this FBR process can effectively remove mono- and dinitrotoluenes from contaminated groundwater, and at appropriate loading rates the FBR process is cost-effective compared to alternative technologies.

INTRODUCTION

The Volunteer Army Ammunition Plant in Chattanooga, TN (VAAP), produced and stored 2,4,6-trinitrotoluene (TNT) until 1977. As a result the groundwater is contaminated with TNT and other nitrotoluenes. At VAAP and other DoD sites nitroaromatic contaminants in the groundwater are the target for remediation. The current treatment strategy involves carbon adsorption. Because of the high cost of carbon regeneration the DoD has identified a need to develop alternative treatment technologies for extracted groundwater at such sites.

This paper summarizes the findings of a demonstration of an aerobic dinitrotoluene (DNT) degradation process in a fluidized bed reactor (FBR) system. The demonstration, conducted over an 11-month period, evaluated the ability of the system to simultaneously degrade 2,4-dinitrotoluene (2,4-DNT) and 2,6-dinitrotoluene (2,6-DNT), as well as the three mononitrotoluenes (MNTs: 2-NT, 3-NT, 4-NT), in groundwater extracted during remediation activities. Field demonstrations of FBR technology for the degradation of chlorinated solvents (e.g., TCE, chlorobenzene) in contaminated groundwater have been successfully conducted (Guarini and Folsom, 1996; Klecka, et al., 1986). This work was the first field demonstration of the FBR technology for treatment of recalcitrant nitroaromatic compounds, such as 2,4-DNT and 2,6-DNT.

Aerobic DNT Biodegradation Process. The aerobic biodegradation strategy involved a mixed culture of microorganisms that has been shown to simultaneously degrade 2,4-DNT and 2,6-DNT in a continuous flow bioreactor (Lendenmann, et al., 1998). Isolated strains of bacteria are capable of utilizing the individual isomers of DNT as growth substrates and the degradation pathways have been determined (Spanggord et al., 1991; Suen and Spain, 1993; Haigler, et al., 1994; Nishino and Spain, 1996). Burkholderia sp. DNT degraded 2,4-DNT at a hydraulic residence time (HRT) of 1 hour in a laboratory-scale fixed bed bioreactor (Heinze et al., 1995).

Although isolated strains were able to use 2,6-DNT as the sole growth substrate, low concentrations of 2,6-DNT inhibited growth of both 2,4-DNT and 2,6-DNT degrading strains (Nishino and Spain, 1996). Degradation of isomeric DNT mixtures was possible in continuous bioreactor systems if the concentration of 2,6-DNT could be maintained at a low steady-state level. The efficacy of the approach was proven in a study which demonstrated simultaneous degradation of 2,4-DNT and 2,6-DNT in a laboratory bench-scale FBR using a mixed culture of bacteria and sand as the support medium (Lendenmann, et al., 1998). Destruction efficiencies of greater than 98 percent for 2,4-DNT and 94 percent for 2,6-DNT were achieved at all loading rates.

FBR Process Description. The FBR system used in the present study coupled the aerobic biodegradation process with a high rate reactor design, using GAC as the support. The basic components of the FBR system include the bioreactor column, the GAC, a suitable biocatalyst, a fluid distribution system in the bottom of the reactor, an oxygenator, a high purity oxygen supply source, feed and influent pumps, a nutrient addition system, and a pH control mechanism. The pilot FBR system (EFB-02, 2" diameter) is designed to operate continuously, 24 hours a day, 7 days a week. The feed pump supplies contaminated groundwater to the system on the suction side of the influent pump. The influent pump circulates the feed water plus recycle water through the oxygenator, where gaseous O_2 is dissolved in the liquid, then into the reactor column in an upward direction fluidizing the bed of GAC media. Microbes attach to the GAC media within the bioreactor and metabolize passing contaminants. A constant liquid level is maintained in the FBR and effluent leaves by gravity flow from the top of the reactor column. Nutrients and pH control chemicals are added in the recycle line. The influent flowrate is automatically controlled in order to achieve stable bed expansion.

EXPERIMENTAL DESIGN AND OPERATION

The field pilot demonstration was designed to meet the following objectives:

1. to determine the feasibility of implementing aerobic, microbial GAC FBR technology in the field to remediate groundwater contaminated with 2,4-DNT and 2,6-DNT, and 2-NT, 3-NT and 4.NT;
2. to obtain removal efficiencies and removal rates for (a) total organics, and (b) nitroaromatics as a function of inlet loading and expanded bed hydraulic retention times;
3. to obtain real-world operating characteristics for groundwater pump-and-treat applications to assess the extent of operator attention required;
4. to obtain data useful for the design of suitable full-scale FBR systems; and
5. to compare the capital and operating costs of the aerobic, microbial FBR technology to those of conventional treatment technologies.

The demonstration consisted of seven evaluation periods (referred to as phases) during the FBR operation and testing. For each evaluation period the conditions were varied to determine the effect on overall nitrotoluene (2,4-DNT, 2,6-DNT, 2,4,6,-TNT, 2-NT, 3-NT and 4-NT) removal. Each operating condition was maintained for a sufficient time to achieve stable effluent quality. The seven phases included five different feed flowrates and two different feed water compositions. The feed water composition was changed by varying the ratio of extracted groundwater from two wells which had different NT concentrations. A comparative analysis of the contaminant removal efficiency measured during each evaluation period was used to determine the optimal loading condition.

Phases 1 through 4 produced the data to evaluate the FBR loading capacity and treatment performance with different feed flowrates. Phase 5 produced the data to evaluate the impact of other FBR operating parameters (for example, the set point for dissolved oxygen concentration) on performance. Phases 6 and 7 produced the data to re-evaluate the performance after activity towards 2,6-DNT greatly improved. Under each operating condition overall DNT, TNT and MNT removal results were determined. Bench-top flask assays were performed at regular intervals to confirm that removal of the nitroaromatic compounds was due to biological activity and not adsorption to the GAC media or by stripping.

RESULTS AND DISCUSSION

Degradation Efficiency of Nitrotoluenes. As the loading to the FBR increased (Phases 1 to 4), the effluent concentrations of the dinitrotoluenes increased and the removal efficiency decreased (Table 1). Conversely, when the loading to the FBR decreased in Phases 5-7, the removal efficiency towards 2,6-DNT dramatically improved at the end of Phase 5 and continued at high levels during Phases 6 and 7. The removal efficiency of the mononitrotoluenes was unaffected by the loading changes over the entire range tested. All the MNTs in the feed were removed below on-site quantification levels under all loading conditions. 2,4 DNT was degraded throughout the study at high efficiency, whereas 2-6 DNT was less effectively degraded at the higher flow rates (Figure 1, "P1" = Phase 1, etc.; "T" - transition period).

The spikes in the effluent concentrations of 2,4-DNT and the MNTs can be explained by transient mechanical problems with the FBR system or by changes in the FBR loading rates. The steady rise in 2,6-DNT effluent concentration during Phases 1-4 appears to be directly related to FBR loading rates. During Phase 5, the declines in 2,6-DNT effluent concentration, such as the one on July 6-7, can be attributed to the sensitivity experiments performed with the FBR. This particular drop was the result of a stoppage in feed flow. In late July, the biodegradation performance towards 2,6-DNT suddenly improved greatly. It seems likely that the bacterial population in the FBR acclimated.

Table 1. Loading and Removal Efficiency of Target Nitrotoluenes

Phase	HRT (min)	NT Load (kg/m^3/d)	COD Load (kg/m^3/d)	[2,4-DNT] (%)	[2,6-DNT] (%)	[2,4,6-TNT] (%)	[2-NT] (%)	[3-NT] (%)	[4-NT] (%)
1	30	1.1	2.2	98	58	73	100	100	100
2	60	0.45	0.9	100	56	67	100	100	100
3	15	1.9	3.6	96	34	40	95	100	100
4	10	3.5	7.6	86	18	33	100	100	100
5	20	2.5	5.0	99	39	46	100	100	100
6	30	1.9	2.9	100	88	54	100	100	100
7	60	1.0	2.2	100	94	58	100	100	100

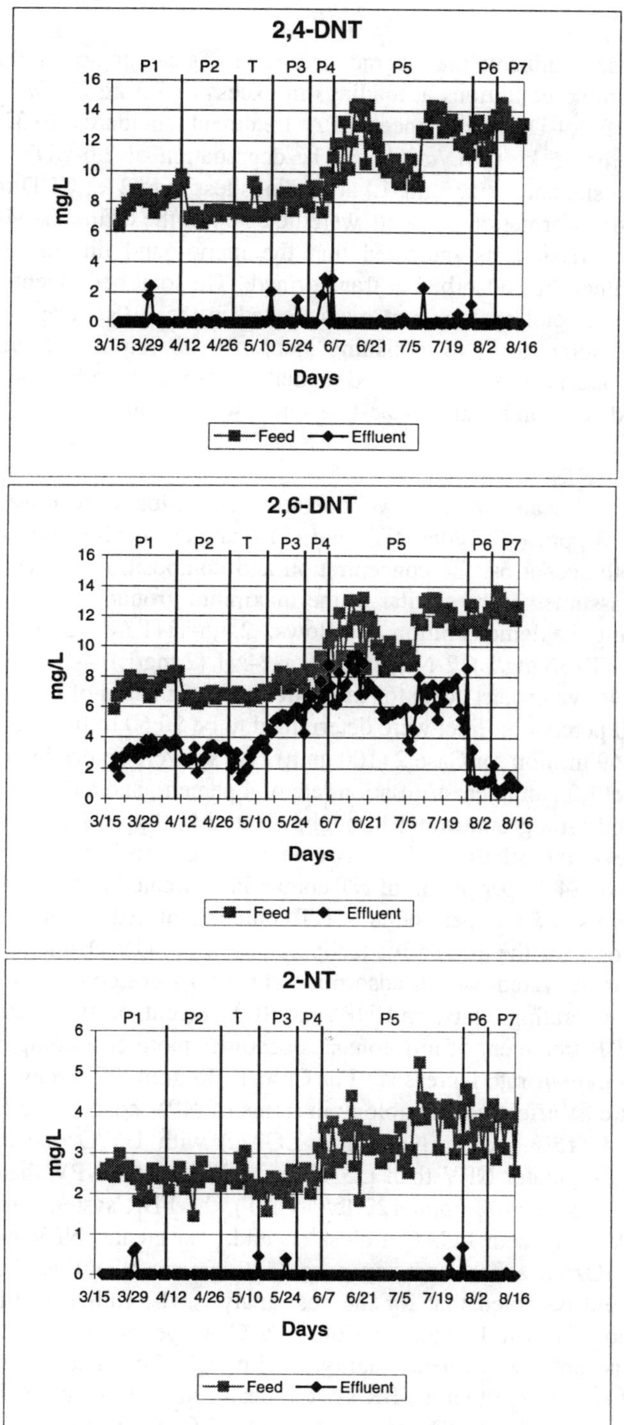

Figure 1. Daily FBR feed and effluent concentrations for 2,4-DNT, 2,6-DNT and 2-NT

The results indicate that degradation of MNTs can be accomplished in the FBR under aerobic conditions at loadings in excess of 7.5 kg COD/m^3/day. The degradation of 2,4-DNT can meet EPA treatment standards (0.32 mg/L) at loadings less than 5 kg COD/m^3/day. The degradation of 2,6-DNT might meet EPA treatment standards (0.55 mg/L) at loadings less than 2 kg COD/m^3/day. No metabolic or transformation products were detected in the effluent. Mass balance and O$_2$ uptake flask tests indicated that the mono- and dinitrotoulenes were mineralized rather than adsorbed or transformed. The total bed inventory was less than 0.1% of the total contaminant mass treated in the FBR. Depending on the FBR loading, some form of polishing step may be required to mitigate the contaminant concentrations to required regulatory levels. A carbon polishing tank or an extended aeration basin may best be suited for this purpose.

COST ANALYSIS

For a full-scale system, two operating scenarios were developed: (1) treatment of 30 gpm (37 lb/day NT) and (2) treatment of 100 gpm (122 lb/day NT). For both scenarios, the concentration and composition of the water to be treated were assumed to be similar to the maximum groundwater characteristics observed during the demonstration as follows: 2,4,6-TNT (4 mg/L); 2,4-DNT (43 mg/L); 2,6-DNT (33 mg/L); 2-NT (10 mg/L); 3-NT (2 mg/L); 4-NT (10 mg/L).

The relative project costs for the FBR system in terms of net present value (NPV), at 100 percent on-line, were determined to be $0.60 million for Case 1 (30 gpm) and $1.49 million for Case 2 (100 gpm). NPVs were determined assuming a 10-year project life, an interest/inflation rate of 4 percent, and a discount rate of 12 percent. Based on a groundwater throughput rate of 30 gpm for Case 1 and 100 gpm for Case 2, the relative unit costs over a 10-year project life are $3.80 per 1,000 gallons or $4.45 per pound of NT contaminant treated for Case 1, and $2.83 per 1,000 gallons or $3.33 per pound of NT contaminant treated for Case 2.

The costs for the alternative technologies of (1) UV/Ozone and (2) liquid phase granular activated carbon adsorption (LGAC) were compared in terms of capital costs, operating costs, and NPV, at 100 percent on-line. Based on the results, the FBR treatment of nitrotoluenes becomes more cost-competitive as the treatment throughput rate increases. For Case 1 (30 gpm, 37 lb/day NT), LGAC was the most cost-effective technology in terms of NPV costs. The FBR system ranked second (15% greater NPV than LGAC) with UV/Ozone as the most expensive (93% greater NPV than LGAC and 67% greater NPV than the FBR). However, for Case 2 (100 gpm, 122 lb/day NT), the FBR system ranked first in terms of NPV costs with LGAC a close second (2% greater NPV cost than the FBR) and UV/Ozone as the most expensive (37% greater NPV than the FBR).

It should be noted that for the cost analysis, the full-scale FBR systems were designed at a low loading rate of 2 kg COD per m^3 per day in order to produce an acceptable effluent quality. Although the calculations were not performed, it is possible that a FBR system designed with a shorter HRT (e.g. a higher loading of 5 kg COD m^3 per day) and LGAC post-treatment to meet required discharge limits, may have an overall system treatment cost lower than that for LGAC alone. The limiting criterion for FBR sizing is the 2,6-DNT concentration. For groundwaters with a greater proportion of 2,4-DNT (compared to 2,6-DNT), higher NT loadings (i.e. shorter HRTs) can be handled by the FBR while satisfying effluent discharge criteria. It is unknown whether similar criteria apply to the alternative technologies (e.g. sizing is based on total organic load or individual species).

Based on this cost analysis, the FBR appears the be the most cost-effective treatment technology when the nitrotoluene removal rate exceeds 120 pounds per day (39 lb/day 2,6-DNT).

CONCLUSIONS

The results of the demonstration support the following conclusions. The FBR effectively removes mono- and dinitrotoluenes from contaminated groundwater with an aerobic biological treatment process. The primary technical factors affecting the processing rate through the FBR are: (1) hydraulic retention time (HRT); and (2) organic load. Based on the system performance at various HRTs and nitrotoluene loads, it has been determined that the optimum FBR organic loading rate is 2 kg COD per day per m^3 of bed ($kg/m^3/day$). At the maximum observed FBR feed concentrations, this translates to an HRT of approximately 90 minutes. A scale-up of the technology should use a loading rate of 2 $kg/m^3/day$ to meet discharge requirements without LGAC polishing. Generally, at very low concentrations LGAC is more cost-effective and at higher concentrations the FBR technology is more economical.

ACKNOWLEDGMENTS

This demonstration was funded by the Department of Defense Strategic Environmental Research and Development Program (SERDP) through Project CU-720, Federal Integrated Biotreatment Research Consortium (Flask to Field).

REFERENCES

Guarini, W., B. Folsom. 1995. "Liquid-phase Bioreactor for Degradation of Trichloroethylene and Benzene," Armstrong Laboratory, Final Report No. AL/EQ-TR-1995-0001.

Haigler, B.E., S.F. Nishino, and J.C. Spain. 1994. "Biodegradation of 4-methyl-5-nitrocatechol by Pseudomonas sp. strain DNT," *J. Bacteriol.* 176:3433-3437.

Heinze, L., M. Brosius, U. Wiesmann. 1995. *Acta Hydrochim. Hydrobiol.* 23:254.

Klecka, G.M., S.G. McDaniel, P.S. Wilson, C.L. Carpenter, J.E. Clark, A. Thomas, and J.C. Spain. 1996. "Field Evaluation of an Granular Activated Carbon Fluid-Bed Bioreactor for Treatment of Chlorobenzene in Groundwater," *Environ. Progress*, 15:93-107.

Lendenman, U., J.C. Spain and B.F. Smets. 1998. "Simultaneous Biodegradation of 2,4-dinitrotoluene and 2,6-dinitro-toluene in a Fluidized Bed Reactor," *Environ Sci. Tech.*, 32:82-87.

Nishino, S.F. and J.C. Spain. 1996. Presentation at the 96th American Society for Microbiology General Meeting, New Orleans, LA.

Spanggord, R. J., J.C. Spain, S. F., Nishino, and K. E. Mortelmans. 1991. "Biodegradation of 2,4-Dinitrotoluene by a Pseudomonas sp.," *Appl. Environ. Microbiol.* 57:3200-3205.

Suen, W.C., and J. C. Spain. 1993. "Cloning and Characterization of Pseudomonas sp. Strain DNT Genes for 2,4-Dinitrotoluene Degradation," *J. Bacteriol.* **175**: 1831-1837.

Sutton, P.M. and P.N. Mishra. 1994. "Activated carbon based biological fluidized bed for contaminated water and wastewater treatment: A state-of-the-art review," *Wat. Sci. Tech.* 29 (10-11): 309-317.

PILOT-SCALE ANAEROBIC BIOSLURRY REMEDIATION OF RDX- AND HMX-CONTAMINATED SOILS

Serge R. Guiot, Chun Fang Shen, Louise Paquet, Jérôme Breton and Jalal Hawari
(Biotechnology Research Institute, NRC, Montreal, Quebec, Canada)
S. Thiboutot and G. Ampleman
(National Defense Canada, DREV, Valcartier, Quebec, Canada)

ABSTRACT: Bioremediation of RDX- and HMX-contaminated soil slurries (40-44% clay soil) was conducted under optimal conditions in two 8 L drum-style reactors and a 250 L pilot-scale reactor. The RDX and HMX content of the soil were 10,355 and 4,720 mg/kg soil respectively in the first case, and 24,060 and 7,860 mg/kg soil respectively in the later case. The results showed that RDX was completely removed after about 45 days of operation, while the HMX removal, which started once RDX was completely removed, took an additional 35 days. HPLC and LC/MS analysis showed that both RDX and HMX were first biotransformed to their corresponding nitroso-metabolites before they were completely degraded. The average RDX removal rate was two times higher in the run with an initial RDX concentration of 24,060 mg/kg than in the run with an initial RDX concentration of 10,400 mg/kg (513 against 213 mg RDX/kg soil.d). This indicates that the explosive mass transfer from the solid to the liquid phases rather than the biodegradation was the limiting step.

INTRODUCTION

Hexahydro-1,3,5-trinitro-1,3,5-triazine (RDX) and octahydro-1,3,5,7-tetranitro-1,3,5,7-tetrazocine (HMX) are important and powerful military explosives. Both RDX and HMX are commonly found together in the soil at sites such as destruction ranges, explosives dumping grounds, industry production and ammunition factories. The relative recalcitrance along with the toxic effects of these explosives lead to concerns regarding their fate in the environment as well as their impact on human health. Biodegradation studies of RDX and/or HMX can be dated back to 1970s (Soli, 1973). McCormick et al. (1981, 1984) demonstrated that RDX and HMX was degraded in anaerobic sewage sludge, although no significant mineralization (only 1.5%) was observed, and reported intermediates such as mono-, di-, and trinitroso-RDX and mono- and dinitroso-HMX due to a sequential reduction of the nitro groups on the parent compounds. Other studies were performed with relatively low concentrations of RDX (< 75 mg/L) in a liquid medium or in a low solid containing slurry (< 5%), and the percentage of RDX mineralized was still less than 10% (Kitts et al., 1994; Binks et al., 1995). Recently however, over 60% mineralization was obtained for both RDX and HMX with anaerobic sludge (Guiot et al., 1997).

Aqueous phase bioslurry reactors are capable of substantially increasing contaminant degradation rates by enhancing process mass-transfer efficiencies. This is conducive to optimal biodegradation in terms of reaction rate and cost-effectiveness (Hampton and Sisk, 1995). Degradation of RDX and/or HMX under anaerobic conditions has also been studied in a soil slurry process (Funk et al., 1993; Kaake et al., 1997). However, the concentrations of RDX and HMX in the contaminated soils were relatively low. To our knowledge, no published results have shown the application of bioslurry systems in the soil remediation of high levels of HMX and RDX contamination.

Objective. The goal of this research was to evaluate the effectiveness of optimal conditions previously studied at bench-scale for the degradation of RDX and HMX in pilot-scale biotreatment of high solid-containing and heavily-contaminated slurries.

MATERIAL AND METHODS

The soil used in this study consisted of 95% clay and 5% sand, as determined according to the method of Nelson and Sommers (1982). The explosive-contaminated soil was homogenized in a six cubic foot peddle-style cement mixer, then mixed with water, and released from large sized particles (> 5 mm). The solids content of the slurries was between 44% and 51% (w/w).

Two 8 L drum-style lab-scale reactors were loaded, each with 5 L (7.2 kg) of the soil slurry, and inoculated with anaerobic sludge, yielding a biomass concentration of 7.9 g VSS/kg slurry. One pilot-scale bioslurry reactor with a volume of 250 litres was constructed by modifying an industrial cement mixer. The reactor was loaded with 168 kg of soil slurry with a 44% solids content, and 20 kg of anaerobic sludge to have a biomass content of 8.3 g VSS/kg slurry. The reactors were operated anaerobically at $35\pm1°C$, with continuous rotation (2 rpm).

RDX, HMX and their metabolites in the soil slurry were extracted with 10 ml of acetonitrile (for a 10 ml aliquot of slurry). The sample was agitated in a Wrist Action Shaker (Burrell Corporation, Pittsburgh, PA) for about 2 hours. Supernatant from the sample was decanted and diluted with a 0.5% calcium chloride solution. The diluted supernatant was filtered through a MILLEX®-HV 0.45 μm filter unit for subsequent analysis by HPLC (Spectra-Physics Model SP 8100) with a Supelcosil CN column and a UV detector. RDX metabolites were identified by comparison with standard compounds. HMX metabolites were identified by a Platform LC/MS using negative electrospray ion source.

RESULTS AND DISCUSSION

The soil slurry (after amendment with anaerobic sludge) had a solids content of 44%. The initial RDX and HMX concentration in the slurry was relatively high : 10,355±720 and 4,720±290 mg/kg dry soil, respectively. Figure 1 showed that RDX was degraded rapidly : about 10 g RDX/kg soil in 45 days. The

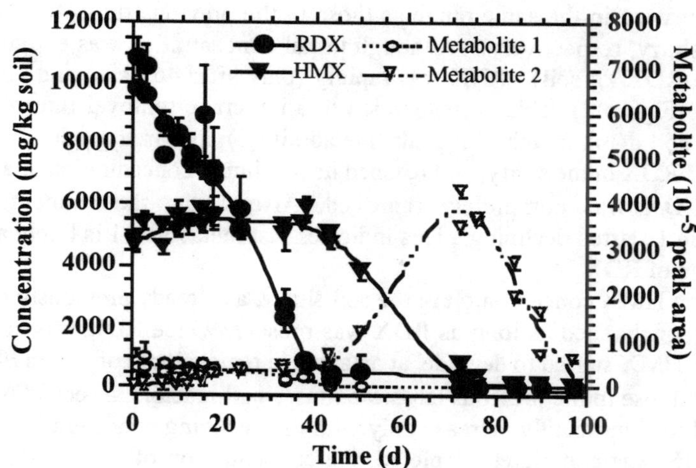

FIGURE 1. RDX and HMX degradation in anaerobic drum-style bioslurry reactors with 44% solids content. Grouped data of two reactors. Metabolites 1 and 2 concentrations expressed as the peak area x 10-5.

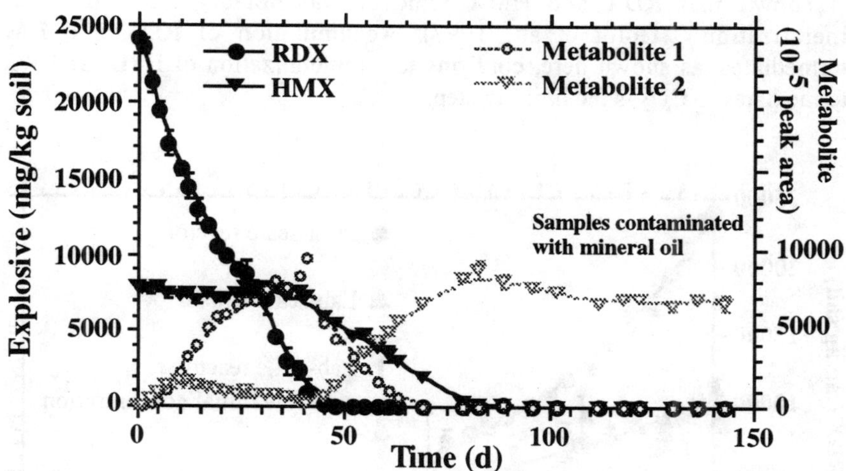

FIGURE 2. RDX and HMX degradation in a pilot-scale bioslurry reactor with 40% solids content. Metabolites 1 and 2 concentrations expressed as the peak area x 10-5.

RDX and HMX removal rates were, on average, 213 and 141 mg/kg soil.d, respectively. Afterwards, a pilot-scale reactor with a volume of 250 litres was operated under a continuous rotating mode (2 rpm) to scale up the experiment from the lab-scale reactors but at much higher RDX and HMX concentrations (24,058 and 7,855 mg/kg soil, respectively). The solid and biomass contents in the soil slurry were in the same range as those in the previous runs (40% and 8.3 g VSS/kg slurry, respectively). Although initial concentration was extremely high (i.e. 24 g RDX/kg soil), RDX was rapidly (only after 46 days) and completely removed (Figure 2). This corresponded to an average removal rate of 513 mg RDX/kg soil.d. A metabolite peak (metabolite 1) appeared gradually with the decline of RDX in the slurry, and reached its maximum concentration at a time just before RDX was completely removed. Afterwards, the concentration of metabolite 1 started declining. This indicates that metabolite 1 is likely a nitroso-metabolite of RDX.

The HMX concentration in the soil slurry, as already previously observed, remained unchanged as long as RDX was present. Once RDX was completely removed, HMX started to degrade, at an average removal rate of 160 mg/kg soil.d, similar to those in the previous lab-scale runs. In all reactors, a second metabolite appeared to significantly increase only with the declining concentration of HMX; once HMX was completely depleted, the concentration of metabolite 2 started declining (Figure 1). This indicates that metabolite 2 originated from the degradation of HMX (likely a nitroso-metabolite of HMX). The relatively high concentration of metabolite 2 long after HMX depletion in the pilot reactor (Figure 2), was explained by a contamination of the samples with mineral oil leachates from the mechanics of the reactor. Previous radiorespirometric assays had shown that RDX and HMX removal was notably faster than their mineralization (Guiot et al., 1997). Accumulation of RDX and HMX intermediates, as shown here, confirms that mineralization of HMX and RDX intermediates to CO_2 is the limiting step.

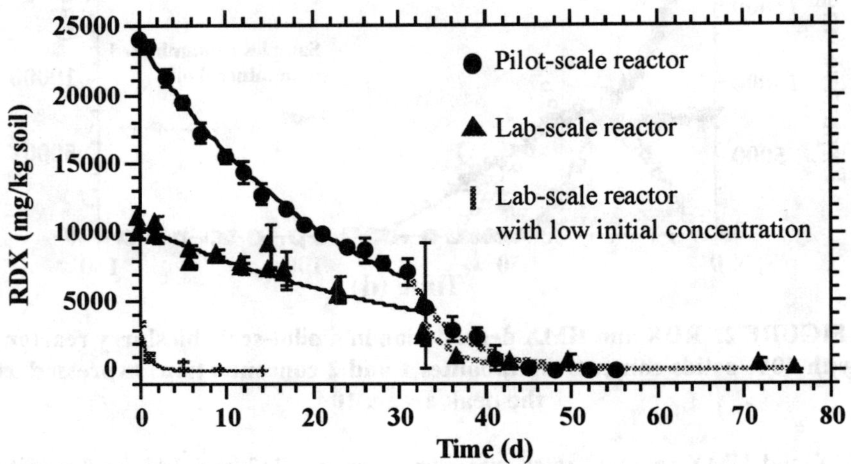

FIGURE 3. RDX removal profiles as two-phase patterns. Curves are first-order fits : C = $C_0.e^{-K_1.t}$; C : RDX residual concentration; C_0 : RDX initial concentration; K_1 = first-order time-constant.

The pilot-scale reactor has seen its average RDX removal rate doubling compared to that of the lab-scale reactors : 513 mg RDX/kg soil.d in the pilot-scale reactor, as compared to 213, in the lab-scale reactors. This change is correlated to the initial overall concentration of RDX : 24 g RDX/kg soil versus 10, respectively. This positive impact of the initial concentration on the removal rate seems to indicate that the solid-to-liquid RDX transfer rate was the limiting step. The increase of the overall concentration from 10 to 24 g RDX/kg soil could not improve the concentration in the liquid phase (limited to the solubility product of RDX which is 46 mg/L), thus could not impact the kinetics of the biodegradation. However, the increase of the overall concentration could improve the concentration gradient between the solid and aqueous phases, and as a result, the rate of RDX transfer from solid to liquid phases. Dependency of the removal rate on the overall concentration is even more evident by showing that data fit first-order kinetics ($dC/dt = -K_1.C$). However, it is somewhat amazing that RDX removal was found to be biphasic (Figure 3). When the residual concentration reached below 6000 mg RDX/kg soil, faster RDX removal seemed to resume to initial rate values, in both cases. This can not be attributed to a time effect, e.g. a lag for adaptation of biomass, since in previous assays within the same drum-type reactor under similar conditions, but with an initial RDX concentration of 3000 mg/kg soil (Shen et al., 1998), RDX removal started immediately at higher rates similar to those of the second phase of the present study, as displayed in Figure 3 for comparison. This could not be attributed either to a negative impact of the metabolites since faster RDX removal resumed while metabolites were still at their highest levels (Figure 2). It is likely that several mass transfer processes (e.g. dissolution from a solid pure RDX to the aqueous phase, sorption from the aqueous phase to the soil particles, desorption from the soil particles to the aqueous phase) are concurrent and affect the removal rate differently. This break in the removal pattern might reflect a change in the weight of one transfer process relative to the others, depending of the effective overall RDX concentration.

CONCLUSIONS

The present results clearly show that the bioslurry approach is likely the most effective way for bioremediating heavily-contaminated soils, which could be treated down to complete depletion, at a rate as high as half a gram of RDX and 0.16 gram HMX per kg soil per day, under optimal conditions.

REFERENCES

Binks, P. R., S. Nicklin, and N. C. Bruce. 1995. "Degradation of hexahydro-1,3,5-trinitro-1,3,5-triazine (RDX) by *Stenotrophomonas maltophilia* PB1." *Appl Environ. Microbiol.* *61*:1318-1322.

Funk, S. B., D. J. Roberts, D. L. Crawford, and R. L. Crawford. 1993. "Initial-phase opimization for bioremediation of munitions compounds-contaminated soils." *Appl. Environ. Microbiol.* *59*:2171-2177.

Guiot, S. R., C. F. Shen, J. A. Hawari, G. Ampleman, and S. Thiboutot. 1997. "Bioremediation of Nitramine Explosive Contaminated Soils." In W. D. Tedder (Ed.), *Proc. American Chemical Society Conference on Emerging Technologies in Hazardous Waste Management IX*, Pittsburgh, PA, Sept. 15-17, pp. 341-344.

Hampton, M.L, and W. E. Sisk, 1995. "Biological treatment of soil contaminated with TNT in a slurry reactor: A pilot/field scale demonstration." In W. D. Tedder (Ed.), *Proc. American Chem. Society Conf. on Emerging Technologies In Hazardous Waste Management VII*, Atlanta, GA, Sept. 17-20, p. 529.

Kaake, R. H., J. Bono, and T. W. Yergovitch. 1997. "Laboratory and field demonstrations of the SABRE™ process at Yorktown naval weapons station." In B. C. Alleman and A. Leeson (Eds.), *In Situ and On-Site Bioremediation*, Vol. 4(2), pp. 61-66. Battelle Press, Columbus, OH.

Kitts, C. L., D. P. Cunningham, and P. J. Unkefer. 1994. "Isolation of three hexahydro-1,3,5-trinitro-1,3,5-triazine-degrading species of the family *Enterobacteriaceae* from nitramine explosive-contaminated soil." *Appl. Environ. Microbiol. 60*:4608-4711.

McCormick, N. G., J. H. Cornell, and A. M. Kaplan. 1981. "Biodegradation of hexahydro-1,3,5-trinitro-1,3,5-triazine." *Appl. Environ. Microbiol. 42*:817-823

McCormick, N. G., J. H. Cornell, and A. M. Kaplan. 1984. "The anaerobic biotransformation of RDX, HMX, and the acetylated derivatives." Tech. Report TR-85/007. U.S. Army Natick Research and Development Lab., Natick, MA.

Nelson, D. W. and L. E. Sommers. 1982. "Total carbon, organic carbon, and organic mater." In A. L. Page, R. H. Miller, and D. R. Keeney (Eds.), *Methods of Soil Analysis, Part 2, Chemical and Microbiological Properties*. American Society of Agronomy and Soil Science Society of America, Madison, WI.

Shen C. F., S. R. Guiot, S. Thiboutot, G. Ampleman and J. A. Hawari. 1998. "Complete biodegradation of RDX and HMX in anoxic soil slurry bioreactors: laboratory and pilot-scale experiments." In *Proc. Sixth Int. FZK/TNO Conference on Contaminated Soil*, 17-21 May, Edinburgh, UK, pp. 513-522. Thomas Telford, London.

Soli, G. 1973. "Microbial degradation of cyclonite (RDX)." NWC TP-5525, AD 762-751. Naval Weapons Support Center, Crane, IN.

OPTIMIZING THE BIOTRANSFORMATION OF RDX AND HMX

Robin L. Autenrieth, Michael D. Jankowski, James S. Bonner, Madhu Kodikanti
Texas A&M University, College Station, TX, USA

ABSTRACT: RDX (hexahydro-1,3,5-trinitro-1,3,5-triazine) and HMX (octahydro-1,3,5,7-hexahydro-1,3,5,7-tetrazocine) are two of the most commonly used high explosives. They are toxic and are present at many present and former munitions sites. Experiments were performed in both suspended growth and slurry reactors at various oxic levels and in sequences of anoxic/oxic time periods in order to maximize RDX and HMX transformation while minimizing the environmental and toxicological impact of the by-products. Aeration reduced the the rate and extent of transformation of the parent compound but enhanced ring-cleavage and appears to produce a more environmentally desirable makeup of by-products.

INTRODUCTION

Most organisms which degrade nitro-explosives start by biochemically altering the nitro groups under anaerobic conditions, usually via reduction, i.e.: $R-NO_2 \rightarrow R-NO \rightarrow R-NHOH \rightarrow R-NH_2$ (Kitts 1994). These reduction reactions are favorable under anaerobic condition. Consequently, anaerobic conditions have been the focus of most of the past work.

McCormick et al. (1981) used a microbial consortia obtained from anaerobic sewage sludge to completely degrade RDX from 50 mg/l within 7 days. The initial intermediates were identified as the mono-, di-, and tri-nitroso derivatives of RDX (MNX, DNX, and TNX, respectively). Other byproducts such as formaldehyde, hydrazine, 1,1-dimethlhydrazine, and 1,2-dimethylhydrazine were also identified. Based on these findings, a pathway for the anaerobic biodegradation of RDX was proposed.

Anaerobic RDX and HMX degradation byproducts such as the nitroso-RDX and nitroso-HMX derivatives, nitrate, nitrite, 1,1-dimethylhydrazine, 1,2-dimethylhydrazine, hydrazine, and formaldehyde, are of concern for their potential toxicity and/or carcinogenicity (Spanggord, 1983a; Young 1997). The few successful aerobic studies have been able to avoid the production of these undesirable intermediates for the most part. Knezovich and Daniels (1991) reported successful RDX biodegradation under aerobic conditions with no potentially carcinogenic byproducts formed. No specific information regarding the extent and extent of degradation and the identification of metabolites was presented. Binks et al. (1995) was able to isolate the bacterium *Stenotrophomonas maltophilia* PB1 which degraded RDX under aerobic conditions from its solubility to nearly zero in 8 days. Only a single metabolite, $C_3H_9N_3O_5$, was formed.

Objective. Experiments designed to maximize the amount of RDX and HMX transformed biotically while minimizing the production of environmentally unfavorable intermediates by varying the oxic conditions in reactors were conducted. Some reactors were also sequenced between oxic states, initially at low dissolved oxygen (DO) and redox levels ("anoxic") as the reduction of the nitro groups would be most feasible under these conditions. Aerated, aerobic ("oxic") stages were then used to promote ring cleavage and further reduce toxicty.

MATERIALS AND METHODS

Cultures. The cultures used in these experiments were developed from soil and/or water samples taken from: an urban playa in Lubbock, TX; a site at the Pantex facility near Amarillo, TX, where explosives were openly burned; and a granular activated carbon-based groundwater treatment system at Pantex.

Reactors. Experiments were performed in duplicate 500mL batch reactors containing 300mL of a mineral salts media. Slurry reactors contained 10% silica (0.5-10 μm) by weight. All reactors contained foam stoppers to allow for a filtered gaseous exchange with the atmosphere. Active aeration was achieved for oxic reactors/stages using compressed air. Biomass was added from an enriched, concentrated stock to an initial level of 500 mg/l based on total solids analysis. Carbon was supplement in the form of a mixture of glucose, succinic acid, and glycerol. Ammonium nitrate was added at 120 mg/l in HMX experiments. After each sampling, reactors received full nitrogen, nutrient, and/or carbon replenishment.

Sets of reactors were set up under the following oxic conditions: (1) all days anoxic, (2) 4 days per sequence (starting with anoxic), (3) 2 days per sequence (starting with anoxic), and (4) all days oxic. Suspended growth experiements were carried out for 8 days and slurry experiments lasted 16 days.

Sample Analysis. Sampling was performed every two days. Samples taken for the analysis of explosives and metabolites were centrifuged at 12,800xg for 15 minutes. The supernatant was withdrawn and injected directly in a Hewlett Packard 1050 HPLC with a photo-diode array detector. RDX analysis was done at 205nm using a Waters Novapak C-8 column (4.6 mm x 150 mm with a 5 μm particle size) with an 82:18 water:isopropanol mobile phase at 0.8 ml/min (Waters Chromotography, Cambridge, Massachusetts, USA). HMX analysis was done at 229nm using a Restek Pinnacle Cyano Amine Column (4.6 mm x 150 mm with a 3 μm particle size) with a 50:50 water: methanol mobile phase at 0.8 ml/min (Restak Corporation, Bellefonte, Pennsylvania, USA). Acetonitrile was added to the remaining pellet for extraction purposes.

Dissolved Oxygen and Redox. Probes were obtained from Microelectrodes, Inc. (Bedford, New Hampshire, USA). D.O. and redox were monitored either continuously (every 1-30 minutes) or semi-continously (4-5 samplings/day) using a PC-based data acquisition system run by VirtualBench software (National Instruments, Austin, Texas, USA).

RESULTS AND DISCUSSION

Suspended Growth - RDX Studies. RDX transformation was highest in anoxic reactors, at a DO below 0.5 mg/l (average = 0.3 mg/l) and a redox average of 117 mV. Transformation after 8 days was slightly higher than the sequencing and oxic reactors. Transformation in the sequencing reactors was lower during oxic phases than anoxic phases (Figure 1). The transformation in the oxic reactors (average DO = 8.4 mg/l, average redox = 325 mV) was lowest of all but not significantly different from the sequencing reactors. Comparsions to a mixture of MNX, DNX and TNX based on retention time and UV absorbance spectrum indicated that RDX was transformed to MNX, but no DNX or TNX was produced. Therefore, the pathway is different from the anaerobic pathway, probably because the conditions were not reduced enough to proceed with the reduction of the nitro groups of MNX. This is particularly important because 1,1-dimethylhydrazine and 1,2-dimethylhydrazine are by-products of DNX metabolism and are therefore will not be produced under the anoxic pathway observed.

While RDX transformation lagged in the oxic phases, these stages produced the highest amount of RDX transformation/least amount of MNX accumulation. The reactors that were under oxic conditions throughout the entire 8 days showed less MNX in chromatograms than the other reactors. These reactors also produced greater quantities of an unidentified metabolite than the other reactors. Based on absorbance comparisons at 205nm and 230nm, this metabolite appears to be a product of ring-cleavage, indicating a probable reduction in the toxicity of the system as compared to the presence of MNX or other intermediates with the triazine ring intact. The sequencing reactors did not end up with significantly lower levels of MNX or higher levels of the ring-cleavage metabolite than the anoxic reactors.

Suspended Growth - HMX Studies. Similar to the RDX studies, HMX transformation was best in the anoxic reactor sets and lowest during oxic phases (Figure 2). However, the oxic reactors produced a metabolite that did not appear in chromatograms from the other reactors. This metabolite appeared to be a product of ring-cleavage. Peak areas of most of the metabolites could not be separated enough for a comparison of peak areas. Extending the sequencing periods to 4 days resulted in an improvement over 2-day sequencing in terms of HMX transformation.

Slurry Phase - RDX Studies. Results paralleled those of suspended growth studies for the most part (Figure 3). Possibly explanations for the observed differences include the use of a different media, the longer experimental duration, or a higher sensitivity due to a higher initial concentration of RDX in these studies (50 mg/l), the differences between reactor sets were more pronounced. Again, the

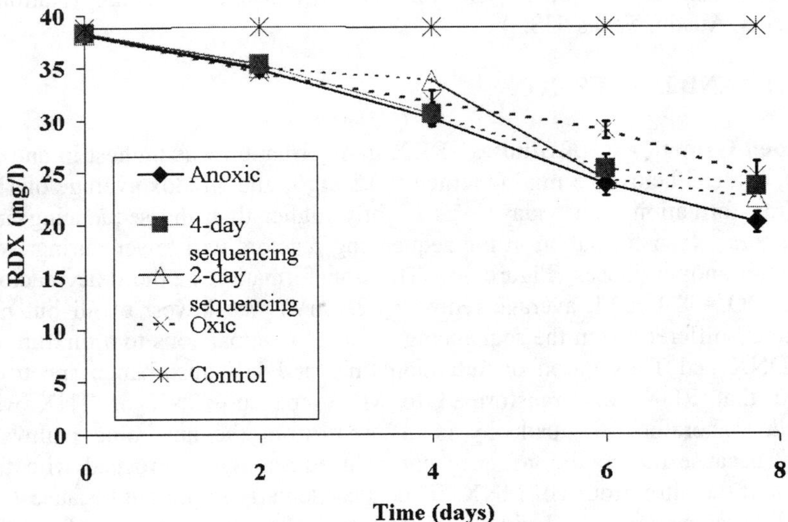

FIGURE 1. Suspended growth RDX transformation over 8-days. Periods of aeration are marked by dotted lines.

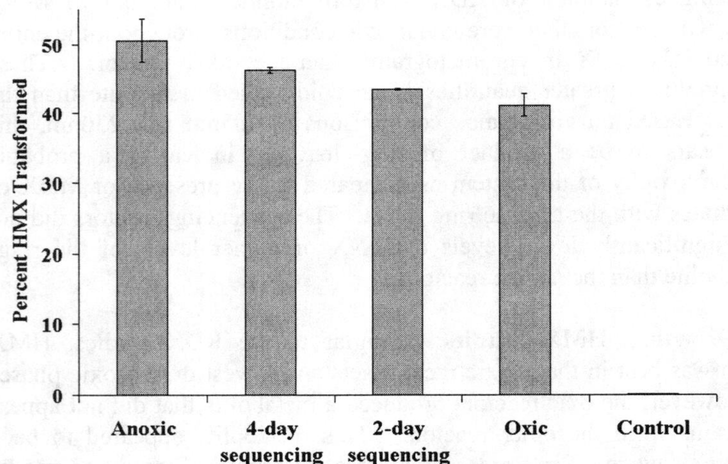

FIGURE 2: Suspended growth HMX transformation over 8-days from an intial HMX concentration of 1.0 mg/l.

anoxic reactors transformed more RDX than the oxic and sequential reactors. However, at the end of 16 days, twice as much MNX existed in the anoxic reactors compared to the other reactors, and levels of the ring-cleavage metabolite were up to ten times lower than those found in the sequential and oxic reactors.

In this case, however, an additional metabolite was eventually formed in the anoxic and sequential reactors that did not appear in the oxic reactors. While there is no evidence that this metabolite is the product of ring-cleavage, its

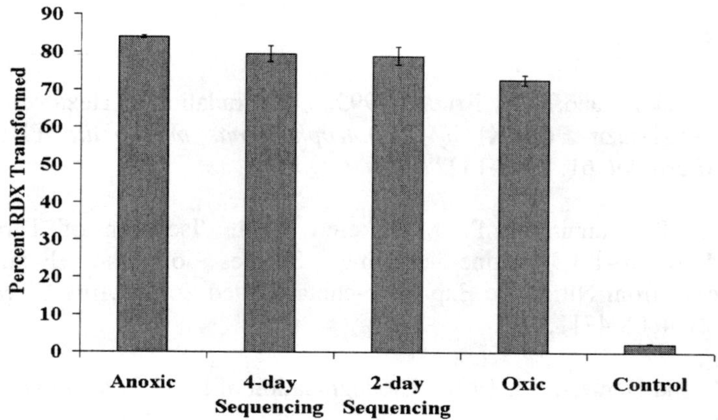

FIGURE 3. Transformation of RDX over 16 days in slurry reactors from an initial RDX concentration of 50 mg/l.

indicates a separate or further metabolic process under anoxic or sequential conditions that may not occur under obligately oxic conditions. The sequential reactors performed closer to the oxic reactors in MNX transformation and ring-cleavaged metabolite production than they did in the suspended studies.

Slurry Phase - HMX Studies. The results of the 16-day slurry experiments matched those of the 8-day suspended growth experiments. 4-day sequencing outperformed 2-day sequencing and worked almost as well as the anoxic reactors at transforming HMX (Figure 4). The ring-cleavage metabolite noted under oxic conditions in the suspended growth experiments was not observed here, but that may be due to a slight alteration of the analytical method made to enhance the sensitivity of HMX analysis.

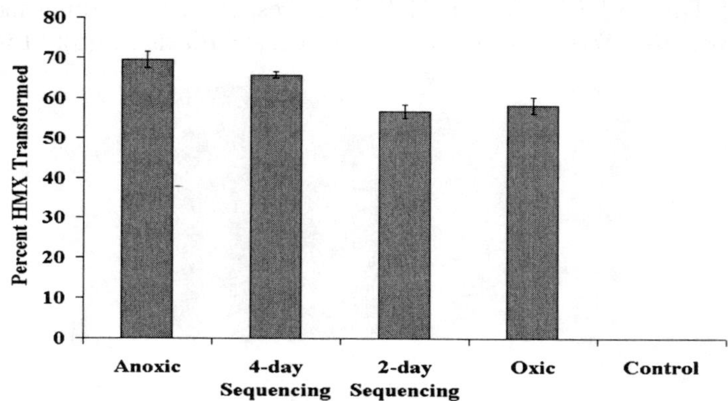

FIGURE 4. Transformation of HMX over 16 days in slurry reactors form an initial HMX concentration of 2.0 mg/l.

REFERENCES

Binks, P. R., S. Nicklin, and N. C. Bruce. 1995. "Degradation of Hexahydro-1,3,5-Trinitro-1,3,5-Triazine (RDX) by *Stenotrophomonas maltophilia* PB1." *Appl. Environ. Microbiol.* 61: 1318-1322.

Kitts, C. L., D. P. CunninghamP. J. Unkefer. 1994. "Isolation of Three Hexahydro-1,3,5-trinitro-1,3,5-triazine-degrading Species of the Family *Enterobacteriaceae* from Nitramine Explosive-contaminated Soil." *Appl. Envir. Microbiol.* 60(12): 4608-4711.

Knezovich, J. P., and Daniels, J. I. 1991. "Biodegradation of HE.*" Water Environ. Technol.* 3(3): 34

McCormick, N. G., J. H. Cornell, A. M. Kaplan. 1981. "Biodegradation of Hexahydro-1,3,5-Trinitro-1,3,5-Triazine." *Appl. Envir. Microbiol.* 42(5): 817-823.

Spanggord, R. J., W. R. Mabey, T. W. Chou, S. Lee, P. L. Alferness, D. S. Tse, T. Mill. 1983a. *Environmental Fate Studies of HMX. Phase II - Detailed Studies.* Contract No. DAMD17-C-2100, US Army Medical Research and Development Command, Fort Detrick, Maryland.

Spanggord, R. J., W. R. Mabey, T. Mill, T. W. Chou, J. H. Smith, S. Lee, and D. Robert. 1983b. *Environmental Fate Studies on Certain Munitions Wastewater Constituents. Phase IV - Lagoon Model Studies.* ADA138550, US Army Medical Research and Development Command, Fort Detrick, Maryland.

Young, D.M., P. J. Unkefer, and K. L. Ogden. 1997. "Biotransformation of Hexahydro-1,3,5-Trinitro-1,3,5-Triazine (RDX) by a Prospective Consortium and Its Most Effective Isolate *Serratia marcescens.* Biotech. and Bioeng. 53(5): 515-522.

MODELING THE MINERALIZATION OF TRINITROTOLUENE AND DINITROTOLUENES IN AEROBIC/ANOXIC BIOFILM

R. Guy Riefler
Barth F. Smets
(University of Connecticut, Storrs, Connecticut)

ABSTRACT:
This study investigates the feasibility of using a single aerobic biofilm reactor, operated such that anoxic zones develop, to mineralize a mixed waste stream of 2,4-dinitrotoluene (24DNT), 2,6-dinitrotoluene (26DNT), and trinitrotoluene (TNT). In the anoxic zone of the biofilm, TNT is reduced to 2-amino-4,6-dinitrotoluene (2A46DNT) and 4-amino-2,6-dinitrotoluene (4A26DNT), and in the aerobic zone, 24DNT, 26DNT, 2A46DNT, and 4A26DNT are mineralized. In this research, a biofilm model was constructed to depict the simultaneous transformation of these nitroaromatic compounds in a fluidized bed biofilm reactor. Parameters for the reactor system and DNT degradation were determined from a fluidized bed biofilm reactor mineralizing 24DNT and 26DNT. TNT transformation parameters were taken from a kinetic analysis of anaerobic TNT reduction rates. Results from the model indicate that this system is capable of mineralizing 24DNT, 26DNT, and TNT simultaneously. However, in order for anoxic zones to develop in a system with such low maximum growth rates, substrate flux must to be increased beyond the treatment capacity of the system. As a result, effluent concentrations for 24DNT, 26DNT, and TNT are high because of the short retention time. These problems might be resolved by using a different reactor design.

INTRODUCTION

Nitroaromatic compounds remain an environmental problem at military waste sites and munitions production facilities because of their suspected toxicity and their resistance to degradation. Because TNT is manufactured by sequential nitration of toluene, TNT is often found together with 26DNT and 24DNT. Under anaerobic conditions, TNT is susceptible to aspecific reductions which can transform the nitro groups into amino groups. This transforms TNT into 2A46DNT and 4A26DNT, and those products into 2,4-diamino-6-nitrotoluene. Complete mineralization of TNT by reduction has proven difficult to implement, however. It has recently been shown that 24DNT and 26DNT can be mineralized in an aerobic fluidized bed biofilm reactor when present as the sole carbon, energy, and nitrogen source (Lendenmann *et al.*, 1998). Treatment of nitroaromatic waste streams or contamination poses a challenge, because TNT and DNTs require different conditions for mineralization.

A fundamental property of biofilms is their diffusional resistance which limits the transport of chemicals into biofilm. It has often been noted that anaerobic zones can form in deep aerobic biofilms due to this diffusional resistance; however, little work has been conducted to make use of this naturally forming stratification. In fact, many pollutants can only be mineralized by sequential exposure to different redox environments. This study investigates the feasibility of using a single aerobic biofilm reactor, operated such that anoxic zones develop, to mineralize a mixed waste stream of 24DNT, 26DNT, and TNT. In the anoxic zone of the biofilm, TNT will be reduced to 2A46DNT and 4A26DNT. Because the enzymes responsible for 24DNT mineralization have a broad substrate range (Suen *et al.*, 1994), it is expected that 2A46DNT and 4A26DNT can be mineralized in the aerobic zone of the biofilm while 24DNT and 26DNT are present to induce expression. Through this redox cycling strategy, it should be possible to mineralize a waste stream of combined TNT, 24DNT, and 26DNT compounds. As the first stage of this research, a rigorous mathematical model of a redox stratified biofilm has been developed.

MATERIALS AND METHODS

AQUASIM 1.0e was used to simultaneous solve partial differential equations describing diffusion, transformation, and growth (Reichert, 1994). Only reaction equations will be presented here, as the complete biofilm model is described elsewhere (Wanner and Reichert, 1996). Four kinetic aerobic equations describe reactions in which O_2 serves as the terminal electron acceptor and 24DNT, 26DNT, 2A46DNT, or 4A26DNT serve as electron donors. These take the form;

$$r_{24D} = k_{24D} X_H \left(\frac{S_{24D}}{K_{24D} + S_{24D}} \right) \left(\frac{S_{O2}}{K_{O2} + S_{O2}} \right) \tag{1}$$

where, r_{24D} is the rate of 24DNT degradation, k_{24D} is the maximum specific degradation rate of 24DNT, S_x is the concentration of the x, K_x is the half maximum degradation constant for x. These reactions are controlled by the concentration of electron donor (S_{24D}) and electron acceptor (S_{O2}). Aerobic endogenous decay has the form;

$$r_{H,aer} = b_H X_H \left(\frac{S_{O2}}{K_{O2} + S_{O2}} \right) \tag{2}$$

where, $r_{H,aer}$ is the aerobic rate of heterotrophic biomass decay and b_H is the linear heterotrophic biomass decay constant.

In the anoxic zone of the fluidized bed reactor, biomass decays with a rate similar to equation 2.

$$r_{H,ano} = b_H X_H \left(\frac{K_{i,O2}}{K_{i,O2} + S_{O2}} \right) \tag{3}$$

where, $r_{H,ano}$ is the anoxic rate of biomass decay and $K_{i,O2}$ is the oxygen inhibition constant. As stored energy sources are consumed by the bacteria in the anoxic zone, some electrons will be aspecifically transferred to TNT, subsequently reducing it to 2A46DNT and 4A26DNT. This loss of TNT is depicted by.

$$r_{TNT} = b_H X_H f \left(\frac{S_{TNT}}{K_{TNT} + S_{TNT}} \right) \left(\frac{K_{i,O2}}{K_{i,O2} + S_{O2}} \right) \qquad (4)$$

where, r_{TNT} is the TNT reduction rate and f is the maximum TNT reduction rate divided by the biomass decay rate.

Because an experimental system to degrade these compounds as described has not yet been constructed, the biofilm model was calibrated in steps with available published data. The fluidized bed biofilm reactor was modeled after the reactor presented by Lendenmann *et al.* (1998) which is a 1.5 L water-jacketed vessel filled with sand as carrier medium with temperature and pH control and aeration provided in the recirculation line. Optimal operation of the reactor was determined by varying the oxygen delivery rate to minimize $S_{TNT}^{2} + S_{24DNT}^{2} + S_{26DNT}^{2} + S_{2A46DNT}^{2} + S_{4A26DNT}^{2}$. Results were determined for influent concentrations of 24DNT = 40 mg/L, 26DNT = 10 mg/L, and TNT = 10 mg/L while the reactor hydraulic retention time (HRT) was varied from 3.60 hr to 0.180 hr.

RESULTS AND DISCUSSION
In the base model, 24DNT and 26DNT mineralization was matched with the results reported from a fluidized bed biofilm reactor (Lendenmann *et al.*, 1998; Smets *et al.*, in press). In this study, steady-state concentrations of 24DNT and 26DNT and biomass per mass of sand were reported for four HRTs: 12.5 hr, 6.3 hr, 3.1 hr, and 1.5 hr. Results were also reported for HRT = 0.75 hr, however because a significant change in cell kinetics was observed, data from this HRT were not used in this study (Smets *et al.*, in press). Using the surface area of sand biomass per mass of sand was converted to biomass per surface area or biofilm density times biofilm thickness. Results for HRT = 12.5 hr and 3.1 hr were used for model calibration and HRT = 6.3 hr and 1.5 hr for model validation. Results are shown in Table 1. Parameters obtained from this calibration were b_H (0.15 1/d) and the oxygen supply rate to the reactor (0.723 g/d for HRT = 3.1hr). In addition, several reported kinetic parameters for DNT degradation (k_{24DNT} = 0.35 COD/COD d, k_{26DNT} = 0.16 COD/COD d, K_{26DNT} = 0.37 mg COD/L) were confirmed, although better results were obtained when the reported value for K_{24DNT} was changed from 0.072 mg COD/L to 0.0072 mg COD/L (Smets *et al.*, in press).

The full model for the fluidized bed reactor transforming 2,4-DNT, 2,6-DNT, and TNT was constructed with the calibrated DNT model and the addition of equations 3 and 4. Values for the kinetic parameters of TNT reduction were obtained from reported values (Admassu *et al.*, 1998). In this study, TNT was reduced by an anaerobic culture with k_{TNT} = 0.014 g TNT/g cells min and K_{TNT} = 2.02 mg TNT/L while the maximum growth rate was 0.029 g cells/g cells min. Using the reported yield (0.17 g cells/g sugar), the maximum growth rate was converted to a maximum sugar consumption rate (0.183 g COD/g cells min) and k_{TNT} was converted to equivalent units (0.0664 g COD/g cells min). Thus, a ratio of TNT reduced per energy consumed was calculated ($f = 0.0664 / 0.183 = 0.36$). In addition, it was assumed that 4A26DNT and 2A46DNT were formed in the ratio of 2/1 (Hawari *et al.*, 1998). The kinetic parameters for 24DNT mineralization were used for 2A46DNT, and the kinetic parameters for 26DNT were used for 4A26DNT.

TABLE 1. Calibration and validation results for DNT mineralization model.

	HRT=12.5 hr	HRT=6.3 hr	HRT=3.1 hr	HRT=1.5 hr
24DNT effluent concentration (mg/L)				
Experimental	0.037	0.038	0.071	0.215
Model	0.018	0.036	0.088	0.254
% Diff	51.3%	5.9%	23.5%	18.0%
26DNT effluent concentration (mg/L)				
Experimental	0.101	0.218	0.164	0.279
Model	0.103	0.106	0.124	0.191
% Diff	2.3%	51.3%	24.5%	31.6%
*Biofilm density * biofilm thickness (g COD/m^2)*				
Experimental	0.191	0.214	0.500	0.700
Model	0.125	0.241	0.497	1.040
% Diff	34.6%	12.6%	6.0%	48.6%

Steady-state effluent concentrations predicted for the fluidized bed biofilm reactor treating 24DNT, 26DNT, and TNT using an oxygen delivery rate that minimizes effluent concentrations for each HRT are shown in Table 2.

TABLE 2. Predicted effluent concentrations for redox stratified fluidized bed biofilm reactor using optimal oxygen delivery rates.

HRT (hr)	3.60	0.720	0.360	0.180
Biofilm thickness (m)	8.35×10^{-5}	3.74×10^{-4}	6.69×10^{-4}	1.10×10^{-3}
Oxygen delivery (g/d)	0.523	2.55	4.97	9.24
Effluent TNT (mg/L)	2.37	2.77	3.88	5.91
Effluent 2A46DNT (mg/L)	0.00492	0.00466	0.00303	0.00196
Effluent 4A26DNT (mg/L)	0.0538	0.0541	0.0288	0.0103
Effluent 24DNT (mg/L)	0.321	1.75	3.99	8.96
Effluent 26DNT (mg/L)	0.407	0.642	1.03	1.99

Using an HRT similar to the experimental system, HRT = 3.6 hr, 76.3% removal of TNT, 99.2% removal of 24DNT, and 95.9% of 26DNT was achieved. However, the oxygen profile for this biofilm (Figure 1) shows that significant redox stratification did not occur. Because the biofilm is thin (83.5 μm) and the oxygen demand is low (low k_m for DNTs), the oxygen concentration at steady-state only varies from 0.125 mg/L at the sand surface and 0.170 mg/L at the biofilm surface. Similar profiles for the nitrotoluenes indicate that the biofilm is largely fully-penetrated but with low oxygen concentrations. Thus, these modeling results are dependent on DNT mineralization rates under oxygen limitation and TNT reduction rates under oxygen inhibition. These two processes are controlled by the parameters K_{O2} = 0.5 mg/L (Horn and Hempel, 1997) and $K_{i,O2}$ = 0.1, which are poorly estimated.

In order to increase the biofilm thickness, substrate fluxes were increased by raising the influent concentrations of 24DNT and 26DNT or by raising the

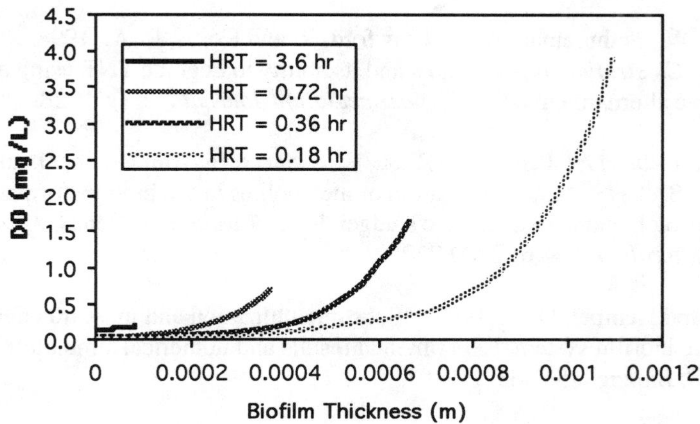

FIGURE 1. Steady-state oxygen concentration in biofilm at different HRTs.

reactor flow rate (decreasing HRT). Higher influent concentrations had little effect on biofilm thickness, but resulted in higher effluent concentrations (data not shown). As shown in Figure 1, lower HRTs led to thicker biofilm formations and development of anoxic zones. However, the lower HRTs also allow less time for treatment resulting in higher effluent concentrations for TNT, 24DNT, and 26DNT (Table 2). This demonstrates that TNT mineralization is possible in this treatment system, although low effluent concentrations may not be achievable.

CONCLUSIONS

In this research, a biofilm model was constructed to depict the simultaneous transformation of 24DNT, 26DNT, TNT, 2A46DNT, and 4A26DNT in a redox-varying fluidized bed biofilm reactor. Parameters for the reactor system and DNT degradation were determined from a fluidized bed biofilm reactor mineralizing 24DNT and 26DNT (Lendenmann *et al.*, 1998). TNT transformation parameters were taken from a kinetic analysis of anaerobic TNT reduction rates (Admassu *et al.*, 1998). Results from the model indicate that this system is capable of mineralizing 24DNT, 26DNT, and TNT. However, in order for anoxic zones to develop in a system with such low maximum growth rates, substrate flux had to be increased beyond the treatment capacity of the system. As a result, effluent concentrations for 24DNT, 26DNT, and TNT are high because of the short retention time. These problems might be resolved using a hollow fiber membrane reactor, rather than a fluidized bed reactor. This reactor design allows the growth of thicker biofilms because of reduced shear which would improve oxygen zonation. In addition, a growth substrate could be supplied to the base of the biofilm which would increase oxygen demand, also improving oxygen zonation, and providing an electron donor to increase TNT reduction rates.

REFERENCES

Admassu, W., Sethuraman, A. V., Crawford, R. and Korus, R. A. 1998. "Growth kinetics of *Clostridium bifermentans* and its ability to degrade TNT using an inexpensive alternative medium." *Bioremediation Journal* . 2(1):17-28.

Hawari, J., Halasz, A., Paquet, L., Zhou, E., Spencer, B., Ampleman, G. and Thiboutot, S. 1998. "Characterization of metabolites in the biotransformation of 2,4,6-trinitrotoluene with anaerobic sludge: Role of triaminotoluene." *Appl. Environ. Microbiol.* 64(6):2200-2206.

Horn, H. and Hempel, D. C. 1997. "Substrate utilization and mass transfer in an autotrophic biofilm system: Experimental results and numerical simulation." *Biotechnol. Bioeng.* 53:363-371.

Lendenmann, U., Spain, J. C. and Smets, B. F. 1998. "Simultaneous biodegradation of 2,4-dinitrotoluene and 2,6-dinitrotoluene in an aerobic fluidized-bed biofilm reactor." *Environ. Sci. Technol.* 32:82-87.

Reichert, P. 1994. "AQUASIM-A tool for simulation and data analysis of aquatic systems." *Wat. Sci. Tech.* 30(2):21-30.

Smets, B. F., Riefler, R. G., Lendenmann, U. and Spain, J. C. in press. "Kinetic analysis of dinitrotoluene biodegradation in an aerobic fluidized bed biofilm reactor." *Biotechnol. Bioeng.*

Suen, W.-C., Haigler, B. E. and Spain, J. C. 1994. "2,4-Dinitrotoluene dioxygenase genes from *Pseudomonas* sp. strain DNT: Homology to naphthalene dioxygenase." *Abstracts of the 94th Annual Meeting of the American Society for Microbiology* . 458.

Wanner, O. and Reichert, P. 1996. "Mathematical modeling of mixed-culture biofilms." *Biotechnol. Bioeng.* 49:172-184.

TECHNOLOGY DEVELOPMENT FOR BIOLOGICAL TREATMENT OF EXPLOSIVES-CONTAMINATED SOILS

Douglas E. Jerger, Ph.D. (IT Corporation, Knoxville, Tennessee)
Patrick Woodhull, P.E. (IT Corporation, Findlay, Ohio)

ABSTRACT

The primary explosive contaminants in soils at DoD sites include TNT, DNT, RDX, HMX and tetryl. Windrow composting and slurry-phase treatments have been applied for full-scale remediation of soils, while other technologies such as soil piles, white rot fungi and phytoremediation have been tested at the pilot-scale. IT Corporation has conducted bench- to full-scale projects applying these technologies and presents a process comparison and process economics based on actual field experience.

INTRODUCTION

Explosives-contaminated soils exist on most DoD sites. The primary explosive contaminants are TNT, DNT, RDX, HMX and tetryl. Windrow composting and slurry-phase treatments have been applied for full-scale remediation at Army and Navy sites. Other technologies such as white-rot fungal treatment, soil piles and phytoremediation are under development.

IT Corporation has conducted bench- to full-scale projects applying these biological treatment technologies for the remediation of explosive-contaminated soils at DoD sites. IT is performing full-scale windrow composting on 20,000 cubic yards of RCRA-listed contaminated soil at Pueblo Chemical Depot (PCD), CO. IT conducted a field pilot test on an anaerobic/aerobic soil pile technology at PCD. Phytoremediation was also evaluated for treatment of TNT-contaminated soils at PCD. In addition to these technologies IT performed full-scale remediation of explosives-contaminated soils at Yorktown Naval Weapons Station using slurry-phase biological treatment.

The benefits and costs for each process based on actual field experience are presented in the following case histories.

CASE HISTORY 1: WINDROW COMPOSTING

Pueblo Chemical Depot (PCD), Colorado is a 35-square mile Army depot which served as an ammunition storage and transfer facility from 1942 to 1988. The site was used from the late 1940s through 1974 for the reclamation of 2,4,6-TNT and Composition B (60 percent RDX and 40 percent TNT) from standard ammunition casings. The operations included high-pressure steam washout, drying, flaking, pelletizing, and packaging of TNT and Composition B.

A soil removal project involving the excavation and staging of explosives-contaminated soils was completed in early 1998. Approximately 21,500 cubic yards (cy) of soil containing explosives contaminants such at TNT, DNT, and TNB were excavated, loaded into dump trucks, and transported to a storage building. The field test was conducted in an enclosed building at PCD.The initial average concentrations for TNT, DNT, TNB were approximately 3,800 milligrams per kilogram (mg/kg), 11 mg/kg, and 35 mg/kg, respectively. RDX concentrations were below the method detection limits (MDLs) 1.2 mg/kg. The mono- and di-amino compounds were generally detected at low concentrations in the soil. The only compound detected in the Toxicity Characteristic Leaching Procedure (TCLP) analysis (EPA Method 1311/8270B) was DNT at a concentration of 0.2 mg/L.

Construction of the windrow compost consisted of blending 15,000 kg of soil with the following amendments at a volume ratio of 30% soil and 70% amendments: 1.4 cy of chicken manure, 4.0 cy of potato waste, 7.0 cy of alfalfa, 7.0 cy of yard waste (wood chips), and 8.7 cy of cow manure.

The analytical results from the sample events indicated rapid degradation of all the target compounds. By Day 15, both DNT and TNB had been degraded below 0.20 mg/kg, and the TNT concentration was 1.2 mg/kg. The mono- and di-amino degradation products, TNT, and DNT all showed greater than 99 percent removal in the first 15 days of treatment. The Day 35 data indicated that the average TNT and DNT concentrations in the windrow compost were 0.4 mg/kg and 0.5 mg/kg, respectively. Both TNB and RDX concentrations were below the reporting limit of 0.2 mg/kg, and all of the mono-and di-amino aromatic compounds were below 0.5 mg/kg. TCLP concentration for DNT and nitrobenzene were below the reporting limits of 0.1 mg/L.

CASE HISTORY 2: SOIL PILE TREATMENT

In support of developing a simple, cost-effective alternative to the site-specific processes currently available, The Waste Management Inc. Bioremediation Process Improvement Team undertook the development of an innovative two-stage static soil (TOSS) process to treat explosives-contaminated soil at fixed-based WMI facilities. The TOSS process is a sequential process that combines an initial anaerobic soil pile stage with an aerobic compost stage A field pilot test was conducted to further develop this process after successful labortory tests with soils from PCD (Green, 1998).

Construction of the TOSS soil pile at PCD consisted of blending 15,000 kg (approximately 12 cubic yards) of soil with the following amendments: 1.25 cy of potato waste, 1.25 cy of sugar beet tailings, 3.5 cy of cow manure and 500 gallons anaerobic digester sludge. The soil was placed on visqueen in 6-foot wide, 4-foot tall, and 45-foot long pile for an initial volume of approximately 20 cubic yards. At the completion of construction and sampling activities, a layer of 6-mil visqueen covered the TOSS pile to minimize water loss. The building was heated with portable propane heaters to maintain an air temperature above 32°F.

A TNT removal efficiency greater than 75 percent was achieved during the first phase of the anaerobic stage operated at a temperature of 12°C for 113 days. The monoamino-dinitrotoluene transformation products increased during day 0 to 50 as the transformation of TNT occurred. The diamino-mononitrotoluene transformation products increased after day 50 as the monoamino-dinitrotoluene transformation products were further reduced.

After 113 days an aliquot of soil was supplemented with anaerobic digester sludge and cow manure and incubated at 35°C in laboratory reactors. An additional 21 days of treatment resulted in an overall TNT reduction greater than 99 percent for the anaerobic stage of the TOSS process. Stage 2 tests were not conducted since the cleanup goals for TNT were achieved in stage one.

The ambient operating temperatures during the demonstration at PCD had a substantial impact on the rates of contaminant degradation. Operation of the TOSS process at higher temperatures was shown to improve treatment kinetics and shortened treatment times.

CASE HISTORY 3: PHYTOREMEDIATION

IT evaluated the efficacy of phytoremediation for the treatment of TNT in soils from PCD. Plants selected using enzyme assay techniques reduced TNT concentrations in explosives-contaminated soils from 2600 mg/kg to 230 mg/kg within eight weeks. Degradation products including monoamino-dinitrotoluene and diamino-mononitrotoluene were rapidly degraded to simple, nonhazardous compounds. The data revealed a first-order reaction rate with a TNT half-life of 15 days using plants in the early stages of growth. As the plants mature the TNT half-life will decrease due to the production of higher concentrations of the nitroreductase enzymes. Therefore, the time required to achieve soil cleanup goals of less than 10 mg/kg would be approximately eight additional weeks.

CASE HISTORY 4: SLURRY PHASE BIOLOGICAL TREATMENT

IT was authorized by the U.S. Navy under the LANTDIV Remedial Action Contract (RAC) to participate in a field pilot test for remediation of explosives-contaminated soils at Yorktown Naval Weapons Station in Virginia. The Comprehensive Long-term Environmental Action Navy (CLEAN) contractor selected the Simplot Anaerobic Biological Remediation (SABRE) process developed by the J.R. Simplot Company for testing.

The field pilot test consisted of excavating and treating 700 cubic yards of soil. The excavated soil was transported, screened and conveyed to a fluidizer for slurry preparation. The slurry was pumped to a gantry which distributed the material into the double-lined bioremediation cell. The TNT concentrations ranged from 18 to 1600 mg/kg and the RDX concentrations ranged from 16 to 850 mg/kg. The treatment goals of 30 mg/kg for TNT and 100 mg/kg for RDX were achieved within 30 days of operation.

Following successful completion of the field test, IT provided full-scale treatment of 1200 cy of explosives-contaminated soils. Explosive concentrations ranged from non-detect to maximum concentrations of 35,000 mg/kg for TNT; 5,400 mg/kg for RDX; 11,000 mg/kg for HMX and 1.2 mg/kg for the amino DNTs.

The full-scale remediation consisted of enlargement of the biotreatment cell to accommodate 1200 cy of soil. IT's application of the SABRE process for explosive-contaminated soils involved flooding the soils with water, adding amendments and maintaining anaerobic conditions. During the treatment process 1200 cubic yards of soil were covered with approximately 2 ft of water. Initial TNT concentrations did not exceed 1,000 mg/kg. The process operating conditions were: pH target range 6.5-7.0, redox potential <-200 mV, and temperatures above 18°C Slurry samples were collected weekly and analyzed for TNT, DNT and degradation intermediates. The biocell was operated for a 30-day period to achieve treatment goals of 30 mg/kg, 8.0 mg/kg, 50 mg/kg and 3900 mg/kg for TNT, DNT, RDX and HMX respectively.

CASE HISTORY 5: WHITE ROT FUNGI

White-rot fungi (WRF) have the ability to degrade a variety of nitroaromatic compounds in both aqueous phase (e.g. red and pink water) and in soil (Lamar, 1998). EarthFax conducted bench-scale tests on TNT-contaminated soil from Mead Army Ammunition Plant. Initial TNT soil concentrations were >1600 mg/kg. The tests evaluated the ability of several different species of WRF to degrade TNT in soils. All of the species of WRF were able to rapidly degrade TNT however differences were observed. A soil TNT concentration of 10 mg/kg (>99% reduction) was achieved in approximately 40 days with the most robust species. Concentrations of 120 mg/kg RDX in soils inoculated with the WRF were treated to less than 20 mg/kg in 30 days.

EarthFax performed bench-scale tests on tetryl-contaminated soil obtained from the Joliet Army Ammunition Plant (JOAAP). After 4 weeks of treatment with the WRF <u>Trametes versicolor</u> an initial concentration of 12,000 mg/kg tetryl was reduced to 4900 mg/kg. The rate of degradation was similar to the rate observed for TNT with WRF indicating that further treatment would occur with time. EarthFax is also participating in a field demonstration project at JOAAP in which 10 tons of soil containing TNT, RDX and HMX have undergone successful treatment using the WRF <u>P. chrysosporium</u>.

PROCESS ECONOMICS AND COMPARISON

The process costs for these technologies were developed for treatment of RCRA hazardous soil at facilities similar to Pueblo Chemical Depot and include construction of RCRA hazardous waste treatment facility with secondary containment, land application of treated soil and more extensive analytical requirements. The soil volume is 10,000 cubic yards or 14,000 tons.

Treatment of explosives-contaminated soils using the TOSS process stage 1 is approximately $110 cubic yard ($78 ton). If stage 2 is necessary the cost increases to $254 cubic yard ($180 ton) (Table 1). The cost for windrow composting for the RCRA hazardous soil is $312 cubic yard ($223 ton). This is similar to the cost reported for the full-scale composting of explosives-contaminated soils of $256 ton at Umatilla Depot in Oregon (Emery and Faessler, 1997).

The estimated process cost for phytoremediation of explosives-contaminated soils in a 1 acre lined treatment cell is $98 cubic yard (Table 1). Process setup costs include plant selection, seedling growth, and on-site planting. Process operating costs for a six-month treatment season include water management, fertilizer application,

TABLE 1. Process Economics for Treatment of Explosives-Contaminated Soils

Task	Slurry-Phase	TOSS	Windrow Compost	Phyto-remediation	White-Rot Fungi
	($/cy)	($/cy)	($/cy)	($/cy)	($/cy)
Setup/teardown	58	44	44	20	40
Stage 1 Operations	200*	44	182	15	100
Stage 1 Sampling and Analysis	18	18	64	15	
Stage 2 Operations		82		15	--
Stage 2 Sampling and Analysis		42		15	--
Admin/Support	22	14	14	10	10
Treated Material Management	100	8	8	8	8
Total	398	252	312	98	158

* Includes technology transfer fee.

TABLE 2. Process Comparison for Treatment of Explosives-Contaminated Soils

PROCESS	ADVANTAGES	DISADVANTAGES
Windrow Composting	Fast, Proven	Cost, Amendments, Curing
Slurry-Phase Treatment	Proven, Treat high explosive concentrations	Cost, Engineering
Soil Pile (TOSS)	Simple, Off site option	Temperature-dependent
Phytoremediation	Cost, No volume increase	Bench-scale testing only
White-Rot Fungi	Cost, Treat high explosive concentrations	Temperature control

process monitoring, sample collection and analyses. The soil would be treated in 3 foot lifts over two growing seasons.

The process economics for the SABRE technology are based on treating 10,000 cubic yards of soil over a 4 yr period. Two soil batches are treated per season in a lagoon reactor without an external heat source. Included in the stage 1 operations cost are soil screeing and reactor loading and unloading. Also included in this category is a technology transfer fee. The treated material management category includes dewatering in a sand drying bed and on site land application of the dewatered soil. Soil treatment costs of $398 per cubic yard include a proprietary inoculum and nutrient amendment, a technology transfer fees, and the specialized mixing system for the reactor (Table 1).

Treatment of the soils by the WRF process involves construction of four forced-aeration biopiles. Each biopile would be approximately 220'X75'X 8'. Treatment of the explosive constituents in the biopiles is anticipated to occur over a 90-day period. The estimated cost for this process is $148/cy (Table 1).

In summary IT has been involved in the on-site application of biological treatment technologies for the remediation of explosives-contaminated soils at DoD sites. IT understands the costs and benefits of each process based on actual field experience (Table 2).

ACKNOWLEDGEMENTS

The authors express their appreciation to Waste Management Inc., Stan Wharry of PCD and Tim Green of IT for their support during the PCD project.

REFERENCES

Emery, D. and D. Faessler, 1997, "First Production Level Bioremediation of Explosives-Contaminated Soils in the United States, " *Bioremediaiton of Surface and Subsurface Contamination*, 829:326-340.

Green, R. B., P. E. Flathman, G. R. Hater, D. E. Jerger, and P. M. Woodhull, 1998, "Sequential Anaerobic-Aerobic Treatment of TNT-Contaminated Soils," *Designing and Applying Treatment technologies: Remediation of Chlorinated and Recalcitrant Compounds*, G. B. Wickramanayake and R. Hinchee (eds.)Battelle Press, Columbus, pp 271-276.

Lamar, R., 1998, "Fungal-Based Bioremediation of Explosives-Contaminated Soils," Earth Fax Development Corporation, Logan, Utah.

PHYTOTOXICITY REDUCTION BY ANAEROBIC-AEROBIC TREATMENT OF EXPLOSIVES-CONTAMINATED SOIL

Paul W. Barnes (Waste Management Inc., Cincinnati, Ohio)
Roger B. Green and Gary R. Hater (Waste Management Inc., Cincinnati, Ohio)

ABSTRACT: Early seedling growth studies were conducted to evaluate residual toxicity to plants following the treatment of explosives-contaminated soil by a proprietary two stage anaerobic-aerobic treatment process. Five species of plants were tested over the range of 0% to 95% of test soils taken prior to and following each stage of treatment. Parameters which indicate toxicity including; time to seedling emergence, germination rate, dry biomass yield, and seedling length were measured for each combination of species, test soil, and dilution. Differences between treatments were evaluated by one-way analysis of variance. While little difference in time to seedling emergence and germination rate were observed between treatments, significant differences in biomass yield and seedling length were evident with treated soil allowing substantially better growth of all species than untreated soil.

INTRODUCTION

Sites at which explosives contamination is typically found are often well suited to restoration as recreation areas, protected habitats or other vegetated natural areas. Since revegetation of these sites is often required after remediation it is important that the ability for treated soil to support plant growth be demonstrated. Reduced plant growth capacity may result not only from the presence of residual contaminants but also from treatment process amendments or altered physical properties such as density or permeability.

30 day early seedling growth studies were used to evaluate the suitability of soil treated by Waste Management's two-stage anaerobic-aerobic treatment process (TOSS process) for revegetation (Green et. al., 1998). Contaminated test soil used for the study was obtained from Joliet Army Ammunition Plant (JAAP) and treated at a Waste Management facility in Louisville, Kentucky.

Objective. The objective of this study was to evaluate the toxicity to common grassland and crop plants resulting from explosives residuals in soil treated through each of the TOSS process phases. The goal was to demonstrate that treated soil would support the growth of native plant species and to quantify the quality of that growth relative to growth in both untreated contaminated soil and clean commercial potting soil. A secondary objective was to evaluate any toxicity which might result from process amendments alone.

MATERIALS AND METHODS

The early seedling growth test procedure was drawn in large part from the methods described by the American Society for Testing and Materials (ASTM) in their document E 1598-94, *Standard Practice for Conducting Early Seedling Growth Tests* (ASTM, 1994). The study consisted of a full-factorial evaluation of all combinations of five test plant species, four soil types, and five dilutions. Test soil was taken from various stages of a pilot demonstration of the T.O.S.S. process conducted by Waste Management at a permitted soil treatment facility in Louisville, Kentucky. Contaminated soil used in the pilot study was obtained from Joliet Army Ammunition Plant in Joliet, Illinois and contained 2,4,6-trinitrotoluene (TNT), octahydro-1,3,5,7-tetranitro-1,3,5,7-tetrazocine (HMX) and hexahydro-1,3,5-trinitro-1,3,5-triazine (RDX) contaminants.

Test Soil Preparation. Four soil types were selected for testing; untreated contaminated soil, soil which had completed phase I anaerobic treatment, soil which had completed phase II aerobic treatment, and clean soil amended with phase I process amendments and treated through phase I. Each soil type was collected from the pilot study at the appropriate time, dried and stored in sealed containers until the toxicity study was begun. In preparation for testing each soil aliquot was homogenized in a 4 cubic foot portable cement mixer and sampled for contaminant analysis. Mean contaminant concentrations with accompanying 95% confidence intervals for triplicate samples are presented in Table 1. The homogenized soil batches were then serially diluted with clean commercial potting soil to proportions of 0%, 25%, 50%, 75% and 100% test soil.

TABLE 1. Contaminant concentrations in undiluted test soil.

Test Soil	TNT(mg/kg)	RDX (mg/kg)	HMX (mg/kg)
Untreated	1333 ± 457	130 ± 23	521 ± 431
Post phase I treatment	38 ± 21	77 ± 4	80 ± 16
Post phase II treatment	$292 \pm 566^{\dagger}$	25 ± 35	19 ± 23
Amended clean soil	ND[††]	ND	ND

[†] mean of triplicate samples with values 3.6, 2.3, and 870 mg/kg
[††] contaminant not detected

Experimental Procedure. Prepared soil was placed in labeled and randomly numbered four-inch diameter by six-inch high plastic pots and watered to slightly less than saturation. Five seeds of the appropriate species were planted in each pot at depths approximately equivalent to twice the seed diameter, evenly spaced around the perimeter. The plant species tested were selected for their germination properties, appropriateness to the final use of the test site, and previous use in accepted toxicity assessments. The species tested (Table 2) consist of three dicotyledons and two monocotyledons.

TABLE 2. Early seedling growth study test plants

Plant Type	Species	Common Name
Dicotyledon	*Desmodium canadense*	Showy Tick-Trefoil
	Trifolium hybridum	Alsike Clover
	Lactuca sativa	Buttercrunch Lettuce
Monocotyledon	*Lolium perenne*	Perennial Pleasure Ryegrass
	Panicum virgatum	Blackwell Switchgrass

Triplicate pots were prepared for each combination of plant species, soil type, and dilution. Prepared pots were placed, by number, in randomized locations under wide spectrum fluorescent grow lamps producing approximately 900 foot-candles at the soil surface 14 hours per day. During the 30 day study, ambient conditions were maintained at 50% relative humidity and $25 \pm 5°C$. Pots were turned 90 degrees in place daily and watered as needed to maintain uniform moisture. After approximately five days it became apparent that soil pots containing 100% test soil were drying quickly at the surface and were unable to accept sufficient water due to high clay content. Soil was removed from these pots, diluted to 95% with commercial potting soil to improve permeability, and replanted.

Measurements. The date of emergence and number of seeds germinating in each pot was recorded daily. After 30 days of growth, each plant was measured for height above the soil surface and harvested by severing at the soil surface. The harvested seedlings were dried at $70 \pm 5° C$ for 48 hours and weighed to determine biomass yield.

RESULTS AND DISCUSSION

Biomass Yield and Seedling Length. While some differences between treatments were observed in mean germination rate and time to seedling emergence, particularly at higher proportions of test soil, more pronounced differences were evident in the growth of the emergent seedling. After 30 days of growth there were marked differences in both biomass yield and length between seedlings grown in any given proportion of contaminated soil and a corresponding proportion of either Phase I or Phase II treated soil. Without exception these differences suggest that plants of all tested species achieved greater length and biomass in the treated soil. In most cases it also appears that Phase II treatment results in increased biomass yield and/or seedling length over plants grown in soil treated only through Phase I.

Figures 1 through 4 depict mean biomass yield and length for the four species germinating in sufficient numbers to allow statistical analysis; *T. hybridum, D. canadense, L. sativa, and L. perenne*, respectively. The fifth tested species (*P. virgatum*) proved to germinate too slowly to provide enough viable seedlings within the time allotted and has therefore been excluded from this discussion. Biomass yield is presented as biomass per seed which accounts for non-germinating plants by assigning them zero mass.

Analysis of Variance. One-way analyses of variance (ANOVA) were used to compare the observed differences between treatments to differences within sets of treatment replicates. An F statistic was computed to compare each treatment to growth in untreated soil. F_{crit} values representing the F statistic predicted at 95% confidence for any two data sets with the same numbers of degrees of freedom were also computed and used to produce the F/F_{crit} ratio. When greater than 1, the

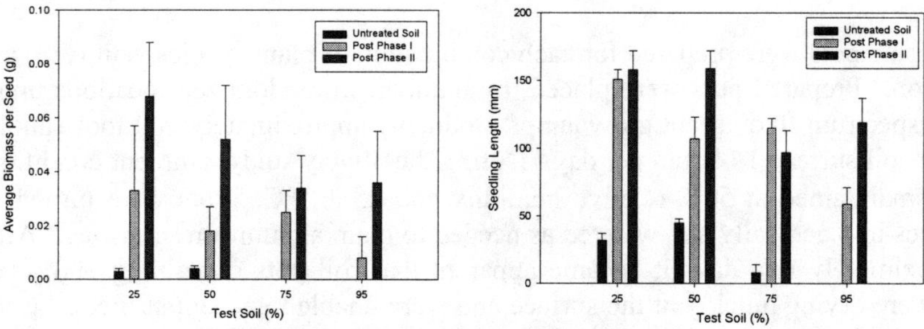

FIGURE 1: Biomass yield and length at 30 days,
T. hybridum **(alsike clover).**

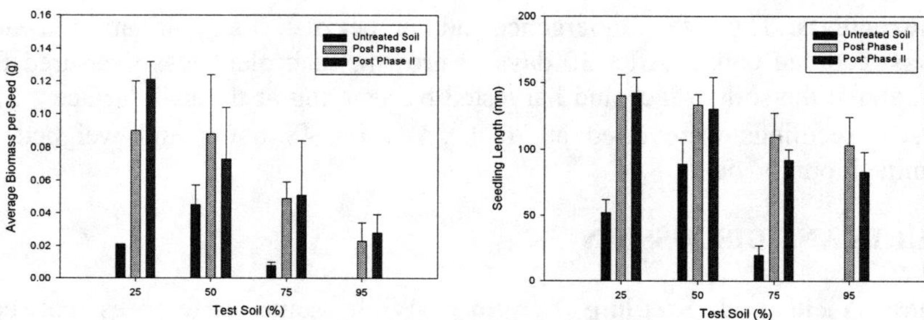

FIGURE 2: Biomass yield and length at 30 days,
D. canadense **(showy tick-trefoil).**

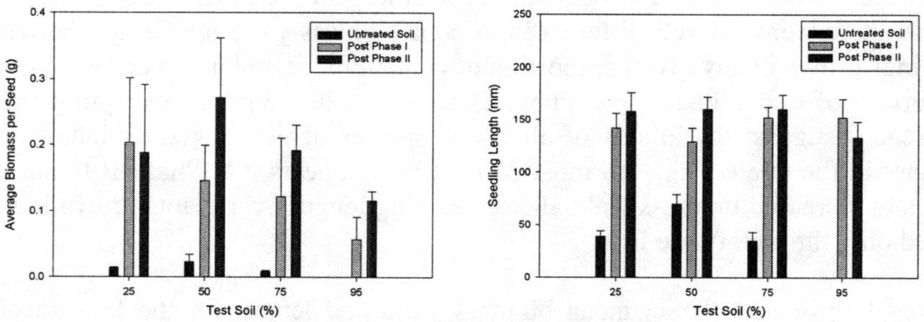

Figure 3: Biomass yield and length at 30 days,
L. sativa **(buttercrunch lettuce).**

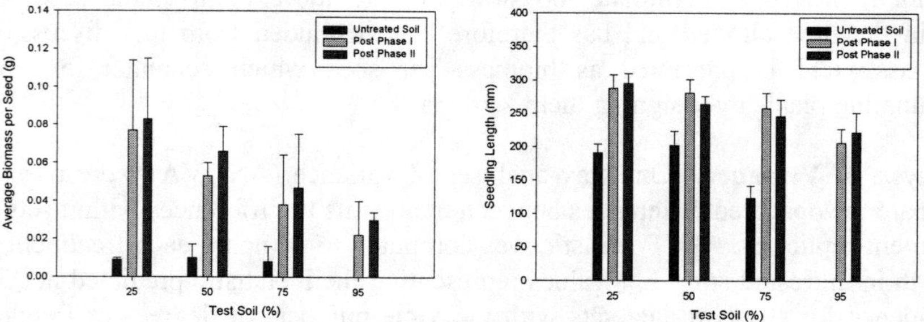

FIGURE 4: Biomass yield and length at 30 days,
L. perenne **(perennial ryegrass).**

F/F$_{crit}$ ratio indicates that the mean values being compared are not likely to be the same at the chosen level of confidence, 95% in this case.

For this discussion, increasing F/F$_{crit}$ values greater than 1 are taken to indicate increasing certainty (above 95%) that that observed differences between treatments are real and cannot be attributed to natural variability. This comparison could not be made for most of the 95% test soil groups because seeds in that proportion of untreated soil failed to germinate in sufficient numbers. Similar comparisons to the commercial potting soil controls also could not be made due to attenuated growth which has been attributed to a verified deficiency in necessary inorganic nutrients. ANOVA results (Tables 3-6) indicate that with few exceptions the differences between treatments far exceeded differences within treatment replicate sets.

CONCLUSIONS

The greater biomass yields and seedling lengths achieved in treated soil indicates that treatment by the T.O.S.S. process results in substantial reduction in phytotoxicity from untreated soil. Additionally, while the majority of contaminant transformation is known to occur in the first (anaerobic) phase of treatment, further reduction in phytotoxicity does occur during the second (aerobic) phase of treatment.

TABLE 3. *T. hybridum*: ANOVA for biomass yield and seedling length.

Test Soil	Untreated (mean)	Post Phase I (mean)	Untreated vs. Phase I (F/F$_{crit}$)	Post Phase II (mean)	Untreated vs. Phase II (F/F$_{crit}$)
Per Seed Biomass Yield					
25%	0.003 g	0.033 g	0.8	0.069 g	5.3
50%	0.004 g	0.018 g	1.2	0.052 g	2.8
75%	0.003 g	0.025 g	4.3	0.034 g	3.9
95%	0.000 g	0.008 g	NA	0.036 g	NA
Seedling Length					
25%	32 mm	151 mm	191.0	158 mm	56.5
50%	45 mm	107 mm	10.1	159 mm	48.8
75%	8 mm	115 mm	76.5	97 mm	15.8
95%		59 mm	NA	119 mm	NA

TABLE 4. *D. canadense*: ANOVA for biomass yield and seedling length.

Test Soil	Untreated (mean)	Post Phase I (mean)	Untreated vs. Phase I (F/F$_{crit}$)	Post Phase II (mean)	Untreated vs. Phase II (F/F$_{crit}$)
Per Seed Biomass Yield					
25%	0.018 g	0.079 g	1.7	0.081 g	1.6
50%	0.043 g	0.064 g	0.1	0.059 g	0.1
75%	0.005 g	0.036 g	3.9	0.048 g	0.9
95%	0.006 g	0.021 g	NA	0.023 g	NA
Seedling Length					
25%	52 mm	140 mm	20.0	142 mm	31.3
50%	89 mm	133 mm	3.4	130 mm	1.7
75%	20 mm	109 mm	15.3	92 mm	41.3
95%	3 mm	103 mm	NA	83 mm	NA

TABLE 5. *L. sativa*: ANOVA for biomass yield and seedling length.

Test Soil	Untreated (mean)	Post Phase I (mean)	Untreated vs. Phase I (F/F_{crit})	Post Phase II (mean)	Untreated vs. Phase II (F/F_{crit})
Per Seed Biomass Yield					
25%	0.013 g	0.184 g	3.2	0.188 g	1.4
50%	0.022 g	0.123 g	4.3	0.223 g	93.0
75%	0.005 g	0.103 g	1.1	0.165 g	13.4
95%	0.000 g	0.049 g	NA	0.108 g	NA
Seedling Length					
25%	39 mm	142 mm	44.5	148 mm	34.2
50%	70 mm	129 mm	13.0	160 mm	31.2
75%	35 mm	152 mm	59.1	160 mm	38.9
95%	12 mm	152 mm	19.7	133 mm	15.5

TABLE 6. *L. perenne*: ANOVA for biomass yield and seedling length.

Test Soil	Untreated (mean)	Post Phase I (mean)	Untreated vs. Phase I (F/F_{crit})	Post Phase II (mean)	Untreated vs. Phase II (F/F_{crit})
Per Seed Biomass Yield					
25%	0.009 g	0.052 g	1.4	0.075 g	7.9
50%	0.010 g	0.050 g	3.9	0.066 g	7.4
75%	0.007 g	0.036 g	0.5	0.047 g	1.0
95%	0.000 g	0.022 g	NA	0.025 g	NA
Seedling Length					
25%	191 mm	288 mm	16.4	295 mm	25.5
50%	202 mm	281 mm	6.8	264 mm	6.0
75%	124 mm	258 mm	18.7	247 mm	13.7
95%	33 mm	206 mm	15.9	222 mm	11.0

REFERENCES

American Society for Testing and Materials, *Standard Practice for Conducting Early Seedling Growth Tests,* Designation: E 1598-94, 1994.

Green, R.B., P.E. Flathman, G.R. Hater, D.E. Jerger, and. P.M. Woodhull. 1998. Sequential Anaerobic-Aerobic Treatment of TNT-Contaminated Soil. In: G.B. Wickramanayake and R.E Hinchee, (eds), *Designing and Applying Treatment Technologies: Remediation of Chlorinated and Recalcitrant Compounds.* Battelle Press, Columbus, Ohio. pp.271-276.

Oak Ridge National Laboratory, *Characterization of Explosives Processing Waste Decomposition Due to Composting*, Project Report, Army Project Order No. 89PP9921, 1994.

PILOT DEMONSTRATION OF THE SEQUENTIAL ANAEROBIC-AEROBIC BIOREMEDIATION OF EXPLOSIVES-CONTAMINATED SOIL

Roger B. Green (Waste Management, Inc. Cincinnati, Ohio)
Paul W. Barnes and Gary R. Hater (Waste Management, Inc., Cincinnati, Ohio)

ABSTRACT: A pilot-scale demonstration of a proprietary two stage anaerobic-aerobic treatment process for the bioremediation of explosives contaminated soil termed the TOSS process is described. Approximately 120 tons of soil contaminated with 2,4,6-trinitrotoluene (TNT), hexahydro-1,3,5-trinitro-1,3,5-triazine (RDX), and octahydro-1,3,5,7-tetranitro-1,3,5,7-tetrazocine (HMX) from the Joliet Army Ammunition Plant (JAAP) was used in the demonstration.

The demonstration was performed at the Waste Management Outer Loop Recycling and Disposal Facility in Louisville, Kentucky. This facility has been granted a permit for the bioremediation of explosives contaminated soil. Major activities performed during the demonstration included: construction and operation of a large scale treatment cell (102 tons), construction and operation of a portable treatment unit (12 tons), and the performance of a phytotoxicity test. This paper discusses results of tests with the large scale treatment cell and the portable treatment unit. Results from these tests showed significant reductions in the soil concentrations of TNT (99.9%), RDX (99.5%), and HMX (98.1%) after sequential anaerobic-aerobic treatment.

INTRODUCTION

The TOSS process is designed to degrade and immobilize explosives compounds through a sequence of reducing (anaerobic) and oxidizing (aerobic) treatment stages (Green et al., 1998). In the first stage of treatment, explosives contaminated soil is combined with an electron donating substrate (starch and sucrose), a microbial inoculum (cow manure and/or anaerobic digester sludge), and water to achieve anaerobic conditions. The resulting mixture is formed into a static pile or placed into a treatment cell and maintained under anaerobic conditions. This first stage of treatment facilitates the chemical reduction of nitroaromatic and nitramine explosives in a solid-phase system. Nitroaromatics are reduced to corresponding amines in multistep process involving nitroso (N=O) and hydroxylamine (HNOH) intermediates (Walsh, 1990). The complete reduction of TNT to triaminotoluene (TAT) via mono- and diaminonitrotoluenes has been reported under anaerobic conditions with redox potentials less than −200 mV (Lenke et al., 1994).

The second stage consists of combining the anaerobically treated soil with composted yard waste and constructing the mixture into a biopile. The biopile is aerated, either through forced air conveyed by perforated piping buried within the pile or, by turning the pile with a compost turner. This approach of combining the anaerobically treated soil with composted yard waste and aerating the mixture is based on the following considerations. Aromatic amines, the reduction products that are formed during the anaerobic stage of treatment, are known to undergo rapid polymerization on exposure to oxygen (Rieger and Knackmuss, 1995). These reduction products are believed to form covalent bonds with a variety of functional groups present in soil organic matter (Major et al., 1997). Recent research suggests that the binding of aromatic amine metabolites to soil humic material is mediated by soil microorganisms (Held et al., 1997). The aerobic stage of treatment exploits these processes by providing a source of aeration, humic-rich organic matter in the form of composted yard waste, and a matrix containing the microorganisms responsible for humification.

MATERIALS AND METHODS

Soil Preparation.   Approximately 120 tons (83 cubic yards) of explosives contaminated soil was transported from the JAAP to the treatment site. This soil was homogenized prior to the start of the demonstration. Homogenization was accomplished by uniformly distributing the individual loads of soil within a windrow using a front-end loader and then making four passes through the windrow with a Scarab® Model 18 windrow turner. After this homogenization step, approximately 12 cubic yards (yd^3) was returned to the stockpile area and reserved for use in: the portable container test (8 yd^3), in a control (3 yd^3), and in amendment optimization and phytotoxicity tests (1 yd^3). The remaining 71 yd^3 of mixed soil was used for the large-scale TOSS demonstration.

Large Scale TOSS Test.   Seventy-one yd^3 of explosives contaminated soil was combined with an inoculum consisting of 15 yd^3 of cow manure and 15 yd^3 of anaerobic digester sludge and a carbon source consisting of 1 ton of macerated potatoes and 1 ton of dextrose. The amendments were blended into the soil by making four passes through the windrow with the windrow turner.

After blending, the soil-amendment mixture was placed into a recompacted clay-lined test cell with a volume of approximately 100 yd^3. Water was added to the mixture during placement to approximately 100% saturation. Soil samples were collected from each 5 yd^3 of material as it was placed in the test cell. After all of the soil was transferred to the test cell, the surface was covered with a spray-on material consisting of recycled paper fiber and polymers. Except for periodic soil sampling, the test cell was left undisturbed during the anaerobic stage of treatment, which lasted for a period of 245 days.

After the anaerobic stage of treatment, the soil in test cell was excavated and combined with approximately 100 yd^3 of composted yard waste to begin the aerobic stage. The anaerobically treated soil was placed in a windrow, composted yard waste was added to the top of the windrow, and the materials were blended together with the windrow turner. The blended material, which was left uncovered, was aerated three times weekly with the windrow turner.

The yard waste compost was produced from a starting mixture of leaves, grass, and wood wastes collected from the Louisville, Kentucky area. Prior to use the yard waste had been composted for three months and contained, on a weight basis, 15.4% total humus, 10.6% humic acid, and 4.8% fulvic acid. The compost was determined to be moderately to well- matured based on testing with the Solvita™ compost maturity test (Woods End Research Laboratory, Inc.)

Portable Container TOSS Test. Eight yd^3 of soil were combined with 1.8 yd^3 of cow manure, 1.8 yd^3 of anaerobic digester sludge, 240 pounds of dextrose, and 240 pounds of macerated potatoes in a water-tight 20 yd^3 steel roll-off waste container. The soil and amendments were mixed together with the bucket of an excavator. Water was added to saturation and uniformly mixed into the soil using the bucket of the excavator. The top surface of the mixture was covered with the spray-on cover previously mentioned. Except for periodic soil sampling, the roll-off container was left undisturbed during the anaerobic stage of treatment, which lasted for a period of 166 days.

The anaerobically treated soil was removed from the roll-off container and combined with 11 yd^3 of composted yard waste using the windrow turner. The resulting material was returned to the roll-off container. This material was periodically aerated by mixing with the bucket of an excavator.

Soil Sampling. Soil samples were collected at a frequency of one sample per 5 yd^3 of material from the anaerobic TOSS cell. Discrete samples were collected from 20 equally spaced locations using a stainless steel hand auger. Soil sampling events were conducted at 0, 14, 28, 47, 68, 81, 95, 116, 138, 165, 180, 202, and 245 days after the start of anaerobic treatment.

Soil samples were collected from the aerobic TOSS windrow at a frequency of one sample per 6 cubic yards of material. Discrete samples were collected from 22 equally spaced locations (11 on each side of the windrow) using a hand trowel. Soil sampling events were conducted at 0 and 23 days after the start of aerobic treatment (246 and 269 days after the start of anaerobic treatment).

Samples of the material in the roll-off container were collected from four equally spaced locations along the long axis of the container. Soil samples were collected at a depth of two to three feet below the top surface of the material with clean shovel. Soil sampling events were conducted at 0, 14, 68, 95, 116, 137, and 166 days after the start of anaerobic treatment, and 0, 14, and 80 days after the start of aerobic treatment (or 166, 180, and 246 days after the start of anaerobic treatment).

Explosives Analysis. Laboratory testing of soil samples for explosives was performed using USEPA SW-846 Method 8330 at a U.S. Army Corps of Engineers certified laboratory. Due to the relatively large amount of organic amendment contained in the samples, samples were milled in a Wiley mill prior to extraction to produce a homogeneous product. In addition, to improve representativeness, the method was modified so that 20 grams of sample was extracted, rather than 2 grams as specified in the method. The transient reduction products of TNT, 2,4-diamino-6-nitrotoluene and 2,6-diamino-4-nitrotoluene, were also quantified in addition to the 14 target analytes normally quantified by Method 8330.

RESULTS

Large Scale TOSS Test. The results of explosives analysis through 269 days of anaerobic-aerobic treatment are summarized in Figure 1. The mean concentration and the 95% confidence interval are presented for the sampling events conducted at the start of the anaerobic stage, completion of the anaerobic stage, and the completion of the aerobic stage.

The removal of TNT and its reduction products; 2-amino-4,6-dinitrotoluene and 4-amino-2,6-dinitrotoluene (collectively aminodinitrotoluenes or ADNTs), and 2,4-diamino-6-nitrotoluene and 2,6-diamino-4-nitrotoluene (collectively diaminonitrotoluenes or DANTs), was essentially complete before terminating the anaerobic stage at 35 weeks of treatment. The anaerobic treatment period was extended to further evaluate the performance of the anaerobic stage for degradation of RDX and HMX.

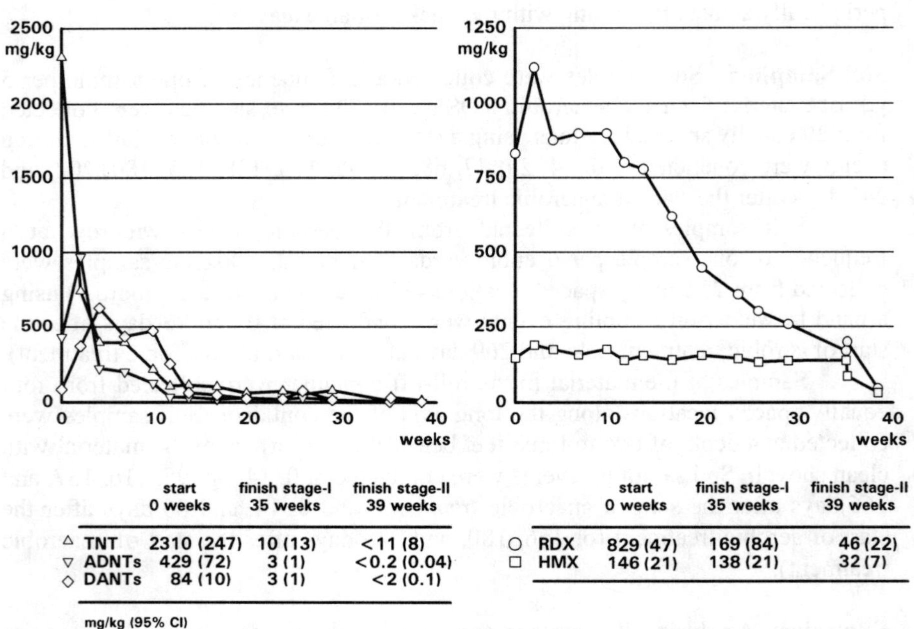

	start 0 weeks	finish stage-I 35 weeks	finish stage-II 39 weeks
△ TNT	2310 (247)	10 (13)	<11 (8)
▽ ADNTs	429 (72)	3 (1)	<0.2 (0.04)
◇ DANTs	84 (10)	3 (1)	<2 (0.1)

	start 0 weeks	finish stage-I 35 weeks	finish stage-II 39 weeks
○ RDX	829 (47)	169 (84)	46 (22)
□ HMX	146 (21)	138 (21)	32 (7)

mg/kg (95% CI)

FIGURE 1: Explosives contaminant results for the large-scale TOSS test.

Portable Container TOSS Test - The results of explosives analysis through 246 days of anaerobic-aerobic treatment are summarized in Figure 2 for the portable container test. The mean concentration and the 95% confidence interval are presented for the sampling events conducted at the completion of the anaerobic stage, and the completion of the aerobic stage. Concentration results for the initial sampling event represent values for composite samples.

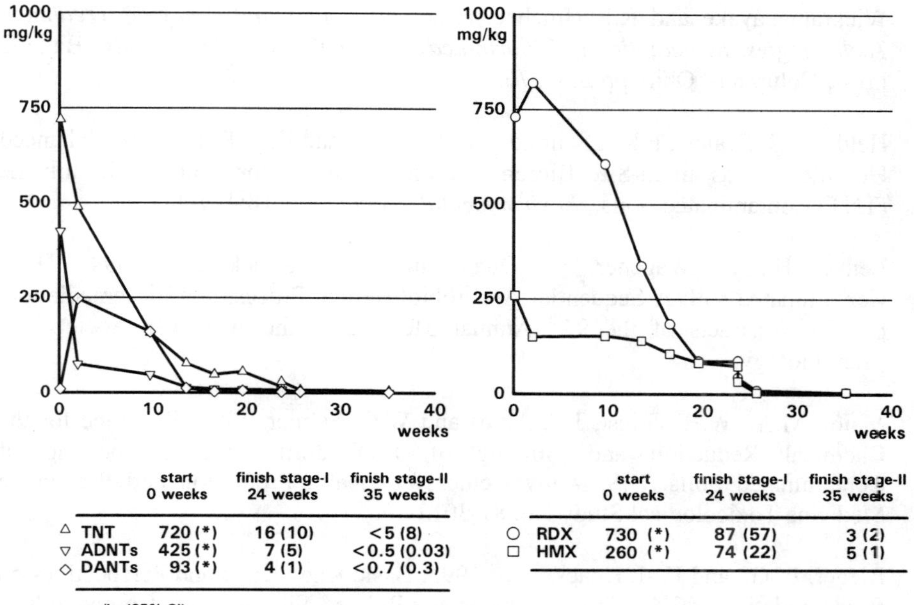

		start 0 weeks	finish stage-I 24 weeks	finish stage-II 35 weeks
△	TNT	720 (*)	16 (10)	<5 (8)
▽	ADNTs	425 (*)	7 (5)	<0.5 (0.03)
◇	DANTs	93 (*)	4 (1)	<0.7 (0.3)

		start 0 weeks	finish stage-I 24 weeks	finish stage-II 35 weeks
○	RDX	730 (*)	87 (57)	3 (2)
□	HMX	260 (*)	74 (22)	5 (1)

mg/kg (95% CI)
(*) composite sample

FIGURE 2: Explosives contaminant concentrations for the portable container TOSS test.

CONCLUSIONS

The data generated from the pilot demonstration provide for the following conclusions:

- The TOSS process was able to degrade TNT, RDX, and HMX to levels which would allow the beneficial reuse of the remediated soil as landfill cover as specified in the explosives bioremediation permit for the Outer Loop Recycling and Disposal Facility;
- The TOSS process was effective when performed in a portable treatment container;
- The combined concentration of TNT and amino intermediates was reduced by >99.5%;
- RDX was not appreciably degraded until the combined concentration of TNT and its reduction intermediates was less than the concentration of RDX;
- Explosives degradation rates for the portable container were significantly higher than rates for the treatment cell, this is most likely explained by a higher average soil temperature in the metal roll-off container versus the treatment cell.

REFERENCES

Green, R.B., P.E. Flathman, G.R. Hater, D.E. Jerger, and. P.M. Woodhull. 1998. Sequential Anaerobic-Aerobic Treatment of TNT-Contaminated Soil. In: G.B.

Wickramanayake and R.E Hinchee, (eds), *Designing and Applying Treatment Technologies: Remediation of Chlorinated and Recalcitrant Compounds*. Battelle Press, Columbus, Ohio. pp.271-276.

Held, T., G. Draude, F.R.J. Schmidt, A. Brokamp, and K.H. Reis. 1997. Enhanced Humification as an In-Situ Bioremediation Technique for 2,4,6-Trinitrotoluene (TNT) Contaminated Soils. *Environmental Technology*, 18:479-487.

Lenke, H., B. Wagener, G. Duan, and H.-J. Knackmuss. 1994. TNT-Contaminated soil: A Sequential Anaerobic/Aerobic Process for Bioremediation. p. 456. Abstracts of the 94th Annual Meeting of the American Society for Microbiology.

Major, M.A., W.H. Griest, J.C. Amos and W.G. Palmer. 1997. Evidence for the Chemical Reduction and Binding of TNT during the Composting of Contaminated Soils. U.S. Army Center for Health Promotion and Preventive Medicine Toxicological Study No. 87-3012-95.

Rieger, P.-G., and H.-J. Knackmuss. 1995. Basic Knowledge and Perspectives on Biodegradation of 2,4,6-Trinitrotoluene and Related Nitroaromatic Compounds in Contaminated Soil. In: J.C. Spain (ed.). *Biodegradation of Nitroaromatic Compounds*. Plenum Press, New York. pp. 1-17.

Walsh, M.E. 1990. Environmental Transformation Products of Nitroaromatics and Nitramines. U.S. Army Corps of Engineers Cold Regions Research and Engineering Laboratory, Special Report 90-2.

COMPOSTING TREATMENT OF EXPLOSIVES-CONTAMINATED SOIL, TOOELE, UTAH

Rosa E. Gwinn, Ph.D., Dames & Moore, 7101 Wisconsin Avenue, Bethesda, MD 20814-4870

ABSTRACT: Dames & Moore conducted a composting treatability study for the treatment of explosives-contaminated soil at Tooele Army Depot (TEAD) in Tooele, Utah (Dames & Moore, 1998). Soil from trinitrotoluene (TNT) washout ponds was mixed with locally available food waste to determine the applicability of composting as a treatment technology. The study included a bench-scale test in which contaminated soil at 20 and 30 percent by volume was mixed with lettuce, barley, onions, and a lettuce and barley mixture. Alfalfa, cow and chicken manure, and wood chips were used in every mixture. The lettuce and barley mixture performed the best, maintaining temperatures conducive to biological degradation of explosives (i.e., 55 to 65°C) for the greatest length of time. Subsequently, contaminated soil was subjected to a pilot-scale test using the lettuce and barley mixture at 20 and 30 percent soil loading rates, again with alfalfa, cow and chicken manure, and wood chips. Thermophilic conditions were maintained for 40 days, with daily monitoring and adjustment of moisture and oxygen. Unique to this treatability study, molasses mixed with water was used in both the bench-scale and pilot-scale tests to add moisture and provide readily available nutrients. The explosives concentrations were reduced to below detection limits (and below cleanup goals) within 10 days.

INTRODUCTION

Biological activity in composting results in the reductive denitrification and mineralization of nitroaromatic compounds, such as trinitrotoluene (TNT) and dinitrotoluene (DNT), and nitramines, such as cyclotrimethylenetrinitramine (RDX) and cyclotetramethylenetetranitramine (HMX). In composting, contaminated soil is mixed with food waste, manure, and bulking agents to enhance the biodegradation of explosives by indigenous microorganisms. The degradation of explosives by native microorganisms proceeds under aerobic, thermophilic conditions (55 to 65°C). To achieve and maintain thermophilic conditions, environmental parameters within the compost must be managed to optimize oxygen availability, moisture, and nutrients. Innovations in amendment mixtures are drawn from Dames & Moore's successful completion of a soil composting treatability study for TNT- and RDX-contaminated soil at Tooele Army Depot (TEAD), Utah.

TEAD is located approximately 40 miles southwest of Salt Lake City in the Great Salt Lake Basin. The Depot has operated as a major munitions storage facility since 1942; it also provided vehicle and equipment maintenance for other installations until 1995. TEAD was placed on the National Priorities List (i.e., designated a Superfund site) in October 1990. In January 1991, TEAD was issued a Resource Conservation and Recovery Act (RCRA) Post Closure Permit by Utah Department of Environmental Quality, which regulates cleanup activities at a number of sites within the Depot. One of these is the TNT Washout Facility.

The TNT Washout Facility was used to decommission explosive munitions. Rinsewater from steam cleaning of munitions casings flowed through a series of settling tanks and filters, and was routed outside into settling tanks. From the settling tanks, the water flowed to several unlined washout ponds. This practice ended in 1986, but the rinsewater left remnant explosives contamination in the soil at the former washout ponds.

Contamination in soil is laterally confined within the area of the eight washout ponds. Concentrations are above risk-based cleanup goals, and require remediation. The maximum concentration of explosives in micrograms per gram ($\mu g/g$) is presented in Table 1.

TABLE 1. Maximum concentration of selected explosives in soil

Explosive Compound	Maximum Concentration ($\mu g/g$)
2,4,6-TNT	2,500
2,4-DNT	5.87
RDX	1,100
HMX	257

Explosives concentrations vary among the ponds, and are heterogeneously distributed within the soil at each pond.

Composting is one of the corrective measures under consideration for cleaning up the washout ponds. Composting control parameters are temperature, oxygen content, moisture content, pH, concentration of organic constituents, and concentration of inorganic constituents (e.g., nitrogen and phosphorus). Typically, composting is successful under aerobic, thermophilic (high-temperature) conditions (55 to 65°C). Mixing and aeration can control temperature and oxygen content. Moisture is controlled through the application of water, which is consumed in the biological activity of the microorganisms. The composition of the selected amendments largely controls the remaining parameters.

The study at TEAD included a bench-scale test of available amendments, and a pilot-scale test of the selected amendment mixtures (Dames & Moore, 1998). The amendments are biodegradable organic matter that act as a food source. These amendments are mixed with manure - which contains the microorganisms, wood chips for bulking, alfalfa for bulking and food, and the contaminated soil. The bulking agents increase the surface area available for microbial growth and enhance airflow into the composting pile or windrow.

BENCH-SCALE TEST

Locally available amendments that had not been tested in previous applications at other sites were evaluated in the bench-scale test. Previous soil composting efforts at such sites as Umatilla Depot Activity in Oregon relied on potato waste as a primary food source, with alfalfa and wood chips added for bulking (Roy F. Weston, 1993). However, because potato waste was not available near TEAD at the time of the treatability study, onions, lettuce, and barley were tested at the bench scale. Although molasses is too fluid to use as a composting substrate, its value as an additive in composting was tested. The local supplies of manure, alfalfa, and wood chips were not subject to testing because previous studies indicate that these do not vary appreciably with geographic location (Roy F. Weston, 1993)

The amendments tested were lettuce, onions, barley, and molasses. Amendments were first analyzed for total organic carbon, total Kjeldahl nitrogen, oxidation/reduction potential (ORP), CO_2 production, and percent solids. A target ratio of carbon to nitrogen in the amendment mixture is 20:1 (Weston, 1993), and amendments-mixed with manure-are all in acceptable range. ORP was positive for all amendments, signifying oxidizing conditions. Relatively higher CO_2 production signifies biological activity; it was highest in the molasses. The pH of all amendments fell in an acceptable range between 4.98 and 6.22, i.e., near neutral. The greatest difference occurred in the percent solids. Barley has very little moisture (90% solid), whereas molasses is almost entirely liquid.

Test beds of approximately 1.5 cubic feet (0.04 cubic meters) were constructed. Three different amendment mixtures at 20 and 30 percent soil loading were evaluated, and duplicate beds were created for each mixture. Twelve test beds and a negative control bed containing only contaminated soil were created. All mixtures contained cow and chicken manure, wood chips, and alfalfa. The amendments for the three mixtures were lettuce, onions, or equal amounts of lettuce and barley. Mixtures were monitored for temperature and moisture daily. If the bed was too warm or too wet, it was mixed to aerate and allow evaporation of water. If the bed was too dry, water was added. If the bed was too cool, water and molasses were added to stimulate microbial activity.

All bed mixtures reached a minimum of 55°C, but only one—the lettuce and barley mix—maintained the temperatures necessary for biodegradation for the entire 24-day test. From the bench-scale test, the resulting optimum amendment mixture was novel—lettuce and barley, with later additions of molasses. Molasses increased moisture and provided a readily available nutrient source. The mixture appeared to provide energy to microorganisms in stages: lettuce was rapidly consumed (10 days), barley provided nutrition slowly and evenly, and molasses acted as a thermal booster. In the small-volume beds established for the bench-scale test, the ability of molasses to boost temperatures was obvious from daily temperature measurements. The application proved so beneficial that molasses was incorporated into the design of the pilot-scale test. The bed with only 20 percent soil loading maintained consistently higher temperatures, but even at 30 percent loading thermophilic conditions prevailed.

PILOT-SCALE TEST

The pilot-scale test treated 6.5 cubic yards (5 cubic meters) of soil in two windrows of approximately 12.5 cubic yards (11.3 cubic meters) each. The optimal mixture of lettuce and barley was used. One windrow had 20 percent soil loading, the other had 30 percent. Windrows were constructed by placing wood chips and alfalfa on the asphalt work pad; these were moistened with water and mixed. Barley, lettuce, chicken manure, and cow manure were added and mixed with the windrow turner. Soil was added and sprayed with water to reduce dust generation during two complete mixings with the windrow turner.

Windrows were monitored for temperature, percent moisture, pH, and oxygen levels over the 40-day test. Temperature was measured daily, while the other parameters were monitored twice a week. Oxygen was consumed rapidly, so the windrows were turned daily for the first 15 days and every other day thereafter. The biological activity consumed water, therefore, water was added daily to maintain the moisture content of the compost. Initially, approximately 35 gallons (132 liters) were added. The volume was reduced to 20 gallons (78 liters) on days when the windrows were turned and 15 gallons (57 liters) on days when they were not turned. On days 13 through 15 and every other day thereafter, molasses was mixed with water at a 1:20 ratio. This technique worked well in the bench-scale test, providing extra energy for microbial activity. The addition of molasses with the water proved ideal in increasing the readily available carbon as well as moisture levels.

The main objective in the treatability study was to achieve and maintain thermophilic conditions. Such conditions were reached within 24 hours and maintained for nearly 40 days (Figure 1).

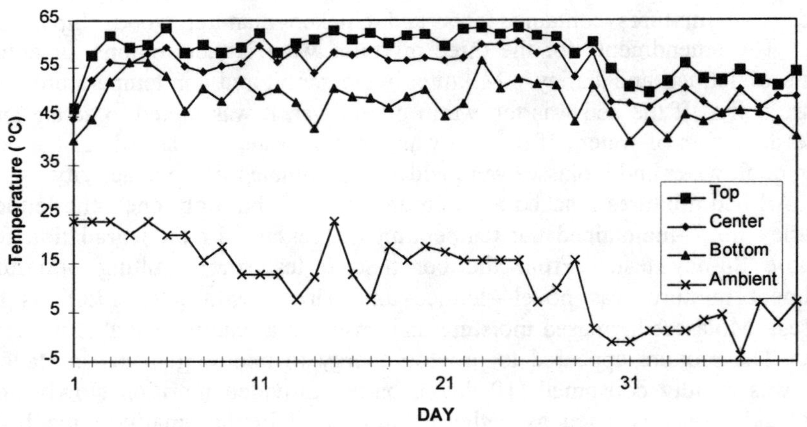

FIGURE 1. Windrow temperatures, 30 percent soil loading.

The temperature results from the windrow with 30 percent soil loading are presented here, because full-scale treatment at a higher loading rate will treat more soil in less time than at the lower loading rate. The temperature data prove that the lettuce and barley mixture (with the manures, alfalfa, and wood chips) provides an excellent environment for biological activity. Thermophilic conditions were maintained for nearly 30 days of the pilot-scale test.

The success of composting at TEAD was also illustrated by the rapid reduction in explosives concentrations to below cleanup goals. Average explosives concentrations in the most soil-rich windrow were reduced to below goals within 10 days (Table 2).

TABLE 2. Concentration of selected explosives in compost (µg/g) over time.

Compound	Day 1	Day 10	Day 20	Day 30	Day 40	Goal
TNT	1,890	22	4.99	0.1	0.1	94
RDX	724	18.3	10.9	3.60	2.96	34
HMX	61.1	3.48	0.87	0.162	4.45	18,000

These sample concentrations represent the average of five samples collected from different locations within the windrow. Daily mixing homogenized the windrow; the samples from Day 1 demonstrated greater variability among the five replicates than on Day 40. Two additional observations support the biological degradation of explosives. First, dilution can be discounted because the decrease in concentrations was too great, and did not occur on Day 1. Second, intermediate breakdown products appeared, then disappeared from the analytical results.

Finally, a toxicity characteristic leachate procedure (TCLP) analysis was performed on the final composted material, and all analytes were below detection limit. Therefore, the composted material is safe for on-site disposal. One year of "curing" is recommended before trying to establish vegetation on the compost.

SUMMARY

The composting treatability study at TEAD met all performance objectives. The bench-scale test identified locally available amendments that can be used to compost soil at the site. The pilot-scale test showed that compost was able to maintain thermophilic conditions for nearly 40 days; 10 days appeared to be a minimum treatment time. A 30 percent soil loading rate can be used, and molasses is an excellent source of food for microbes. Total composting costs from this study were approximately $230 per cubic yard ($300 per cubic meter), which is equivalent to $153 per ton ($168 per metric ton) of treated soil. Based on the success of the treatability study and its low relative cost, composting is a likely candidate for full-scale treatment of the contaminated soil at the Washout Ponds at TEAD.

REFERENCES

Dames & Moore, 1998. *Technical Report: Soil Composting Treatability Study, TNT Washout Facility (SWMU 10), Tooele Army Depot, Tooele, Utah*. Prepared under Contract No. DACA31-94-D-0060 for Tooele Army Depot, Utah.

Roy F. Weston, Inc., 1993. *Windrow Composting Demonstration for Explosives-Contaminated Soils at the Umatilla Depot Activity, Hermiston, Oregon*. Prepared under Contract Number DACA31-91-D-0079 for the US Army Environmental Center, Aberdeen Proving Ground, Maryland.

COST-EFFECTIVE EXPLOSIVE COMPOUNDS BIOREMEDIATION BY WINDROW COMPOSTING HAWTHORNE ARMY DEPOT

Richard Brunner (Tetra Tech, Inc., San Francisco, California)
Dennis Potter (US Army Corps of Engineers, Sacramento, California)

ABSTRACT: Windrow composting is an effective method for mitigating explosive compounds in soil. Past engineering practices made treating soil moderately high but cost-competitive. During a pilot study at HWAD, explosive compounds in soil were remediated successfully by combining standard composting techniques with cost-saving practices and by using local resources. At the same time, on-site workers and off-site receptors and the environment were protected. The soil was treated using less expensive environmental controls and local low-cost additives and the premixing process was eliminated, thereby minimizing materials handling. To keep costs down, the windrows were composted on a compacted soil pad, remediation took place outside of an enclosed building, and water was applied to the windrows to control dust and to maintain a proper bioremediation environment. Monitoring data showed no impact to the treatment area either during or after the remediation treatment. Low-cost additives included recycled wood from HWAD that normally would be considered solid waste. All of the explosive compounds were remediated successfully to less than the treatment goal in approximately three weeks. Treated compounds included the common explosives TNT, RDX, and HMX and the less common ammonium picrate or "yellow-D," which had not previously been bioremediated by composting.

INTRODUCTION

HWAD is in the west-central part of Nevada, approximately 140 miles southeast of Reno. The past method of disposing of explosives-laden wastewater ("pink water") was to transfer the wastewater away from the facilities through a series of settling tanks and drain lines to unlined catchment pits. The explosives in the pits would accumulate as the wastewater evaporated and percolated into the soil. Remedial investigations of the pits showed that most of the explosives-ladened soil in these pits was at a depth of less than five feet below the ground surface (bgs). Based on these investigations, there is approximately 60,000 cubic yards of explosives-ladened soil at 21 sites requiring remediation. Bioremediation by windrow composting was selected as the remedial practice for these soils based primarily on the successful remediation of explosives-ladened soils at the Louisiana Army Ammunition Plant[1] and the Umatilla Army Depot[2].

PILOT STUDY PARAMETERS AND GOALS

A full-scale production pilot study of windrow composting was initiated to determine the feasibility and cost to remediate explosives ladened soils under HWAD's site conditions. To develop parameters and goals for this pilot study, a team was formed of personnel from the HWAD Environmental Operations Group, Day Zimmermann Hawthorne Corporation (the HWAD operating contractor), the Army Industrial Operations Command, the Army Environmental Center, the US Army Corps of Engineers, Sacramento District, and Tetra Tech, Inc., the Corps' remediation contractor.

The key conditions developed by the HWAD Team are the following:

- Six catchment pits and one disposal trench in the vicinity of Building 101-41 at HWAD were selected as the source of the explosives-ladened soil for this pilot study;

- A minimum of 2,000 cubic yards of soil would be remediated to show that the treatment technique could be implemented on a larger scale, ultimately to treat 60,000 cubic yards of explosives-ladened soil;

- The primary chemicals of concern were known to be 2,4,6-trinitrotoluene (TNT), cyclotrimethylenetrinitramine (RDX), and octahydro-1,3,5,7-tetranitro-1,3,5,7-tetrazocine (HMX);

- Soils that contained yellow-D, which had not been remediated previously by this technique, would be included in the pilot study;

- Total concentrations of explosives in the soils were known to be as high as six percent (60,000 mg/kg) and would likely be found at concentrations greater than 12 percent (120,000 mg/kg);

- The explosives-ladened soil is primarily sand and silty sand with low organic content and little clay;

- Excavation goals were set to remove soils with concentrations of explosives greater than 100 mg/kg; and

- Health-based treatment goals were set for each explosive chemical, ranging from 4 mg/kg for 1,3,5-trinitrobenzene (TNB) to 4,000 mg/kg for HMX.

PILOT STUDY TREATMENT

The pilot study treatment was bioremediation composting in active windrows. This process degrades the explosive chemicals in the soil by mixing the soil with bulking and nutrient additives and maintaining a favorable environment for thermophilic bacterial activity. Under these conditions, the bacteria degrade the explosive chemicals to nontoxic compounds in a humic soil media. The compost recipes that were used at HWAD were modified from the previously successful bioremediation compost projects by adjusting the recipes for the cost-effective available additives. The percentages of the additives were varied to assess the most cost-effective recipe while maintaining the remedial efficiency. Table 1 lists the additives and their percentages for each recipe.

Bioremediation Results. Based on the previous treatment studies, the bioremediation at HWAD was conducted for 28 days. It was found that the thermophilic conditions in the compost windrows was reached within two days, and all of the explosives were degraded to concentrations less than their respective treatment goals in less than 28 days. Table 2 lists the maximum TNT, RDX, and yellow-D concentrations per recipe throughout the pilot study as indicator chemicals of the degradation.

TABLE 1. Recipe types and volumes of additives in the Bioremediation Pilot Study of Explosives-ladened Soils at HWAD.

	Explosives – impacted Soil (Type)	Hay	Wood Chips	Manure	Potatoes & Potato Waste	Cubic Yards of Compost
Recipe 1	30% (TNT)	20%	20%	20%	10%	2,230 cy
Recipe 2	30% (TNT)	15%	20%	15%	20%	1,930 cy
Recipe 3	30% (TNT)	10%	25%	25%	10%	1,900 cy
Recipe 4	30% (TNT)	20%	20%	10%	20%	2,020 cy
Recipe 5	30% (Yellow-D)	15%	20%	15%	20%	1,380 cy
Total Volume	2,840 cy	1,535 cy	1,990 cy	1,620 cy	1,480 cy	9,460 cy

TABLE 2. Maximum explosives concentrations in the Bioremediation Pilot Study at HWAD.

	Recipe 1		Recipe 2		Recipe 3		Recipe 4		Recipe 5
Treatment Days	TNT (mg/kg)	RDX (mg/kg)	TNT (mg/kg)	RDX (mg/kg)	TNT (mg/kg)	RDX (mg/kg)	TNT (mg/kg)	RDX (mg/kg)	Yellow-D (mg/kg)
In Situ*	120,000	60,000	120,000	60,000	120,000	60,000	120,000	60,000	10,000
Stockpile	5,720	1,950	5,720	1,950	5,720	1,950	5,720	1,950	2,900
Day 2	70	275	65	95	30	560	400	530	7,570
Day 5	50	62	120	130	500	55	17	57	531
Day 10	500	600	9	34	14	600	10	23	1,600
Day 16	11	60	10	17	12	19	2,000	17	0.4
Day 28	5	25	<1	2	<1	4	2	9	0.4
Goals	233	64	233	64	233	64	233	64	7

*In situ concentrations estimated from field screening kits and field observations.

The cost to remediate 2,840 cubic yards of explosives-ladened soil was less than $250 per cubic yard. The success of this pilot study showed that a large-scale bioremediation program could be conducted at an average rate of approximately 75 cubic yards per day. Costs could be reduced further by using the economy of scale and a more efficient analytical program.

COST-EFFECTIVE INNOVATIONS

The key cost savings for the HWAD bioremediation pilot study was to implement several innovative techniques to the proven compost process, without affecting the environment or the effectiveness of the remediation but still maintaining a safe environment for on-site and off-site receptors. These innovations and the rationale for using them are listed below.

Environmental Controls. Previous windrow composting projects were conducted on impermeable pads inside a temporary structure to control the environment. Traditionally, a capital expenditure of approximately $200,000 would be needed to construct such a facility. This facility would allow approximately 400 cubic yards of compost to be treated at one time. The 60,000 cubic yards of affected soil to be treated eventually will result in approximately 200,000 cubic yards of compost.

If the compost were treated inside a structure in 400 cubic yard increments for one month per treatment, the project would take approximately 33 years. Although larger or multiple structures could reduce this schedule, the HWAD team proposed eliminating this environmental control. Instead, the HWAD composting pilot study was conducted on a compacted soil pad in the open environment. The pad for this pilot study was approximately 325 feet long by 570 feet wide and covered four acres. This area allowed for the simultaneous construction of two windbreak windrows and 26 compost windrows 325 feet long. These windrows were six feet high and 12 feet wide and were spaced 10 feet apart. The treatment pad was constructed by grubbing a treatment area and compacting the soil in the pad while increasing its moisture content. A one-dimensional vadose zone transport model was used to demonstrate that the treatment would have minimal impact on the treatment pad soil. At the end of the treatment, 100 samples were collected from the pad to confirm that the treatment did not affect the soil beneath the compost windrows. Only four of these samples contained explosives concentrations greater than 1 mg/kg, with a maximum concentration of RDX at 18 mg/kg.

Windbreak windrows were constructed at the ends of the treatment pad, and the pad was oriented so that the windrows were perpendicular to the prevailing wind direction. An extensive moisture control program was implemented to assure a favorable environment for the bioremediation treatment. Through these controls, a favorable environment for the remediation was maintained, and the compost, which contained explosives, was not released to on-site or off-site receptors or to the environment.

The success of these innovative environmental controls allowed for approximately 9,500 cubic yards of compost to be treated cost-effectively over six weeks. The cost-efficiency of this pilot study includes the $200,000 saved from not having to construct an impermeable pad and structure and the labor and equipment costs saved by treating 9,500 cubic yards of compost in six weeks versus treating the same volume of compost in 24 400 cubic yard patches over 48 to 72 weeks.

Minimal Material Handling. The HWAD team proposed minimizing one of the material handling tasks by eliminating premixing the explosives-ladened soil with the additives. The objective was motivated by the large volume of soil to be treated. Each constituent in the recipe was measured at its appropriate percentage based on a ten cubic yard patch. Each patch was loaded into a 12-yard dump truck and was transported to the treatment pad and placed in a windrow. The compost material was mixed and aerated three times during the first day of treatment. It was found that the windrows reached the optimum thermophilic conditions of greater than 50° Celsius within two days of treatment and showed no adverse affects from the treatment process. With proper dust control measures, this process had no impact on the potential on-site or off-site receptors or the environment.

Low-cost Additives. A site-specific cost-effective additive was a supply of wood chips from the base. Over the years, HWAD had accumulated thousands of wood pallets and ammunition shipping creates, which were considered solid waste and were scheduled for landfill disposal. A large grinder was used to recycle this waste into wood chips for the compost treatment. Most of the wood had been treated with a preservative containing pentachlorophenols (PCPs), but analytical data from wipe samples, wood chip samples, and compost samples showed that the low levels of PCPs (less than 100 mg/kg) in the wood did not interfere with the bioremediation treatment. In fact, the PCPs were degraded along with the explosives to less than the typical laboratory quantitation limit of 0.1 mg/kg.

SUMMARY

The successful pilot study of bioremediating 2,840 cubic yards of explosives-ladened soil by windrow composting at HWAD implemented several procedures to cost effectively treat large volumes of material. The study demonstrated the procedures that can be implemented at HWAD to treat up to 60,000 cubic yards of soil. These procedures included treating the compost windrows on a compacted soil pad and constructing windbreak windrows and implementing a dust control program to maintain the moisture level of the compost and to eliminate fugitive dust emissions. It also demonstrated that materials handling time can be reduced by eliminating the premixing task during windrow construction and by using local cost-effective additives, such as wood waste from the base that normally would be disposed of in a landfill. The treatment costs for this pilot study were approximately $250 per cubic yard of soil but likely will be reduced for the full-scale treatment by an economy of scale and a more cost-effective analytical program. With the proper teaming and cooperative effort, these and other cost-effective procedures can be used at other sites with similar conditions to achieve more cost-effective remediation.

REFERENCES

Roy F. Weston, Inc. 1988. *Field Demonstration – Composting of Explosives-contaminated Sediments at the Louisiana Army Ammunition Plant.* US Army Toxic and Hazardous Materials Agency Technical Report AMXTH-IR-TE-88242.

Roy F. Weston, Inc. 1993. *Windrow Composting Demonstration for Explosives-contaminated Soils at Umatilla Depot Activity, Hermiston, Oregon.* US Army Environmental Center Technical Report CETHA-TS-CR-93043.

COMPOSTING OF EXPLOSIVE-CONTAMINATED SOILS: A PILOT- AND FULL-SCALE CASE HISTORY

Patrick Woodhull, P.E. - IT Corp.; Findlay, OH
Douglas Jerger, Ph.D. - IT Corp.; Knoxville, TN
Paul Barnes - IT Corp.; Pueblo, CO
Rod Staponski, P.E. - Earth Tech; Englewood, CO
Stan Wharry - U.S. Army; Pueblo Chemical Depot, CO

ABSTRACT: IT Corporation and Earth Tech completed a pilot-scale demonstration of windrow composting technology on explosives-contaminated soils at the Pueblo Chemical Depot, Colorado. The field pilot study was conducted on-site under a RCRA Treatability Study Exemption. The soil for the pilot study was selected from the stockpile to obtain a representative concentration of contaminants: 3,000 mg/kg TNT, 10 mg/kg DNT, and 40 mg/kg TNB. Amendments, a mixture of cow manure, chicken manure, chopped alfalfa, potato waste, and yard waste (wood chips and grass clippings), were blended to a final volume ratio of 70% amendments and 30% soil. After 15 days of treatment, DNT and TNB concentrations were degraded below 0.20 mg/kg and the average TNT concentration was 1.2 mg/kg. Additional sampling and analysis documented the degradation of DNT below the RCRA TCLP criteria (<0.13 mg/l). Based on the success of the field pilot study, windrow composting was selected for the full-scale remediation of the stockpiled soil. Full-scale production will begin in February 1999 and require approximately 20 months to complete.

INTRODUCTION

IT Corporation (formerly OHM Remediation Services Corp.) and Earth Tech (formerly Rust Environment & Infrastructure [RUST]) were contracted by the U.S. Army Corps of Engineers (USACE) to remediate approximately 21,500 cubic yards (cy) of explosive-contaminated soils at the Pueblo Chemical Depot (PCD). The remediation was part of a voluntary, RCRA-corrective action initiated by PCD as a part of the restoration program at the facility (RUST, 1997).

The contaminated soil was part of a Solid Waste Management Unit that encompassed the former Trinitrotoluene (TNT) Washout Facility and Discharge System. The contaminants of concern identified for the soils included TNT, 2,4- and 2,6-dinitrotoluene (DNT), 1,3,5-trinitrobenzene (TNB), and royal demolition explosive (RDX). The contaminated soils were excavated, transported and stockpiled in a permitted containment building at PCD. The soil was managed as a hazardous waste due to the potential for the soil to fail the TCLP test for 2,4-DNT (RCRA waste code D030). An Interim Corrective Measure Study, using RCRA-specific criteria, identified windrow composting as the recommended treatment technology. Prior to full-scale treatment a field pilot study was completed to demonstrate that composting technology could reduce the contaminated concentrations in the soils to the risk-based treatment criteria (WM BioPIT, 1997).

MATERIALS AND METHODS

The pilot test was conducted in an existing building at PCD. To prepare the building for use, the existing equipment was removed and the concrete floor patched and sealed. The building was insulated and heated with temporary propane heaters to ensure the amendments were not frozen prior to windrow construction. Two rolling panel doors at each end of the building allowed equipment access during construction and operation.

Pilot Test Soil. Based on excavation sampling, field screening analysis, and RCRA Facility Investigation data, the soil was selected from the stockpile to obtain representative concentrations of contaminants: 3,000 mg/kg TNT, 10 mg/kg DNT, and 40 mg/kg TNB. RDX was not selected as a target contaminant for the pilot test due to the low concentrations detected in the soils. The contaminated soil for the pilot test was removed from the excavation stockpile with a backhoe, placed in a 5 cubic yard (cy) dump truck, weighed, and transported to the treatment building. The soil was homogenized with a backhoe and placed in a 4-foot high pile for initial sample collection. Twelve core samples were collected and shipped off-site for analysis by USEPA Method 8330, Explosives and TCLP analysis (USEPA SW-846 Method 1311/8270B).

Compost Amendment Selection. The amendments for the compost were selected based on the blends utilized at previous windrow composting projects (Umatilla Depot Activity, OR) and were blended with the soil at a volume ratio of 30 percent soil and 70 percent amendments (Weston, 1993). In addition, a benchscale compost self-heating test was conducted following USACE guidelines (Woods End Research Laboratory and Weston, 1994) to confirm that the selected compost mixture could achieve and maintain the desired temperature increase.

The amendment blend contained 30% cow manure, 4% chicken manure, 26% alfalfa, 14% shredded potatoes, and 26% wood chips. The cow manure and alfalfa were obtained from a local feed lot, the chicken manure was trucked from an egg production farm near Colorado Springs, CO, the potatoes were purchased whole and shredded on-site. Shredded, composted, yard waste (wood, bark, and grass clippings) from Waste Management Inc.'s Midway Landfill was used as the source of wood chips.

Windrow Compost Equipment. Mixing and aeration of the compost were performed daily using a commercial compost turner, Resource Recovery Systems of Nebraska, Inc.'s Model KW614 (Sterling, CO). The KW614 was used for composting operations at Umatilla and had previously passed a detailed Hazard Safety Review performed by Allegany Ballistics Laboratory for the U.S. Army. The self-propelled compost turner is approximately 24 feet wide, 12 feet long, and 12.4 feet high. The mixing and aeration action is creased by a horizontal, 16-inch diameter rotating drum with a series of fixed flails. As the machine moves length-wise through the windrow, the compost was lifted, mixed, and reformed into a windrow.

Windrow Compost Construction. Construction of the windrow consisted of blending 15,000 kilograms (approximately 12 cy or 16.5 tons) of soil with the following amendments: 1.4 cy of chicken manure, 4.0 cy of potato waste, 7.0 cy of alfalfa, 7.0 cy of yard waste (wood chips), and 8.7 cy of cow manure. The volume of the amendments was determined with the front bucket of a CASE 580 backhoe. The amendments and soil were mixed for approximately 15 minutes using the backhoe bucket and 10 passes with the KW614. After the initial mixing was complete, the compost was shaped into a windrow in the center of the building. The final windrow dimensions were approximately 10 feet wide at the base, 5 feet high, and 36 feet long for an initial volume of approximately 40 cy.

Windrow Compost Monitoring. Field monitoring of the process parameters was conducted to evaluate the performance of the technology. Daily temperature measurements were taken at the top, middle, and bottom of the windrow at three locations along the windrow length using a Reotemp Compost Thermometer (Forestry Suppliers, Inc.). Approximately once per week, random discrete samples were collected for on-site pH measurement using a standard field pH meter. Three random, discrete samples were collected periodically for on-site moisture content analysis.

Samples for explosives analysis at an off-site laboratory were collected at Days 0, 5, 11, 15, and 35. The samples were collected from the desired depth using a non-sparking shovel to dig into the windrow. The compost was transferred from the shovel to an aluminum mixing pan. The material in the pan was thoroughly homogenized by hand and placed in an appropriate sample container and labeled for analysis. The samples were preserved on ice at 1 to 4°C for overnight shipment. The shovel and mixing pans were decontaminated between sample collection.

The initial (T=0) and final (T=35) sample events consisted of three depth samples (top, middle, and bottom) collected at six points equally spaced along the length of the windrow. Two duplicate samples were collected for a total of 20 samples for both the initial and final sample event. The intermediate (T=5, 11, and 15) samples were collected from the entire depth of the windrow (surface to 4 feet) at six locations spaced along the windrow length.

RESULTS AND DISCUSSION
Soil Stockpile. The initial soil stockpile (prior to amendment addition) concentrations were near the representative range for the pilot study. The initial average concentrations for TNT, DNT, and TNB were approximately 3,800 mg/kg, 11 mg/kg, and 35 mg/kg, respectively. RDX concentrations were below the method detection limits of 10 mg/kg and 1.2 mg/kg for both laboratories. The mono- and di-amino compounds were generally detected at low concentrations in the soil. The only compound detected in the TCLP analysis was DNT at a concentration of 0.2 mg/L. Based on the analytical results, the soil was considered representative of the excavation stockpile and acceptable for use in the pilot test.

Windrow Compost Monitoring.   The initial windrow temperature ranged from 64°F to 67°F and the windrow reached thermophilic conditions with temperatures ranging from 112°F to 140°F with the first five days of operation.   This temperature increase occurred despite an ambient temperature in the treatment building of 35°F to 55°F. In general, the windrow temperature remained above 104°F until mixing was discontinued on Day 35. The windrow temperatures during the pilot test are summarized in Figure 1.

The pH of the compost ranged from 8.6 standard pH units (SU) to 9.0 SU during the pilot test. The pH as within the expected range based on the UMDA data (Weston, 1993) and no attempt was made to adjust the pH of the compost during the pilot test.

The moisture content of the compost ranged from an initial value of approximately 34 percent and declined gradually to approximately 15 percent by Day 35. The initial moisture holding capacity was 40 percent, corresponding to a target moisture content of 24 to 32 percent. Based on visual observations of the compost immediately after water addition, the moisture holding capacity of the compost decreased over the course of the test; however, based on visual observations and handling of the compost during operation, the relative percent saturation remained consistent.

Table 1 presents the average concentration for each of the explosives and intermediate degradation products for the samples collected during the pilot test. The initial stockpile concentrations were within the expected range and the initial windrow concentrations reflect the impact of the addition of amendments to the soil. The mono- and di-amino compounds present in the initial windrow samples are an indication of the rapid reduction of the TNT nitro-groups occurring in the compost.

The analytical results from the intermediate sample events (Day 5, Day 11, and Day 15) show rapid reduction in all of the target compounds. By Day 15, both DNT and TNB have been degraded below 0.20 mg/kg and the average TNT concentration was 1.2 mg/kg. The mono- and di-amino compounds, TNT, and DNT all showed a greater than 99% removal in the first 15 days of treatment.

The data from the final sampling event (Day 35) indicated the TNT and DNT concentrations were 0.4 mg/kg, and 0.5 mg/kg, respectively. Both TNB and RDX were below the analytical reporting limit of 0.2 mg/kg, and all mono- and di-amino compounds were below 0.5 mg/kg. Three samples were also collected from the windrow and analyzed off-site for TCLP following USEPA Method 1311/8270B procedures. The analytical results showed that DNT was less than the reporting limit of 0.10 m/L, which is below the RCRA toxicity characteristic concentration of 0.13 mg/L.

In addition to the chemical analyses for explosives, the treated material was analyzed for nutrients to assess the fertilizer value and to determine the potential leachability of nitrates from the compost.   The analytical results confirmed that the compost material contained nutrients (nitrogen, phosphorous, potassium) and the compost material was acceptable for use as a soil conditioner (RUST, 1998).

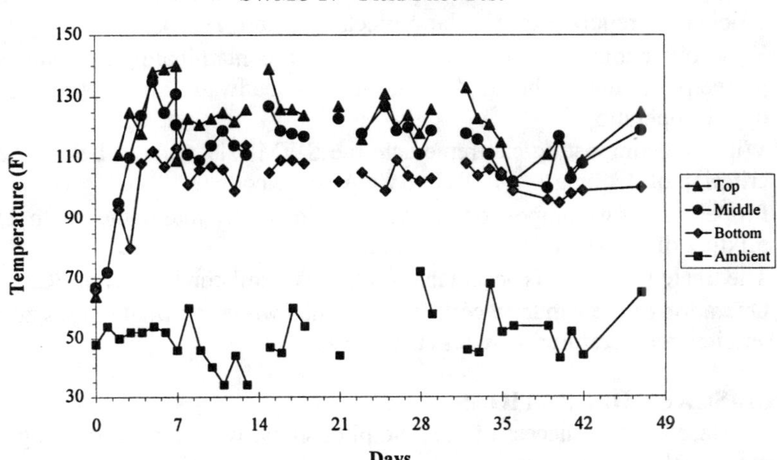

**Figure 1. Windrow Operating Temperature
SWMU 17 - Field Pilot Test**

Table 1.   Average Explosives and Degradation Product Concentrations In Samples Collected During the Pilot Test

Sample	TNT	DNT	TNB	RDX	4AMDNT	2AMDNT	2,4DANT	2,6DANT
Stockpile	4100	11	37	<10	12 (a)	NA	<22	<6
Day 0	1380	4.0	5.1	0.7	225	131	53	8.8
Day 5	34.3	1.0	0.4	2.1	12	4	5.6	1.9
Day 11	51.1	0.5	0.4	1.3	12	8.0	2.3	1.0
Day 15	1.2	<0.2	<0.2	<1.0	0.9	0.3	0.3	0.2
Day 35	0.36	0.5	<0.2	<0.2	0.4	<0.2	0.2	0.3

(a) The total of both isomers is reported as 4AMDNT.

Results are in units of mg/kg (dry-weight basis).
NA = Not Analyzed

4AMDNT = 4-amino-2,6-dinitrotoluene
2AMDNT = 2-amino-2,4-dinitrotoluene
2,4DANT = 2,4-diamino-6-nitrotoluene
2,6DANT = 2,6-diamino-4-nitrotoluene

CONCLUSIONS

Based on the data generated and observations made from the operation of the windrow compost during the pilot test, the following conclusions are drawn:

- Removal efficiencies of the target compounds were comparable to removal efficiencies reported for similar projects (Weston, 1993).
- Windrow composting degraded the target contaminants to concentrations protective of human health, ecology, and groundwater after approximately 15 days of operation.
- Windrow composting can remediate the SWMU 17 soils to below the TCLP criterion of 0.13 mg/L for DNT (RCRA waste code D030).
- Proper windrow compost operating conditions were maintained with a volume mixture of 30% soil and 70% amendments.
- The treated compost is acceptable for use as a soil conditioner at PCD.
- Operation of the windrow compost remained within optimal ranges even when ambient temperatures were less than 40°F.

FULL-SCALE OPERATIONS

Based on the success of the field pilot study, windrow composting was selected for the full-scale remediation of the stockpiled soil. An existing 180 foot wide by 500 foot long warehouse adjacent to the containment building was converted into a permitted treatment facility. During full-scale remediation the facility will contain ten 200-foot long windrows at an average treatment time of 22 days. An on-site, USACE-certified analytical laboratory will provide process monitoring data, confirmation sampling, and target contaminant analysis. At the completion of treatment the compost will be placed in the Wildlife Management Area at PCD as a soil amendment and conditioner. Full-scale production will begin in February 1999 and require approximately 20 months to complete.

REFERENCES

Roy F. Weston, Inc. 1993, Windrow Composting Demonstration of Explosives-Contaminated Soils at the Umatilla Depot Activity Hermison, OR. Report No. CETHA-TS-CR-93043. U.S. Army Environmental Center, Aberdeen, MD.

RUST. 1997. Final (Revision 3) SWMU 17 Interim Corrective Measure Soil Removal Project Work Plan. Prepared for USACE Omaha District.

RUST. 1998. Draft Final Field Pilot Test Report, Windrow Composting; SWMU 17, Pueblo Chemical Depot. Prepared for USACE Omaha District.

Woods End Research Laboratory, Inc., and Roy F. Weston, Inc. 1994. Amendment Selection and Process Monitoring for Contaminated Soil Composting. Report No. ENAEC-TS-CR-93019. U.S. Army Environmental Center, Aberdeen, MD.

WM BioPIT. 1997. Draft Final SWMU 17 Interim Corrective Measure Field Pilot Test Work Plan. Prepared for USACE Omaha District.

BIODEGRADATION OF DDT: THE ROLE OF *Pleurotus* sp., A LIGNICOLOUS FUNGUS

Anne A. Osano (University of Nairobi, Kenya)
George M. Siboe, John O. Ochanda and Kokwaro J.O. (University of Nairobi, Kenya)

ABSTRACT: A laboratory investigation on the role of the white rot fungus *Pleurotus luteoalbus* in the biodegradation of DDT was carried out. DDT and many persistent environmental pollutants are known to be toxic and/or carcinogenic to animals and/or humans. Ligninolytic enzyme production media which was incorporated with ^{14}C DDT was inoculated with the fungus. ^{14}C CO_2 which was used to estimate DDT biodegradation indicated that 69% of ^{14}C DDT was metabolized either to ^{14}C CO_2 or DDT metabolites, during the thirty day incubation. Although not all ^{14}C DDT was metabolized to ^{14}C CO_2, the metabolites can now be easily broken down by other microorganisms.

INTRODUCTION:

DDT (1,1,1-trichloro-2, 2-bis (4-chlorophenyl) ethane) and many toxic or carcinogenic organohalides persist in the environment because micro-organisms are unable to degrade them and when they do so, like the bacterial degradation of DDT, the rates are very slow (Alexander *et al.*, 1981; Anderson *et al.*, 1976).

Recent work has showed that a number of white rot fungi can mineralize not only DDT but also other persistent environmental pollutants and that the metabolic pathway is different from that of the bacterial biodegradation, (Suba-Rao and Alexander 1985; Bumpus *et al.*, 1985, 1987).

Most studies on ligninolytic system have centred on *Phanerochaete chrysosporium* despite the fact that the oxidative extracellular enzyme system is probably not identical in all white rot fungi, resulting in a difference in efficiency (Hattaka and Uusi, 1983, Ander *et al.*, 1980). For comparable efficiency in biodegradation of DDT and other organo-pollutants, it is therefore important to evaluate the system in a number of white rot fungi. *Pleurotus luteoalbus* releases a ligninolytic enzyme of optimal pH value of 4. The enzyme is substrate dependent with a Km value of 0.25 µmoles and V max of 0.357 µmoles (Osano *et al.*, 1995).

Objectives. The overall objective was to study the biodegradation of DDT by *Pleurotus luteoalbus*, while specific objectives as the study of CO_2 production from biodegradation of DDT and metabolite formation in the biodegradation process.

MATERIALS AND METHODS

Pleurotus luteoalbus was cultured from collections from Karura Forest, Kenya. The fungus was maintained on 2% malt extract agar.

Chemicals. All chemicals were donations from Pesticide Laboratory, Chemistry Department, University of Nairobi and included the following:
- ^{14}C P, P^1 DDT,
- non-labeled 98% P, P^1 DDT, DDE and 99% DDD,
- ^{14}C cocktail as a trapping agent,
- Na$_2$SO$_4$, acetonitrile, acetone, toluene, n-hexane, florosil, methanol.

Metabolite determination. 10 ml of enzyme culture media was partitioned three times into 15 ml hexane. The combined hexane layer was concentrated to 5 ml in a rotary evaporator.

Clean-up procedure. This was done by the use of Florisil Column (60 - 100 mesh). Tripple distilled hexane, double distilled acetone and the concentrated enzyme culture media were cleaned. The hexane and the acetone were first cleaned and then used in the clean up of the metabolite extract.

Gas chromatography. A gas chromatograph (GC) (Model Packard 428) equipped with an electron capture detector, with detector current of 4 was used in this study. DDT, DDT metabolites were on a glass column (2 m x 2 in.) packed with G.P. 1.5% SP-2250/1.95% SP-2401 on 100/120 supelcoport. The temperature of the detector was 250 ºC, of the oven was 210 ºC and of injection point was 230 ºC. Nitrogen, white spot was used as the carrier gas. The flow rate was 50 - 60 ml/min, the chart speed being 10 mm/min. and the attenuation was 128.

Culture conditions and analytic procedure. *Pleurotus luteoalbus* was incubated at 21 ºC in liquid medium. 15 g of malt, 2 g of yeast in 1,000 ml of water formed the malt yeast broth, the media used for enzyme production (Niku-Paavola *et al.*, 1988). Each 250 ml of culture flask was equipped with a gas exchange manifold and contained 10 ml of the malt yeast broth. The incubated medium also contained 1.25 n mole of [^{14}C] DDT (50 n Ci) which was added in 10 μl of acetone. The culture was sealed and incubated for 30 days. During the first three days of incubation, cultures were allowed to grow in air. After 3 days and at 3 day intervals thereafter, cultures were flushed with oxygen. Evolution of ^{14}CO$_2$ from [^{14}C] DDT was assayed at three day intervals as described by Bumpus et al. (1985). The ^{14}CO$_2$ produced was trapped in Carbon ^{14}C cocktail for Harvey Biological oxidizer (scintillation fluid), and was subsequently assayed at three day intervals using a scintillation counter [Tri-carb, model 1000 TR from Packard Canberra Co., USA]. DDT disappearance and metabolite formation were assayed by Gas Chromatography (GC) analysis. The mycelia was collected, dried and dry weight taken.

RESULTS

Figure 1 shows the biodegradation of DDT to carbon dioxide (CO$_2$) by *Pleurotus luteoalbus*. During 30 days of incubation, 3% of the initial DDT was mineralized to [^{14}C] CO$_2$.

Five metabolites and of course DDT were observed during the same experiment. This is illustrated in Figure 2. These metabolites are 1,1-dichloro-2,2,-bis(4-chlorophenol)

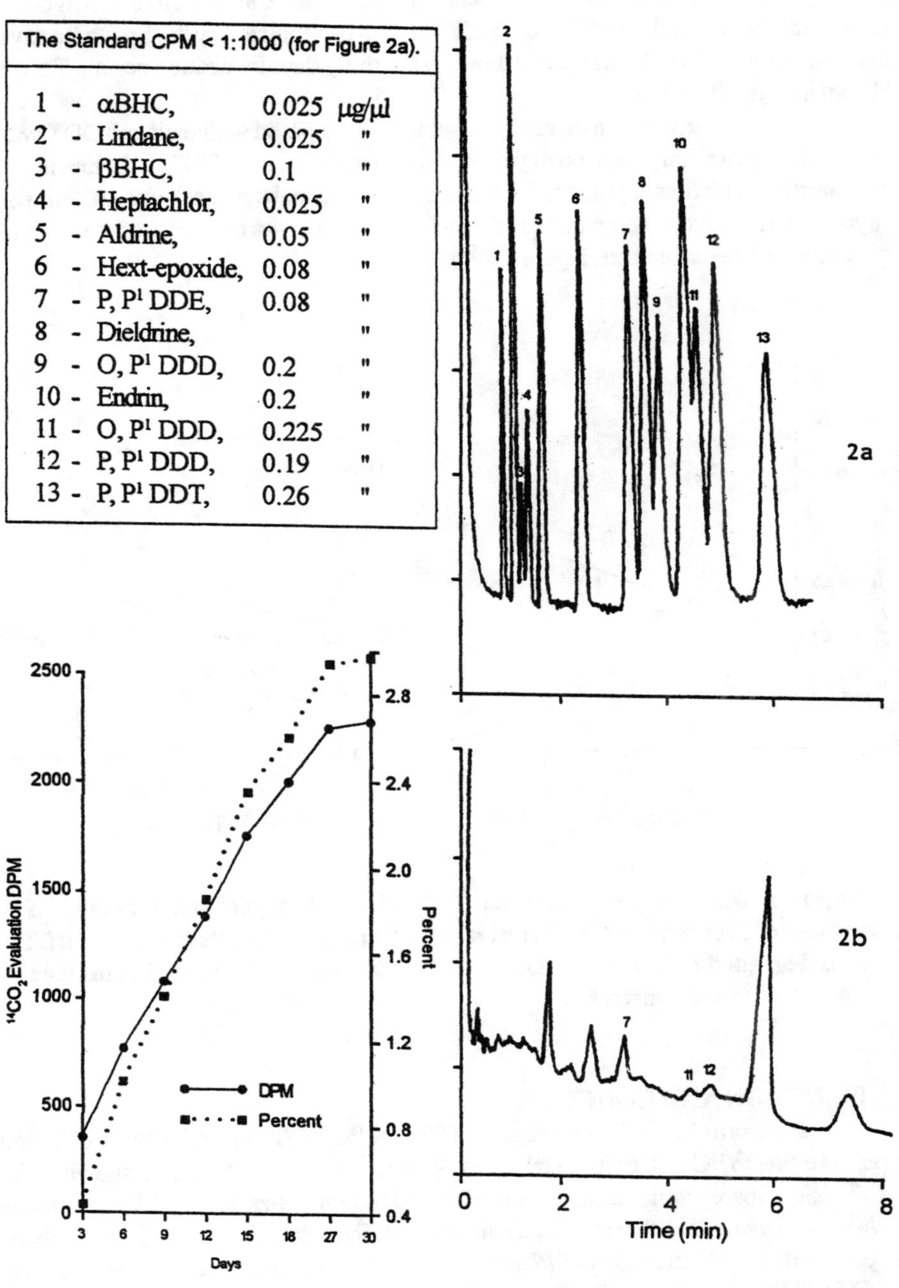

Figure 1. $^{14}CO_2$ evaluation from DDT in cultures of *Pleurotus luteoalbus*. The values are corrected for the control.

Figure 2 (a). Shows the Standards for GC, and (b), Gas Chromatography analysis of DDT and their metabolites in cultures of *Pleurotus luteoalbus* respectively.

ethane (DDD); DDE; and three other unidentified metabolites. Calculations from gas chromatography analysis showed that 68% of DDT was metabolized by the first three days and a maximum of 69% was broken down by the thirty days incubation period. This is shown in Figure 3.

Mycelial production on malt as a substrate compared with malt and DDT was similar, being 0.060 mg and 0.66 mg respectively. The presence of DDT in the media did not interfere with fungal growth. DDT alone was a poor substrate as only 0.029 mg of mycelia was produced compared to other substrates. This was only 50% of the amount produced in other substrates. Figure 4 shows this.

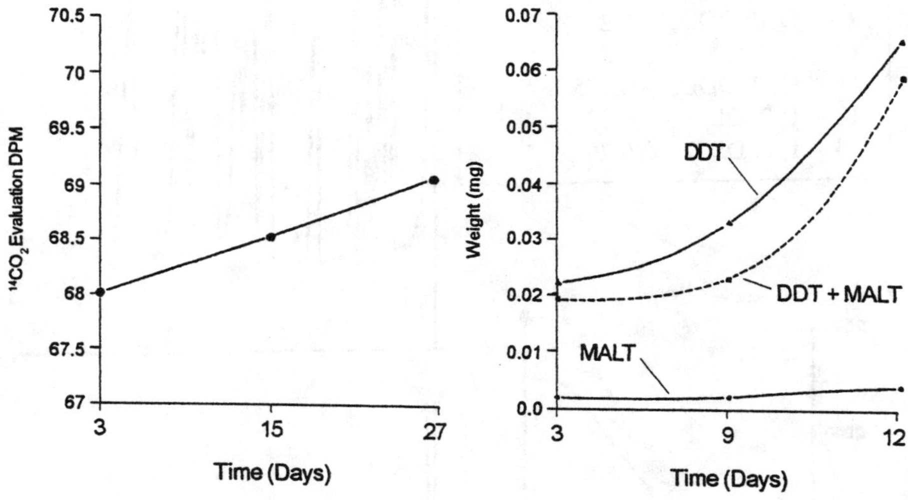

Figure 3. DDT disapearence from cultures of *Pleurotus luteoalbus* as it is degraded to CO and DDT metabolites.

Figure 4. Mycelial dry weights of *Pleurotus luteoalbus* utilising DDT substrates as compared to malt extract.

DISCUSSION AND CONCLUSION

Pleurotus luteoalbus mineralized 69% of ^{14}C DDT during 30 day incubation period either to ^{14}C CO_2 or its metabolites. 3% was metabolised to CO_2 as shown in Fig 1. This compares with earlier report of 4 - 13% mineralization by *Phanerochaete chrysosporium* during the same incubation period. Even though only 3% is metabolized completely to CO_2, the ability of *Pleurotus luteoalbus* to catalyze the initial oxidation of DDT will have a profound effect upon the molecule to undergo subsequent oxidative degradation by other fungi and indeed other microorganisms in the soil environment. It has been shown that some eight species of fungi did not catabolize DDT but were able to degrade DDT metabolites (Suba-Rao and Alexander, 1985). Upto 69% of DDT was metabolised either to CO_2 or metabolites in the 30 days of incubation.

Six metabolites including DDD, DDE and DDT were formed during the biodegradation of DDT by *P. luteoalbus*. Three could not be identified due to lack of standards. DDD seems to disappear from the culture starting at high levels of 0.1236 ng/μl in 3 days but ending up as 0.0173 ng/μl in day 30. According to Bumpus and Aust (1987), metabolic pathway, the produced DDD is used to form FW-152 and hence its disappearance. This is therefore the most likely pathway for DDT biodegradation, as opposed to that of Suba-Rao and Alexander (1985). Even though the presence of dicofol was not confirmed, it has been shown to undergo zinc catalyzed thermolytic decomposition to form DDE (Mickinny, 1974). It is therefore possible that the DDE observed resulted from this reaction. More work is however necessary to confirm this.

ACKNOWLEDGMENTS:

We acknowledge with much appreciation Prof. S.O. Wandiga in whose pesticide laboratory all this work was done and who donated all the chemicals necessary. Special thanks go to Mr. J. M. Makopa, L. Mbuvi, S. Njogu, and C. Ochieng' for the technical advice and help. We are grateful to J. Shilavula for typing the work and L. Wasonga for typesetting it.

REFERENCES

Alexander, M. 1981. *Science* **211**: 132.

Ander,P., Hatakka, A.& Eriksson, K.E. (1980). *Arch. Microbiol.* **125**: 189 - 202.

Anderson, D.W. and J.J. Hickey. *Environ. Pollut.* **10**: 183, 1976.

Bumpus, J.A., M. Tien, D. Wright and S.D. Aust. 1985. Oxidation of persistent environmental pollutants by a white rot fungus. *Science* **228**: 1434 - 1436.

Bumpus, J.A. and S.A. Aust. 1987. Biodegradation of DDT (1, 1, 1, - trichloro - 2, 2 - bis (4 - chlorophenyl) ethane by the white rot fungus *Phanerochaete chrysosporium. Applied and Environmental Microbiology* **53**: 20001 - 20007.

Hatakka, A.I. and Uusi-Rauva, A.K. 1983. *Eur. J. Appl. Microbiol. Biotechnol.* **17**: 235 - 242.

Niku-Paavola, M., M., Karhunen E., Salola, P. and Raumio V. 1988. Ligninolytic enzymes of the white rot fungus *Phlebia radiata*. Biochem. J. **254**: 877-864.

McKinney, J.D., E.O. Oswald, S.M. de Paul Palaszek, and B.J. Corbett. 1974. Characterization of chlorinated hydrocarbon pesticide metabolites of the DDT and polycyclodiene types by electron impact and chemical ionization mass spectrometry, p. 5 - 32. In R. Hague and F.J. Biros (ed.). Mass spectrometry and NMR spectroscopy in pesticide chemistry. Plenum Publishing Corp., New York.

Niku-Paavola, M. Karhunen E., Salola, P. and Raunio V. (1988). Ligninolytic enzymes of the white rot fungus *Phlebia radiata. Biochem. J.* **254**: 877 - 884.

Osano A.A., Siboe, G.M., Ochanda, J.O. and Kokwaro J.O. 1995. Biodegradation properties of white rot fungi in Karura Forest, Kenya. Workshop in Fungal Diversity in Sub-Saharan Ecosystems and their role in rural development held in Dar-es-Salaam, Tanzania.

Suba-Rao, R.V. and M. Alexander. 1985. Bacterial and fungal cometabolism of 1, 1, 1 - trichloro - 2, 2- bis (4 - chlorophenyl) ethane. (DDT) and its breakdown products. *Appl. Environ. Microbiol.* **49**: 509 - 516.

IN-SITU BIOREMEDIATION OF CREOSOTE AND PENTACHLOROPHENOL-CONTAMINATED GROUNDWATER AT A WOOD-TREATING SITE

Abdolhamid Borazjani, Susan V. Diehl, David A. Strobel, and Louis Wasson
(Forest Products Laboratory, Mississippi State University, Mississippi State, MS)

ABSTRACT: The objectives of this multi-year project were to investigate *in situ* biological remediation for the clean-up of organic wood preservatives in groundwater at a wood treating site. A suitable site was selected at a south Mississippi location where creosote components and pentachlorophenol (PCP) were present in both the groundwater and subsoil. Three temporary recharge wells were established in the upgrade and in the same screening zone as the recovery well. A bacterial culture, previously isolated and known to efficiently degrade PCP and polycyclic aromatic hydrocarbons (PAHs) (*Arthrobacter sp.*), oxygen (supplied as 0.005% hydrogen peroxide), and nutrients, all of which were suspended in 500 gallons of water, were introduced into each recharge well. Each recharge well was charged bimonthly for 1993, monthly for 1994-1995, and annually for 1996, 1997, and 1998. Water samples from the recovery well were taken on selected dates. Analyses of the results from the recovery well has shown a significant degradation of total selected PAHs. PAH concentration was reduced from 31.84 mg/L to 4.89 mg/L in 18 months. No significant biodegradation of PCP or tetrachlorophenol was observed. In comparison with the initial results, bacterial populations have shown an inconsistent increase during each year except 1997-1998.

INTRODUCTION

Contaminated groundwater has been found in all fifty states. Some of this water is contaminated with chlorinated organic pollutants such as pentachlorophenol (PCP) and polycyclic aromatic hydrocarbons (PAHs). The wood treating industry in Mississippi has been in operation for over one hundred years. Groundwater contamination by wood preservatives has been the result of leaching of these compounds from the soil at waste disposal sites into groundwater reserves.

In-situ bioremediation is the biological treatment of soil and groundwater without removal from the contaminated site. This technology involves the use of microorganisms which can degrade the contaminating compounds in both soil and water. In-situ bioremediation requires an understanding of microbiological processes relative to biodegradation of the target contaminants as well as the soil's physical and chemical effects on these microbial processes. Often these systems utilize aerobic metabolism and thus involve the addition of oxygen, usually as air, and inorganic nitrogen and phosphorus.

Several methods are available to remove organic pollutants from groundwater (Borow, 1989; CAA Bioremediation Systems, 1988; Campbell et al., 1989; Heyse et al., 1986; Looney et al., 1992; Taddeo et al., 1989; TeKrony and Ahlert, 1992;

Yare et al., 1987; Yare et al., 1989). Two of the most common methods are filtration and biological treatment. Both of these methods are effective in certain situations for groundwater cleanup of organic contaminants. Filtration involves the pumping of groundwater through carbon filters to remove contaminating organics. The cost for filtration of groundwater containing wood treating chemicals averages $1.25-$5.25 per 1000 gallons depending on influent concentration and type of carbon used. This method is labor intensive and the spent carbon used for filtration has to be disposed of in ways other than incineration. This only relocates the contaminants. Biological treatment (pump and treat) involves pumping the contaminated water into bioreactors where cleanup is carried out by means of microorganisms at a cost of $1.00 per 1000 gallons. Both bacteria and fungi have been shown to be important in the bioremediation processes.

In-situ bioremediation utilizes attempts to develop large populations of naturally occurring microorganisms that can detoxify parent compounds of interest to products that are no longer hazardous. Success of this technology, which is even less labor intensive and more cost effective than pump and treat methods, leads to a better and more permanent cleanup level.

The objective of this research was to investigate the use of an in-situ biological technique for the cleanup of organic wood preservatives in groundwater at a wood treating site in southern Mississippi. This involved the naturally occurring microorganisms as well as a bacteria species previously isolated at our laboratory capable of degrading PCP and PAHs at a rapid rate. Since this particular bioremediation process is an aerobic process, the use of added oxygen and micronutrients were also investigated as possible ways to enhance the process. Analytical techniques applied to water samples obtained from wells already in place were used to monitor the treatment effectiveness.

MATERIALS AND METHODS

A suitable site was selected at a south Mississippi location where creosote and PCP are present in both groundwater and subsoil. Three temporary recharge wells were established in the upgrade and in the same zone as the recovery well (WC-5), 60 feet north of WC-5. The three wells were located with 15-foot spacing along a line approximately perpendicular to the direction of groundwater flow. The approximate locations of these wells, pumping wells (WCP-1, WCP-2, WCP-5), and treatment systems are shown in Figure 1. The recharge wells were installed to a depth of approximately 40 feet below land surface (ft bls) and screened from approximately 35-40 ft bls in the same zone screened by the recovery well (WC-5). It should be noted that if constituents occur at greater depths, they may not be impacted by this experiment. The construction of the recharge wells were established by drilling a borehole to a depth of 40 ft bls with a rotary bit. The diameter of the borehole was between 6 and 8 inches. Soil samples derived from the drilling were inspected and later were analyzed to provide information concerning contamination of subsoil. The wells were assembled from 2-inch diameter Schedule 80 PVC casing, a 10-foot long, 2-inch diameter Schedule 80 PVC well screen with 0.020 inch (20-slot) slot width, and a PVC bottom well cap. The wells were placed vertically in the center of the borehole. The casings were extended 1 foot above the ground surface.

A filter pack consisting of 20-40 grade silica sand was poured into the boreholes annuluses to a level at least 3 feet above the top of the screen. Approximately 15 pounds of 1/2-inch diameter bentonite pellets were poured into the boreholes annuluses with water and were allowed to hydrate for one hour. A cement slurry consisting of Portland cement and water was poured into the borehole annuluses until the ground surface was reached. The casings were capped with PVC caps. After allowing the cement slurry to set for 48 hours, the recharge wells were pumped to withdraw fine-grained particulates from well casings and filter packs.

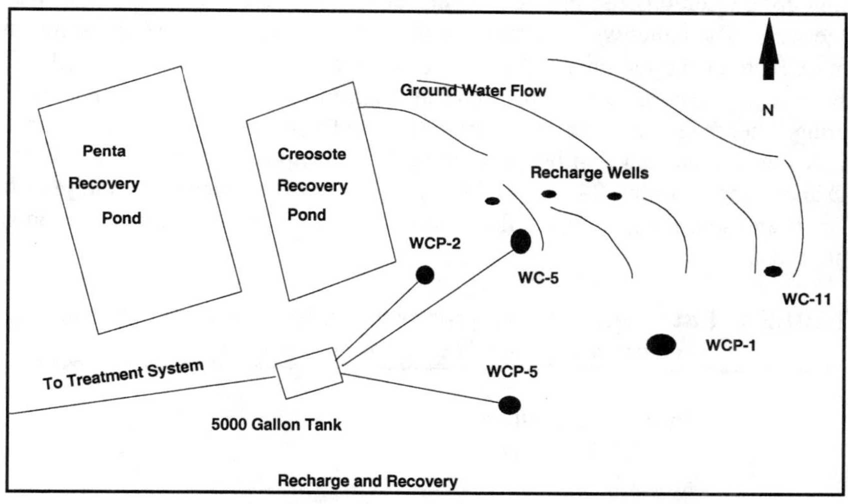

FIGURE 1. Diagram of study site. The three temporary recharge wells are in upgrade of the recovery well (WC-5).

A bacteria culture, previously isolated and known to efficiently degrade PCP and PAHs (*Arthrobacter sp.*) (Walker, 1992), oxygen (supplied as 0.005% hydrogen peroxide), and nutrients (as 0.040% nitrogen, 0.080% phosphorus, and 0.040% potassium), all of which were suspended in 500 gallons of water, were introduced into each recharge well. Hydrogen peroxide was used as an oxygen source because it can supply a much higher concentration of oxygen than can be achieved by dissolving air or even pure oxygen in water. This is important because the limitation of oxygen that can be supplied by aeration was the most common problem previously encountered in in-situ bioremediation work (Heyse et al., 1986). Although hydrogen peroxide is cytotoxic at high concentrations, studies indicate it can be added to groundwater in concentrations up to 100 mg/L without adverse effects, assuming proper formulation and control (Heyse et al., 1986). This suspension was infused into the subsoil and groundwater, which flows downgrade to the recovery well (WC-5) and pumping wells (WCP-1, WCP-2, WCP-5) (Figure 1).

The charging system required for each well is by gravity feed and does not require pressurization. Each recharge well has been charged bimonthly for 1993, monthly for 1994-1995 and annually for 1996, 1997, and 1998. Water samples from the recovery well have been taken on selected dates.

Special care was exercised to prevent contamination of the groundwater during sampling. Water contained within and adjacent to the well casing can potentially reflect chemical interaction with the atmosphere (by diffusion of gases down the casing) or the well construction materials (through prolonged residence adjacent to the casing). This water was removed from the well prior to sample collection. A bailer was used to extract water samples from the well. Much care was taken during insertion of sampling equipment to prevent undue disturbance of water in the well. The bailer was lowered into the water gently to prevent splashing and then extracted gently to prevent creation of an excessive vacuum in the well. The sample was poured directly into a one-liter teflon capped amber bottle. While pouring water from the bailer, the water was carefully poured down the inside of the sample bottle to prevent significant aeration of the sample. The sample was kept at 4°C and shipped within 24 hours of sampling. These samples were analyzed for selected creosote components, chlorinated phenols (Table 1) and microorganism populations.

TABLE 1. List of selected polycyclic aromatic hydrocarbons (PAHs) and chlorinated phenols monitored in this study.

Benzo(b) fluoranthene
Benzo(k) fluoranthene
Carbazole
Chrysene
Cresols
Fluoranthene
Naphthalene
Phenanthrene
Pentachlorophenol (PCP)
2,3,4,6-Tetrachlorophenol (TeCP)

EPA Method 3520 was used for extraction of groundwater (US EPA, 1986). PAHs and PCP were analyzed using EPA Methods 8100 and 8140, respectively (US EPA, 1986). The dilution plate method was used to count bacterial colonies. The media used were nutrient agar, nutrient agar amended with 5 mg/L of technical grade pentachlorophenol (Vulcan Chemical Company, Wichita, Kansas) and nutrient agar amended with 20mg/L of whole creosote. The nutrient agar was autoclaved for 20 minutes at 15 psi and 121°C and then cooled to 55°C. Both creosote and PCP dissolved thoroughly in methyl alcohol were added to the cooled nutrient agar. The number of colonies recovered on nutrient agar, nutrient agar amended with PCP, and nutrient agar amended with creosote represents the approximate number of colonies for total bacteria, PCP acclimated bacteria, and creosote acclimated bacteria, respectively.

RESULTS AND DISCUSSION

Concentration levels of total selected PAHs, tetrachlorophenol, and PCP for each sampling date are given in Table 2. A significant biodegradation of total selected PAHs has been occurring since this study began in September 1993. PAH concentration has been reduced from 31.84 to 4.89 mg/L in 18 months. With the exception of samples collected during 1997, this reduction has been gradual and consistent. Tetrachlorophenol degradation was very slow at first, but no detectable amount of this compound was found in the 1998 samples. PCP degradation has been inconsistent, but overall PCP concentration has been reduced from 0.607 mg/L to around 0.450 mg/L during the last 6 years.

TABLE 2. Concentration of total selected PAHs and chlorinated phenols from the recovery well (WC-5) taken since September 1993.

Date Sample Taken	Total Selected (PAHs mg/L)	TeCP (mg/L)	PCP (mg/L)
4-28-93 (before treatment)	31.840	.081	.607
9-13-93 (one month after first injection)	10.460	.047	.159
3-07-94	10.850	.052	.401
6-06-94	18.266	.089	.596
9-01-94	11.202	.065	.369
12-08-94	8.37	ND*	.275
1-26-95	4.89	ND	.426
1995	4.16	.045	.486
1996	3.92	.029	.413
1997	4.26	.033	.428
1998	3.28	ND	.450
Detection limit	---	.020	.020

* ND = Non-detect.
** Each figure for 1995 and 1996 represents an average of four sampling dates and 1997 and 1998 are for two sampling dates.

The bacterial population determined on selected media are given in Table 3. In comparison with the initial results, bacterial populations have shown an inconsistent increase. More consistent increases began after June 1994, when recharge well injections were changed from a bimonthly to a monthly basis. The bacterial counts have been declining significantly in the past two years. This drop could be attributed to the decrease in injection frequencies.

Overall results from this six-year project are very promising if the current pattern continues. We believe that this technique will result in enhanced biodegradation of creosote and chlorinated phenols in both subsoil and groundwater.

TABLE 3. Bacterial populations from the recovery well (WC-5) isolated on selected media since September 1993.

Date Sample Taken	Nutrient Agar (Colony/mL)	Creosote Amended Nutrient Agar (Colony/mL)	PCP Amended Nutrient Agar (Colony/mL)
4-28-93	1500	900	100
9-13-93	800	700	900
2-01-94	115,800	114,400	136,800
3-07-94	16,000	16,000	13,000
6-06-94	5,000	4,000	4,000
10-04-94	78,700	78,200	67,900
12-15-94	107,000	90,000	72,000
1995	127,000	103,000	105,000
1996	285,000	295,000	297,000
1997	82,000	21,000	63,000
1998	18,000	22,000	17,000

Each figure for 1995, 1996, 1997, and 1998 represents an average of 4, 4, 2, and 2 sampling dates, respectively.

REFERENCES

Borow, H.S. 1989. *Biological Cleanup of Extensive Pesticide Contamination in Soil and Groundwater.* Proc., Hazardous Mat. Control Res. Inst., Biotreatment, Nov., Washington, DC. pp. 51-56.

CAA Bioremediation Systems. 1988. *Feasibility of Biodegradation of Waste Liquid Coal Tar.* U.S. EPA, SBIR Report, Contract No. 68-02-4500, April.

Campbell, J.R., J.K. Fu and R. O'Tool. 1989. *Biodegradation of PCP Contaminated Soils Using In-situ Subsurface Bioreclamation.* Proc., Hazardous Mat. Control Res. Inst., Biotreatment. Nov., Washington, DC. pp. 17-27.

Heyse, E., S.C. James and R. Wetzel. 1986. "In-situ Aerobic Biodegradation of Aquifer Contaminants at Kelly Air Force Base." *Environ. Progress.* Vol. 5, No. 3. pp. 207-211.

Looney, B.B., D.S. Kahack, T.C. Hazen and J.C. Corey. 1992. *Environmental Restoration Using Horizontal Wells: A Field Demonstration.* Am. Chem. Soc. Symp. of Emerging Tech. for Hazardous Waste Management, Sept., Atlanta, GA. pp. 551-552.

Taddeo, A., M. Findlay, M. Dooley-Danna and S. Foge. 1989. *Field Demonstration of a Forced Aeration Composting Treatment for Coal Tar.* Proc., Hazardous Mat. Control Res. Inst., Biotreatment, Nov., Washington, DC. pp. 57-62.

TeKrony, M.C. and R.C. Ahlert. 1992. *Pilot-scale Study of In-situ Steam Extraction for the Recovery of Polychlorinated Aliphatic Hydrocarbons from the Vadose Zone.* Am. Chem. Soc. Symp. of Emerging Tech. for Hazardous Waste Management, Sept., Atlanta, GA. pp. 21-25.

US EPA. 1986. *Test Methods for Evaluation of Solid Waste Physical/Chemical Methods.* SW-846. Office of Solid Waste and Emergency Response, Washington, DC.

Walker, K. 1992. *Biodegradation of Selected Polycyclic Aromatic Hydrocarbons and Pentachlorophenol in Highly Contaminated Wastewater.* M.S. Thesis. Mississippi State University, MS. 80 pp.

Yare, B.S., D. Ross and D.W. Ashcom. 1987. *Pilot-scale Bioremediation at the Brio Refining Superfund Site: Superfund.* 1987 Proc. of Hazmat, Nov., Washington, DC. pp. 315-319.

Yare, B.S., W.J. Adams and E.G. Valines. 1989. *Pilot-scale Bioremediation of Chlorinated Solvent and PNA-containing Soils.* Proc., Hazardous Mat. Control Res. Inst., Biotreatment, Nov., Washington, DC. pp. 47-50.

NATURAL ATTENUATION OF PENTACHLOROPHENOL AT A WOOD TREATMENT FACILITY

Shawn R.T. Severn, Ph.D., Christina Noftsker (Exponent Environmental Group, Bellevue, Washington)
J. Stephen Barnett (Exponent Environmental Group, Lake Oswego, Oregon)
RueAnn Thomas (International Paper, Memphis, Tennessee)

ABSTRACT: Review of historical data and results of a field investigation indicate natural attenuation of pentachlorophenol (PCP) is occurring in groundwater at a former wood treatment facility. High total organic carbon levels from a leaking sewage pipe and/or petroleum product releases have contributed to anaerobic conditions, resulting in a pattern of reductive dechlorination of PCP, evidenced by the presence of tetra-, tri-, and dichlorophenol (TeCP, TCP, DCP). The presence of methane, a negative oxidation reduction potential (ORP), and decreases in PCP concentrations over time, along with corresponding increases in PCP dechlorination products, provide indirect and direct evidence that anaerobic biodegradation is occurring. Furthermore, biodegradation of PCP and less chlorinated phenols under aerobic conditions appears to be responsible for decreasing contaminant concentrations on the plume margins, suggesting sequential anaerobic-aerobic processes may be responsible for PCP attenuation. The results of solute transport modeling using site data indicate that biodegradation is a major component in the natural attenuation of PCP. These results are evaluated to provide the first two lines of evidence required by the U.S. Environmental Protection Agency (EPA) to support natural attenuation. The first line of evidence is a decrease in contaminant concentration along the flow path; the second line of evidence is loss of contaminant mass supported by geochemical and biological decay rate data.

INTRODUCTION

PCP biodegrades under both anaerobic and aerobic conditions via pathways similar to those observed in chlorinated solvent biodegradation. Under anaerobic conditions, PCP can undergo reductive dechlorination to less chlorinated daughter products: TeCP, TCP, DCP, and monochlorophenol (MCP) (Figures 1 and 2; Mikesell and Boyd 1986). Reductive dechlorination has been observed under very anaerobic conditions, mainly in methanogenic environments (Nicholson et al. 1992). The less chlorinated phenols (MCP and DCP) are generally more susceptible to aerobic biodegradation than the higher chlorinated phenols (Banerjee et al. 1984). However, unlike fully chlorinated solvents such as tetrachloroethene, PCP can be biodegraded under aerobic conditions (Steiert et al. 1987).

Natural attenuation is evaluated using a "lines of evidence" approach. The three lines of evidence recommended in U.S. EPA (1997) are: 1) reduction of

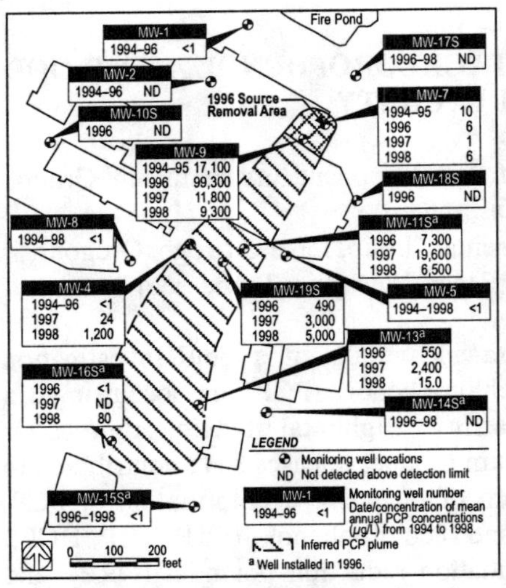

Figure 1. Mean PCP concentrations in groundwater

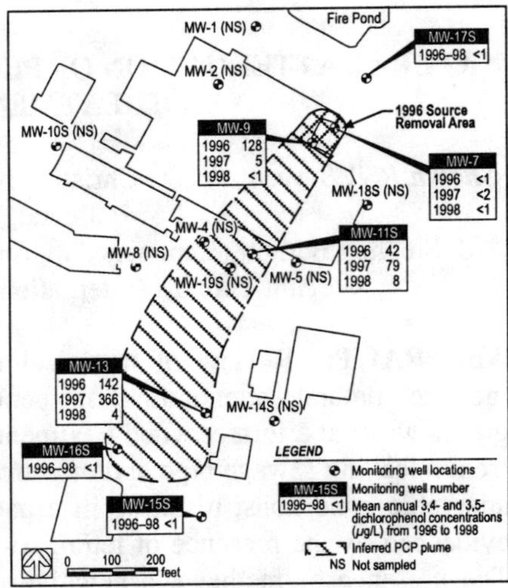

Figure 2. Mean DCP concentrations in groundwater

contaminant concentration along the flow path, 2) documented loss of contaminant mass supported by geochemical and biological decay rate data, and 3) microbiological laboratory data supporting degradation and decay rates. The third line of evidence is usually only required when the first two lines are insufficient in duration and quality to clearly establish the occurrence of natural attenuation. The first two lines of evidence for this site and a three-dimensional solute transport model developed from data obtained at the site are discussed below.

BACKGROUND

The site is an approximately 12-acre (0.4-hectare) former wood processing and treatment facility in northern California that operated from the early 1900s to the 1980s. A spray booth and dip tank, located in the northeastern corner of the site were used to apply PCP, mixed in a carrier solution of mineral spirits or equivalent diesel-range hydrocarbons. The results of site investigations performed in the 1990s have identified the presence of PCP contamination in the spray booth and dip tank area.

Topography at the site slopes to the south toward a creek located approximately 2,000 ft (610 m) downgradient from the site. Groundwater is present at depths ranging from 0.5 to 15 ft (0.15 to 4.6 m) below ground surface, depending on the location and time of year. The shallow and deep water-bearing zones are separated by a clay aquitard. The shallow zone consists of an unconsolidated silty to well-graded sand to sandy gravel unit that varies in thickness from approximately 2 to 20 ft (0.61 to 6.1 m) beneath the site. Groundwater flow in the shallow zone is toward the south-southwest. Field data suggest an interstitial groundwater velocity of 650 ft/year (198 m/year).

PCP concentrations in groundwater have been monitored since 1990. The PCP plume extends from the dip tank and spray booth area to approximately 600 ft (183 m) downgradient (Figure 1). In August 1996, the dip tank and PCP-affected soil in the vadose zone of the source area were removed. Excavation of the source area extended approximately 5 ft (1.5 m) below ground surface, to the depth of the shallow water table. Concurrent with the source removal, a sewer line in the vicinity of the source area was repaired. Post-excavation soil samples from the base of the excavation indicate PCP concentrations up to 2,600 mg/kg remain in saturated soils. The oil-based carrier has been detected in soil and groundwater, and nonaqueous-phase liquid (NAPL) has been historically observed in groundwater in the source area.

Groundwater was sampled quarterly using low-flow sampling methods from 13 of 19 existing wells located adjacent to, upgradient, and downgradient of the source area (Figure 1). Temperature, pH, oxidation reduction potential (ORP), dissolved oxygen, and specific conductivity were measured with a flow-through cell during sample collection.

Groundwater samples from 13 wells were analyzed for PCP and TeCP; samples from 7 wells were analyzed for PCP degradation products, including 2,4,6-TCP, 3,4-DCP, and 3,5-DCP by EPA Method 8151 Modified. Groundwater samples collected from the same 7 wells were submitted for general chemistry analysis (ammonia, nitrate, chloride, total organic carbon, total iron, sulfate, and sulfide). Methane was determined by EPA Method RSK 175.

An analytical Horizontal Plan Source (HPS) model (Galya 1987) was also used to simulate the three-dimensional migration of PCP in groundwater. The HPS model allows for an areal source located at the groundwater surface, and is considered appropriate to represent a PCP source in the excavation area. Because source removal resulted in PCP concentrations that were not representative of steady-state conditions, the solute transport modeling was conducted with data acquired prior to the source removal.

RESULTS AND DISCUSSION

First Line of Evidence. The source area, where NAPL has been observed, is considered to extend from MW-7 downgradient toward MW-11S. Prior to source removal in the former dip tank area (MW-9), a reduction in PCP concentrations along the downgradient flow path was evident (Figure 1). Concentrations decreased from an average of 17,100 μg/L at MW-9 to 550 μg/L at MW-13, located approximately 600 ft (183 m) downgradient. Concentrations of PCP in wells MW-7, MW-9, MW-11S, and MW-13 generally decreased over time throughout 1996 and the first quarter of 1997. In late 1997, a marked increase in the PCP concentrations was detected in wells within the plume (MW-9, MW-11S, and MW-13). This is attributed to source removal in August 1996, which appears to have mobilized residual PCP, and increased the loading of particulate and colloidal material in the shallow groundwater system. The rapid rate of migration of the high concentration spike of PCP agrees with the results of other researchers who found high concentrations of PCP (greater than 20,000 μg/L) in a carrier solution travel essentially unretarded at up to 80 percent of groundwater velocity (R = 1.2) (Davis 1994). Following the PCP spike observed in 1997, PCP concentrations in downgradient wells have decreased by more than a factor of 10 during 1998 (MW-11S to MW-13 and MW-16S) (Figure 1). In addition, the 1998 data indicate the plume axis has shifted 50 to 100 ft (15 to 30 m) to the west, resulting in a broader plume and PCP detections in MW-16S (Figure 1).

Second Line of Evidence. Since 1996, oxygen has been depleted relative to background levels (>4 mg/L) in wells located within the plume and upgradient (average dissolved oxygen ranged from 0.5 to 1.4 mg/L). Wells MW-4 and MW-5, transgradient of the plume, are generally aerobic (dissolved oxygen ranged from 1 to 5 mg/L) (Figure 3). Similarly, ORP in wells located in the plume and in well MW-17S indicate reducing conditions (ORP ranged from −35 to −116 mV); however, transgradient wells are oxidizing (ORP = 50–250 mV). The anaerobic conditions observed in upgradient well MW-17S are likely due to a leaking clay tile sewer line, a large portion of which was repaired in August 1996. Prior to source removal, ORP in the downgradient core of the plume (wells MW-9, MW-11S, and MW-13) ranged from −114 mV to 210 mV, with correspondingly low dissolved oxygen of <1 mg/L. Methane has been detected consistently at concentrations ranging from 80 to 338 µg/L in wells where DCP also has been detected up to 2,600 µg/L in MW-17S. Following the repair of the sewer lines and source removal, dissolved oxygen and ORP appear to be increasing and becoming more positive in the downgradient core of the plume. Sulfate concentrations are approximately 1 mg/L in MW-9 and MW-11S, and increase to 4–5 mg/L in MW-13. These data suggest methanogenic and sulfate-reducing bacteria are active in the source area. Nitrate is depleted and dissolved iron concentrations are elevated in samples taken downgradient of the source area indicating a change to less reducing conditions. At the leading edge of the plume in MW-15S, aerobic conditions are reestablished (dissolved oxygen >2.5 mg/L; ORP >200 mV) (Figure 3 [ORP data not shown]). While methanogenic and sulfate-reducing bacteria appear to play a role in PCP dechlorination in the area between MW-9 and MW-13, aerobic degradation may be limiting migration of lower concentrations of PCP found in the oxidized zone.

Daughter products of PCP dechlorination have been regularly detected in wells downgradient of the source (Figure 2). Prior to source removal, maximum DCP concentrations ranged from 120 µg/L to 140 µg/L in MW-13 and MW-9. However, since 1997, the diminished source of electron donors as a result of the repaired sewer line and removal of hydrocarbons from the source area appears to have slowed dechlorination in the plume. During the three 1998 quarterly sampling periods, mean DCP concentrations in MW-9, MW-11S, and MW-13 were less than 10 µg/L. Nevertheless, continued detections of DCP are clear indications of dechlorination in the plume area.

Anaerobic and aerobic biodegradation of PCP results in the release of chloride ions. Chloride data, while limited, show increasing concentrations along the downgradient flow path, with elevated concentrations found in well MW-15S downgradient of the PCP plume terminus (Figure 4). The decrease in PCP

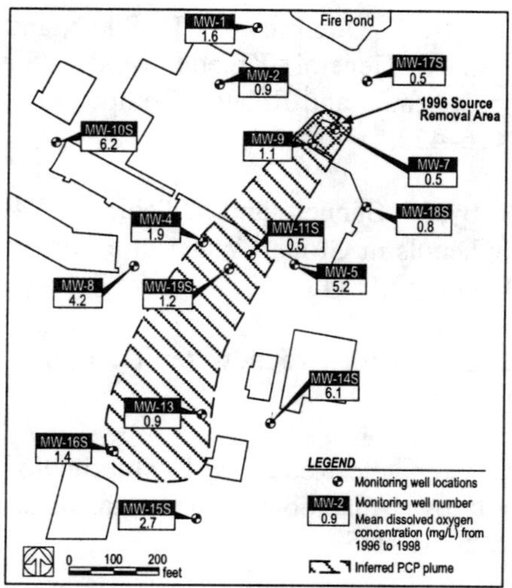

Figure 3. Mean dissolved oxygen concentrations in groundwater

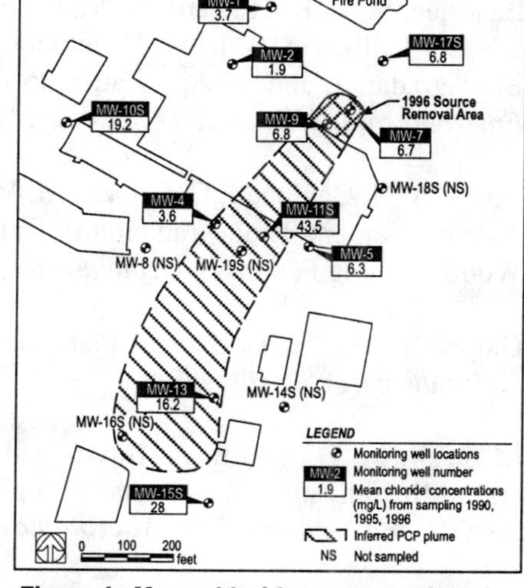

Figure 4. Mean chloride concentrations in groundwater

between wells MW-11S and MW-13 is 3 to 160 times greater than the decrease in chloride concentrations over the same distance (Figures 1 and 4). This supports the argument that biodegradation, and not dilution alone, is responsible for PCP attenuation at the site.

To further evaluate observed conditions, three-dimensional modeling using the HPS solute transport model (Galya 1987) was performed. To calibrate the solute transport model, approximately 100 simulations were performed with reasonable combinations of observed PCP concentrations in downgradient wells. Modeling results indicated that observed results could not be reproduced without ongoing biodegradation. A decay rate of $0.8 \, \text{yr}^{-1}$ was required to reproduce observed data. A PCP degradation rate of $0.6 \, \text{yr}^{-1}$ was also calculated using chloride as a conservative tracer (wells MW-11S, MW-13, MW-15S).

CONCLUSIONS

The results of this study support the lines of evidence approach used to evaluate natural attenuation of PCP. The first line of evidence is supported by historical groundwater data at the site that demonstrate the trend of decreasing PCP concentration along the flow path. The second line of evidence is supported by the presence of PCP daughter products, methane, negative Eh, low dissolved oxygen, and chloride distribution. Intrinsic biodegradation of PCP appears to be occurring at the site through both anaerobic and aerobic processes. Data collection will assist in further evaluation of the role of biological and physical mechanisms (e.g., aerobic vs. anaerobic degradation and sorption) responsible for PCP natural attenuation at this site.

REFERENCES

Banerjee, S., P. H. Howard, A. M. Rosenberg, A. E. Dombrowski, H. Sikka, and D. L. Tullis. 1984. "Development of a General Kinetic Model for Biodegradation and its Application to Chlorophenols and Related Compounds." *Environmental Science & Technology*, 18: 416–422.

Davis, A., J. Campbell, C. Gilbert, M. V. Ruby, M. Bennett, and S. Tobin. 1994. "Attenuation and Biodegradation of Chlorophenols in Ground Water at a Former Wood Treating Facility." *Groundwater*, March-April: 248–257.

Galya. 1987. "A Horizontal Plane Source Model for Groundwater Transport." *Groundwater*, 25(6)733–739.

Mikesell, M. D., and S. A. Boyd. 1986. "Complete Reductive Dechlorination and Mineralization of Pentachlorophenol by Anaerobic Microorganisms." *Applied and Environmental Microbiology*, 52: 861–865.

Nicholson, D. K., S. L. Woods, J. D. Istok, and D. C. Peek. 1992. "Reductive Dechlorination of Chlorophenols by a PCP-acclimated Methanogenic Consortium." *Applied and Environmental Microbiology*, 58: 2280–2286.

Steiert, J. G., J. J. Pignatello, and R. L. Crawford. 1987. "Degradation of Chlorinated Phenols by a Pentachlorophenol Degrading Bacterium." *Applied and Environmental Microbiology*, 53: 907–910.

U.S. EPA. 1997. *Use of Monitored Natural Attenuation at Superfund, RCRA Corrective Action, and Underground Storage Tank Sites.* EPA 9200.4–17. U.S. Environmental Protection Agency, Office of Solid Waste and Emergency Response, Washington, DC.

ACTIVATION OF AN INDIGENOUS MICROBIAL CONSORTIUM FOR BIOAUGMENTATION OF SOIL CONTAMINATED WITH WOOD-PRESERVATION COMPOUNDS

Valérie Bécaert, Maude Beaulieu[1], Josée Gagnon, Richard Villemur[1],
Louise Deschênes, Réjean Samson
NSERC Industrial Chair for site Bioremediation, École Polytechnique, Montréal
(Québec) Canada
[1]INRS, Institut Armand-Frappier, Laval, (Québec) Canada

ABSTRACT: The concept of activated soil is based on the cultivation of selected biomass from a fraction of contaminated soil in bioreactors. A pentachlorophenol (PCP) based wood-preserving mixture (WPM) was used as substrate for soil activation in order to acclimate the microbial flora to several contaminants. Experiments were conducted in two 8 liter fed-batch 10%(w/v) slurry reactors, one being fed with technical grade PCP, the other with PCP based WPM. PCP, C_{10}-C_{50}, pH, microbial density activities and DNA were monitored during the 60 day activation period. The PCP analysis from reactor samples showed that maximal PCP degradation rates were similar in both reactors, going from 19 to 132 mg/L·d in the soil activated with technical grade PCP, and from 22 to 98 mg/L·d in the WPM activated soil. Mass balance indicates that 99.9% of PCP and 99.0% of C_{10}-C_{50} added to the slurries were degraded. The bacterial counts showed that the concentration of PCP-degrading bacteria was not influenced by the presence of the other components in the WPM. The gene encoding PCP-4-monooxygenase (*pcpB*) was detected in both activated soils and amplification of 16S ribosomal genes by polymerase chain reaction and their analysis by single strand conformational polymorphism revealed that the biodiversity dropped dramatically during activation.

INTRODUCTION

Pentachlorophenol (PCP) is a xenobiotic compound used in the wood preserving industry for its biocidal properties and is a significant source of contamination for soil and groundwater. Commercialy, it is used dissolved in diesel oil containing a lot of impurities such as chlorophenols, chloro-dioxins and chloro-furans.

Traditional bioremediation techniques of PCP contaminated soils are costly and often require long treatment periods. Furthermore, PCP-degrading bacteria are very few if present at all in contaminated soils. The concept of activated soil is based on the enrichment of an indigenous microbial consortium, in a fed-batch bioreactor, fed with a targeted contaminant. This process increases in a short period of time the density of the micro-organisms desired (Otte et al., 1994). Once produced, the biomass is mixed with the contaminated soil to reduce the lag time and overall remediation time.

Previous results, using technical grade PCP (NaPCP) as activating solution, showed that this approach can be considered as a promising technology (Barbeau et al., 1997). However, commercial PCP is often accompanied by other contaminants such as diesel oil, dioxins and furans and other chlorophenols. These compounds also need to be remediated.

Objectives: The main objective of this study is to demonstrate that the activation of an indigenous microbial consortium from soil using a wood preserving mixture allows the acclimatization of the microbial flora to the major compounds contained in soil contaminated with wood-preserving substance . Therefore, two consortiums were compared; one using the technical grade PCP activation solution as previously done, the other using a commercial PCP based wood preservation mixture containing diesel oil, dioxins and furans and other impurities.

MATERIALS AND METHODS

Soil characterization. Wood preserving mixture contaminated soil was obtained from a soil treatment company. Soil was sieved (2mm) and kept at 4°C in amber-colored 1-liter bottle until used. Soil was characterized for the following physical and chemical parameters: granulometry (ASTM D422), pH (ASTM D4972), chloride ions (ionic HPLC Dionex DX500) and residual PCP concentration (EPA 625). Soil samples were also characterized to determine the PCP degrading activity by indigenous biomass in radio labeled microcosms by mineralization studies (Barbeau et al., 1997).

Soil activation. The production of the two activated biomass was accomplished by using two 8-L stainless steel fed-batch completely mixed soil slurry reactors. An aeration system provided 0.2 vvm of air and an agitation system (290 RPM) maintained soil particles in suspension. The bioreactors were covered with stainless steel caps, equipped with condensers to reduce evaporation, and were kept at room temperature.

The growth medium was composed of 10% (w/v) contaminated soil (8 kg of soil in 8 L double-buffered mineral salt medium (Greer et al., 1990)). Reagent grade NaPCP (purity of 90%) previously dissolved in 0,25N NaOH (purity of 95-100%) and wood preserving mixture diluted in 0,25N NaOH were spiked in increasing concentrations (50, 100, 100, 100, 200, 200 and 250 mg/L) in two similar reactors. The spikes were added when residual PCP concentration in the aqueous phase was below the detection limit.

Residual PCP concentration were monitored by HPLC (Spectra-system P4000, spectra-focus UV3000). Chloride ions and pH were monitored twice each day using potentiometry (Orion perpHect electrode and Acumet electrode n° 13-620-527 respectively). Petroleum hydrocarbon concentrations (C_{10}-C_{50}, ministère de l'environnement et de la faune du Québec method 410-HYD.1.0), were also monitored at strategic moments. Temperature and dissolved oxygen were maintained constant.

PCP and hexadecane degradation activity were also determined using radio labeled microcosms ([U-^{14}C]PCP specific activity 10.4/mmol, purity >98% and [1-^{14}C]hexadecane specific activity 4,1mCi/mmol, purity >98%) by mineralization studies with 10 mL slurry samples before each PCP spike (Barbeau et al, 1997).

The number of heterotrophic, hydrocarbonoclast and PCP-degrading bacteria were evaluated before each new spike using the MPN method. The growth medium for the heterotrophic bacterial count was nutrient broth (BBL). Hexadecane (purity 99%, Sigma-Aldrich Canada LTD) was chosen as the representative of hydrocarbon and was added to a mineral solution (Mills et al., 1978). An unbuffered mineral solution (MSM) containing 25 mg L^{-1} of PCP was used as growth medium for PCP degrading bacteria.

PCR and biodiversity analysis The polymerase chain reaction (PCR) amplifications were done with total DNA extracted from slurries. Two pairs of primer were used: the universal eubacterial primers and the primers for *pcpB* gene. Amplifications were accomplished in 50 μL reaction volume with 10 mM Tris-HCl pH 9.0, 1.5 mM MgCl$_2$, 50 mM KCl, 200 μM dNTP, 10 pmol of primer, 2.5 U of Taq DNA polymerase. Amplifications were done at 80°C for 3-5 min in which the bacterial DNA (100pg-10ng) was added, then 94°C-5min, 55°C-5min, 30 cycles at 72°C-2min, 94°C-40 sec, 55°C-1min and finally72°C for 10min. 10μL of pcpB PCR products were run on a 1.5% agarose gel with 2μL of loading buffer (glycerol 9%, EDTA 5mM, bromophenol blue 0.02% and xylene cyanol 0.02%) at 80 mV for 45 min. Biodiversity was evaluated using the analysis by single strand conformational polymorphism (SSCP) on the16S ribosomal genes amplified by PCR (Bassam et al., 1991).

RESULTS AND DISCUSSION

Soil characterization. The soil used for the production of the bacterial consortium was a slightly acidic sand. Its PCP concentration was 8mg/kg. The results from the PCP mineralization experiments with initial soil demonstrated a good biodegradation potential since 80% of the [^{14}C]PCP was recovered in [^{14}C]CO$_2$ after only 15 days of incubation compared to 1% for the sterilized soil.

Soil activation. During activation, the degradation rates increased in both reactors They were comparable throughout the activation process (figure 1). Maximal PCP degradation rates (max. slopes) increased from 19 to 132 mg L^{-1}d^{-1} in the reactor fed with technical grade PCP and from 22 to 98 mg L^{-1}d^{-1} when WPM was used as substrate. These results are comparable with Barbeau et al (1997) who reached 90 mg PCP L^{-1}d^{-1}.

However, the latent periods observed in the reactor using technical grade PCP were not present in the reactor using the WPM. The partitioning of PCP between water and petroleum hydrocarbon seems to have eliminated latent periods by reducing the PCP available in the aqueous phase and slowly releasing it with PCP and hydrocarbon degradation. This phenomenon could also influence the PCP toxicity threshold by decreasing the availability of the contaminant for microorganisms.

Mass balance indicates that 99.9% of the PCP added to the slurries were degraded. The PCP sorption on the soil particles was negligible (0.1% of the PCP added). Similar results have been observed in previous soil activation. Indeed, Otte et al (1994) observed a 99.8% PCP degradation and Barbeau et al. (1997) a 99.7% PCP degradation.

The pH was closely monitored during the 60 days of activation in both. Figure 1 shows that pH did not fluctuate below 7 in either reactor. PCP being a weak acid with a pKa of 4,74 in water, when pH decreases below 7, the portion of non-ionized PCP (almost not soluble) increased (Arcand et al., 1995).

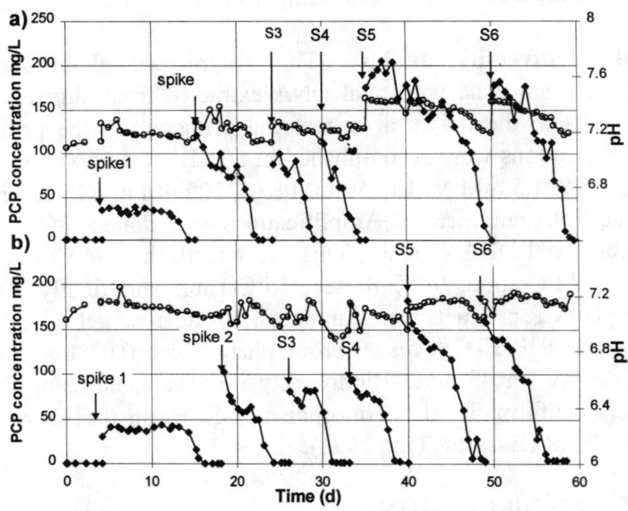

FIGURE 1 PCP degradation in reactor using a) technical grade PCP, b) wood-preserving mixture. (◆ PCP concentration, ○ pH)

C_{10}-C_{50} analysis in the reactor fed with wood-preserving mixture shows the biodegradation of petroleum hydrocarbon. However, sampling was difficult since the hydrocarbons were not soluble in water and therefore no degradation rate was evaluated. Furthermore, mass balance indicates that 99.0% of the C_{10}-C_{50} added to the slurries were degraded.

Microcosm experiments using radio-labeled hexadecane shows a higher level of hexadecane mineralization during activation, increasing from 12% to 60% in the reactor fed with WPM compared to 5% to 22% in the reactor fed with technical grade PCPMicrocosm experiments using ^{14}C-PCP were conducted on samples taken from both reactors. Figure 2 shows three mineralization curves for each reactor. The curves named Time 0 represent the slurries initial state, spike 1 is a sample taken after degradation of the first PCP spike (i.e. when no residual PCP is detected in slurries) and spike 4 was taken after the fourth PCP spike.

The results clearly show an increase in mineralization activity during the acclimation of the microbial flora in both reactors. Indeed, the short latency period as well as the high maximal mineralization rate are the two main characteristics of a very active biomass (Millette et al, 1995). The maximal mineralization rate for spike 4 was similar in both reactors indicating comparable performances.

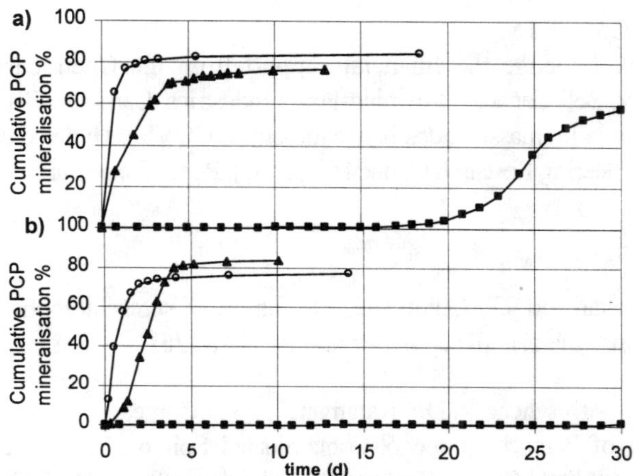

FIGURE 2 ^{14}**C-PCP mineralization in a) NaPCP activated soil, b) wood-preserving mixture activated soil. (■ Time 0, ▲ spike 1, ○ spike 4)**

The bacterial growth in the technical grade PCP activated soil varied from 10^5 to 10^7 cells/ml for the heterotrophic bacteria, 10^2 to 10^6 cells/ml for the PCP degrading bacteria and 10^2 to 10^3 cells/ml for the hydrocarbonoclasts. For the soil activated with the WPM, the total heterotrophic, the PCP degrading and the hydrocarbonoclasts microbial populations increased respectively from 10^3 to 10^8 cells/ml, 10^2 to 10^6 cells/ml and 10^2 to 10^7 cells/ml. These results show that the concentration of PCP degrading bacteria is not influenced by the presence of the other components in the WPM, the PCP-degrader density being the same in both reactors. Furthermore, the use of technical grade PCP seemed to favor the growth of PCP-degrading bacteria since density of total heterotrophic and PCP-degraders were similar

PCR and biodiversity analysis. The enzyme PCP-4-monooxygenase is necessary for the first step in the aerobic degradation of PCP (Orser et al. 1994). The gene encoding PCP-4-monooxygenase is named *pcpB*. It can be detected in soils by extracting total DNA and using the polymerase chain reaction. Detecting the gene means that the microbial flora has the potential to degrade PCP. The gene was strongly detected in both reactors but only faintly visible in the initial contaminated soil. To evaluate biodiversity in the reactors, the amplification of 16S ribosomal genes by PCR and their analysis by SSCP was performed. The results showed an important reduction in biodiversity in the reactor activated with technical grade PCP. Furthermore, there was a marked difference in microbial diversity between the two reactors. The reactor fed with WPM demonstrated a greater biodiversity than the one fed with technical grade PCP. These results corroborate those obtained by bacterial counts since the density of heterotrophic bacteria in the soil activated with wood-preserving substances is higher then the density of PCP-degrading bacteria.

AKNOWLEDGMENTS

The authors acknowledge the financial support from the Chair partners: Alcan, Bodycote/Analex, Bell Canada, Browning-Ferris Industries, Cambior, Centre québécois de valorisation de la biomasse et des biotechnologies (CQVB), Hydro-Québec, Natural Science and Engineering Research Council (NSERC), Pétro-Canada and SNC-Lavalin.

REFERENCES

Arcand, Y., J. Hawari and S.R. Guiot. 1995 " Solubility of Pentachlorophenol in Aqueous Solutions : the pH effect." *Water Res. 29* : 131-136.

Barbeau, C., L. Deschênes, D. Karamanev, Y. Comeau, R. Samson. 1997. "Bioremediation of Pentachlorophenol-Contaminated Soil by Bioaugmentation using Activated Soil." *Applied Microbiology and Biotechnology.48*(6) : 745-752.

Bassam, B., J. , G. Caetano-Anollés and P.M. Gresshoff. 1991. " Fast and Sensitive Silver Staining of DNA in Polyacrylamide Gels." *Analytical Biochestry, 196* : 80-83.

Greer, C.W., J. Hawari, and R. Samson. 1990. " Influence of Environmental Factors on 2,4-dichlorophenoxyacetic Acid Degradation by *Pseudomonas cepacia* Isolated from Peat." *Arch. Microbiol 154*: 317-322.

Millette, D., J.F. Barker, Y. Comeau, B.J. Butler, E.O. Frind, B. Clément and R Samson. 1995. "Substrate Interaction during Aerobic Biodegradation of Creosote-Related Compounds: A Factorial Batch Experiment." *Environmental Science and Technology. 29*(8): 1944-1952.

Mills, A.L., C. Breuil and R. R. Colwell. 1978 " Enumeration of Petroleum-Degrading Marine and Estuarine Microorganisms by the Most Probable Number method." *Can. J. Microbiol., 24* : 552-557.

Orser, C.S. and C.C. lange. 1994. "Molecular analysis of Pentachlorophenol Degradation." *Biodegradation, 9*: 237-288.

Otte, M.-P., J. Gagnon, Y. Comeau, N. Matte, C.W. Greer and R. Samson. 1994. "Activation of an Indigenous Microbial Consortium for Bioaugmentation of Pentachlorophenol/Creosote Contaminated Soils." *Applied Microbiology and Biotechnology. 40* : 926-932.

BIOTIC AND ABIOTIC CONTRIBUTIONS TO REDUCTIVE TRANSFORMATION OF ORGANIC POLLUTANTS

E. Erin Mack (National Research Council), James W. Beck, and W. Jack Jones. (United States Environmental Protection Agency, Athens, GA, USA)

ABSTRACT: The relative contributions of biotic and abiotic reductive transformation processes were probed in two anoxic freshwater sediments by following the transformation of nitrobenzene and 2,4 dichlorophenol (compounds with reducible functional groups of different one electron reduction potential). The sediments differed in their ambient concentrations of iron, organic matter and *in situ* redox potential but both sediments were competent for transformation of the test compounds. The transformation of nitrobenzene is known to be mediated by both biotic and abiotic processes. Of the two sediments tested, nitrobenzene reduction in one sediment was found to be predominantly carried out by abiotic pathways while in the other sediment biotic pathways dominated. Results suggest that the abiotic reduction of nitrobenzene is dependent upon specific pools of reduced iron present in the sediment. In comparison, reductive dechlorination of 2,4 dichlorophenol, a process known to occur via strictly biotic pathways, was limited by availability of organic carbon in one sediment and redox level in the other.

INTRODUCTION

Reductive transformations are important processes for transformation of organic pollutants in anoxic environments. Reductive dechlorination of a pollutant often results in a product that is less toxic, more mobile, and more available to (aerobic) degradation than the parent compound. It is well known that there are both biotic and abiotic reactions that contribute to environmental reductive transformations. For example, specific bacteria transform organic pollutants directly using the pollutants as terminal electron acceptors for the conservation of energy or as co-metabolites. In contrast, reduced iron minerals serve as a reductant for a number of organic pollutants either directly (Klausen et al, 1995) or via a bulk electron donor such as organic matter (Dunnivant, Schwarzenbach, and Macalady, 1992). Although this latter transformation reaction is considered to be abiotic, reduced iron in the environment is present as the result of microbial reduction of iron minerals. In such a case, reductive transformation of the pollutant may be tightly coupled to the physiology of iron reducing bacteria. The relative contributions of both biotic and abiotic processes are important for determining (1) the capacity of a specific environment for transformation of organic pollutants and (2) the products of these transformation reactions.

In this investigation we examined both abiotic and biotic processes for reductive transformation of organic chemicals with reducible functional groups of differing one electron reduction potentials. Transformations of nitrobenzene and 2,4 dichlorophenol were followed in slurries of sediments from two different ecosystems. Nitrobenzene is known to be reduced both biotically and abiotically (Haderlein and Schwarzenbach, 1995), and reduced iron is known to be an important reductant for environmental nitrobenzene reduction (Klausen et al, 1995). In contrast, dechlorination of 2,4 dichlorophenol does not occur in biologically inhibited systems (Liu and Jones, 1995) and is unlikely to be transformed by reduced iron in an abiotic reaction. The abiotic contributions to reductive transformation of nitrobenzene and 2,4 dichlorophenol were identified by comparing rates of degradation in live slurries and in controls that had been treated with the protonophore 3,3',4',5-tetrachlorosalicylanilide (TCS) or by autoclaving. TCS inhibits microbial iron reduction (Kostka and Nealson, 1995) and was used as an alternate means of evaluating non-biologically mediated transformations. For nitrobenzene, a compound for which abiotic paths of degradation are important, the rates of reduction were measured over a range of temperatures at, above and below which biological processes were expected to be active. Because 2,4 dichlorophenol is reportedly degraded strictly through biological pathways, the effects of redox and added organic carbon, were examined.

MATERIALS AND METHODS

Sediments were collected from Cherokee Park Pond (high clay and iron content, 3% organic matter) and Oconee River (low clay and iron content, low organic matter) as surface grabs of bulk sediment. All further manipulations of the sediment samples (sieving and slurry preparation) were carried out in a N_2 filled glove bag. Sediment slurries (sediment diluted 10 or 20% wt/vol with N_2-sparged site water) were prepared and allowed to settle for 30 seconds; supernatant was decanted and aliquots placed into serum vials that were sealed with a butyl rubber stopper and an aluminum crimp seal. The settling step removed a sand fraction of the sediment that contributes little to overall reductive transformation of nitrobenzene and 2,4 dichlorophenol (data not shown). Nitrobenzene, 2,4 dichlorophenol, and 4-chlorophenol were added to the slurries from anoxic aqueous stocks. TCS was added from an acetone stock and sulfide was added from a neutral pH stock; final concentrations were 1 and 1.5 mM respectively. Organic carbon was added as a mixture of lactate, pyruvate, glucose and yeast extract such that the final concentrations were 5, 2, 0.1 mM and 0.01% respectively. The 2,4 dichlorophenol spiked slurries were incubated statically at 25°C. The nitrobenzene spiked slurries were continuously mixed during incubation at 4, 10, 25 and 70°C in order to provide adequate mass transfer of reactants. The slurries were sampled with a N_2-flushed syringe, preserved in an equal volume of acetonitrile and frozen until analysis.

Analysis for nitrobenzene and 2,4 dichlorophenol was carried out on a Hewlett Packard 1050 Series HPLC equipped with a multi-wavelength detector. Nitrobenzene was eluted isocratically with 60% methanol, 38% water, and 2% glacial acetic acid (1.5 ml/min) on a Waters C_{18} - microbondpak column. Nitrobenzene and chlorinated phenols were quantitated at 260 and 280 nm respectively.

Reduced iron in 0.85N HCl extracts of sediment slurries was measured by the ferrozine assay (Kostka and Luther, 1994) on a Hewlett Packard 8453 UV/VIS spectrophotometer. Head space methane concentrations were measured on a Hewlett Packard 5890 GC equipped with FID.

RESULTS AND DISCUSSION

Experimental conditions of autoclaving, TCS amendment, and incubation over a range of temperatures were used to evaluate biotic and abiotic reduction of nitrobenzene to aniline in Cherokee Park and Oconee River sediment slurries. The rates of nitrobenzene reduction in slurries from both sites increased with increasing temperature (Fig. 1). Previous experiments exhibited negligible loss of nitrobenzene from sterile water controls (data not shown). The contribution of biotic pathways to nitrobenzene reduction (defined as the amount of reduction sensitive to autoclaving and TCS) was greatest at 25°C (Fig. 2), the most microbially permissive temperature tested, while abiotic pathways dominated nitrobenzene reduction at 70°C in both sediments. However, nitrobenzene reduction was more sensitive to TCS amendment and autoclaving in the Oconee River sediment slurries incubated at 70°C (Figures 1 and 2) as compared to Cherokee Park sediments. This may be due to the more positive initial redox level of Oconee River sediments causing them to be more sensitive to oxidation during autoclaving and/or interaction of TCS with the comparably smaller pools of reactive iron.

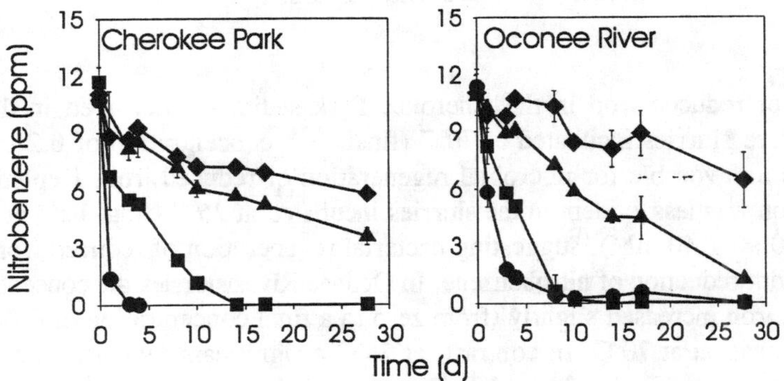

Figure 1: Rates of nitrobenzene reduction in live, unamended Cherokee Park and Oconee River sediment slurries (10% wt/vol) incubated at 70, 25, 10, and 4°C. Each point represents the average of three replicates. Error bars = standard deviation.
—●— 70°C; —■— 25°C; —▲— 10°C; —◆— 4°C.

These results suggest that in general, abiotic reduction of nitrobenzene (mediated by pools of reduced iron minerals) is relatively more important in the Cherokee Park sediments while in Oconee River sediments the reduction of nitrobenzene is tightly coupled to microbial activity. This conclusion is further supported by the observed changes in the reduced iron pools of the slurries. Initially, reduced iron concentrations were 1.36 mM in the Cherokee Park sediments but were below detection limits in the Oconee River sediments. A large

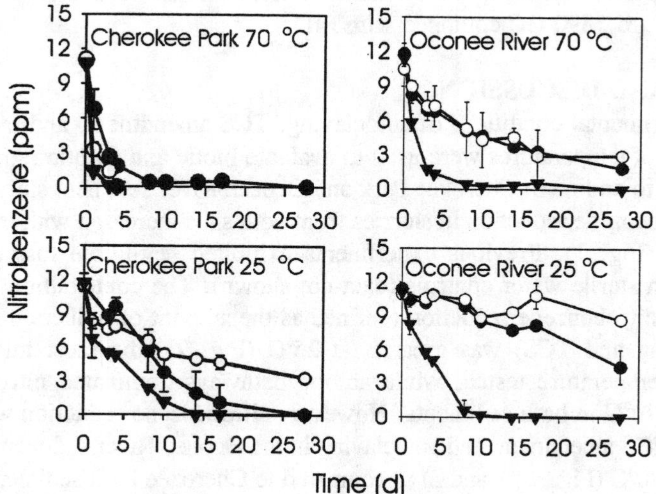

Figure 2: Rates of nitrobenzene reduction in Cherokee Park and Oconee River sediment slurries (10% wt/vol) incubated at 70 and 25°C. Each point represents the average of three replicates. Error bars = standard deviation. —●— Autoclaved; —○— 1mM TCS; —▼— Live, Unamended.

depletion of reduced iron in the Cherokee Park sediments occurred in the un-amended live slurries incubated at 70°C (final Fe^{2+} concentration of 0.26 mM), conditions unfavorable for microbial regeneration of reduced iron. Depletion of reduced iron was less evident in the slurries incubated at 25°C (final Fe^{2+} concentration of 0.90-1.40 mM), suggesting bacterial regeneration of reduced iron concomitant with reduction of nitrobenzene. In Oconee River slurries the concentration of reduced iron increased slightly (from zero to a final concentration of 0.06mM) during incubation at 70°C. In contrast, at 25°C a significant increase in reduced iron was observed (final 0.38 mM Fe^{2+}) as a probable result of active microbial reduction of iron. Compared to reducing sediments like Cherokee Park, oxidizing sediments such as Oconee River require active microbial populations to provide reducing conditions: pools of reductant, such as reduced iron, sufficient for trans-formation of nitrobenzene. Consequently, nitrobenzene reduction in the more oxi-dizing Oconee River sediments is tightly coupled to microbial activity. In contrast the reducing Cherokee Park sediments have a pool of reductant sufficient for abiotic nitrobenzene reduction. However, maintenance of a pool of reduced iron was possible only under conditions favoring microbial iron reduction as evidenced by the significant loss of reduced iron in Cherokee Park sediments incubated at 70°C.

Reductive dechlorination of 2,4 dichlorophenol to 4 chlorophenol was also investigated in Cherokee Park and Oconee River sediments. Reductive dechlorina-tion of 2,4 dichlorophenol is inhibited by autoclaving (this study, Fig. 3 and Liu

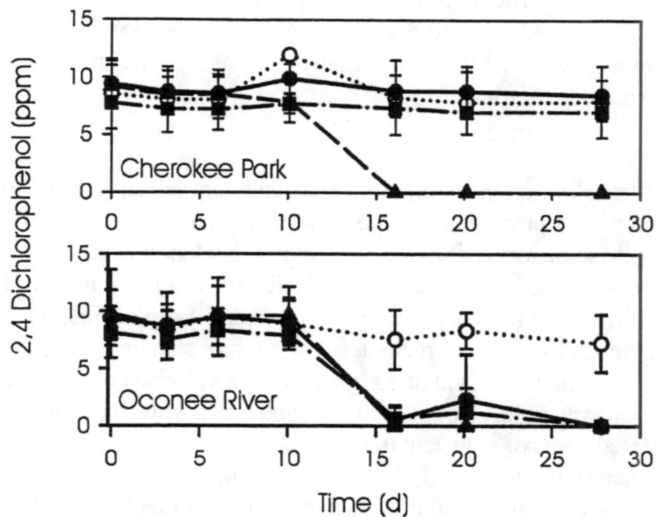

Figure 3: Profile of dechlorination of 2,4 dichlorophenol in Cherokee Park and Oconee River sediment slurries (10 and 20% wt/vol, respectively) with and without amendments. Each point represents the average of three replicates. Error bars = standard deviation. ○ Autoclaved; ● Unamended; ■ 1.25 mM Sulfide; ▲ Organic Carbon.

and Jones, 1995) and addition of TCS (data not shown), indicating that this transformation is mediated by direct microbial processes. Sediment slurries were amended with sulfide or organic carbon to determine if microbial dechlorination of 2,4 dichlorophenol was limited by redox level and/or by availability of a suitable electron donor.

In both Cherokee Park and Oconee River sediments, the 2,4 dichlorophenol was completely dechlorinated to 4 chlorophenol. Further dechlorination to phenol was not observed. Dechlorination of added 4 chlorophenol was also not observed in parallel slurries in which 2,4 dichlorophenol was ortho dechlorinated in unamended or sulfide or organic carbon amended conditions (data not shown). 2,4 dichlorophenol and 4 chlorophenol were not dechlorinated in sterile water controls in the presence or absence of added organic carbon, or sulfide (data not shown). In Cherokee Park sediments, 2,4 dichlorophenol was dechlorinated only when amended with organic carbon (Fig. 3). This result suggests that dechlorination of 2,4 dichlorophenol is limited by available electron donor in Cherokee Park sediments. In contrast, Oconee River sediments dechlorinated 2,4 dichlorophenol under all conditions tested with the exception of the autoclaved control (Fig. 3). This result suggests that dechlorination of 2,4 dichlorophenol in Oconee River sediments is not limited by available electron donor. Final head space concentrations of methane in live slurries were higher than in sterile water and autoclaved controls. Slurries amended with organic carbon produced the most methane. The lag period prior to dechlorination is probably representative of the time needed for the sediment to reach a sufficiently low redox potential to support dechlorination. The inability of either sediment to dechlorinate 4-chlorophenol under any of the tested conditions suggests that the *in situ* microbial populations are not competent for 4-chlorophenol degradation.

CONCLUSIONS

Environmental transformations of organic pollutants are controlled by numerous physical, chemical and physiological site conditions. This study showed that the contributions of these factors to the overall rate of degradation of nitrobenzene and chlorinated phenols varied with the site and pollutant in question. Nitrobenzene reduction was carried out largely through abiotic pathways in sediments that were poised at a sufficiently low enough redox level to produce standing pools of reduced iron minerals. In sediments poised at a higher (more oxidizing) redox level, nitrobenzene reduction was tightly coupled to microbial processes that lowered redox levels and increased pools of reductant (reduced iron minerals). Loss of 2,4 dichlorophenol dechlorination in biologically inhibited controls showed that this process is not dependent upon a standing pool of reductant in the sediment but is mediated by active microbes. Understanding the interacting effects of environmental conditions, microbial activities, and physicochemical properties of organic pollutants is important in reducing the uncertainty in fate, exposure, and assessment models for contaminated ecosystems.

REFERENCES

Dunnivant, F. M., R. P. Schwarzenbach, D. L. Macalady. 1992. "Reduction of Substituted Nitrobenzenes in Aqueous solutions Containing Natural Organic Matter." *Environ. Sci. Technol.* 26:2133-2141

Haderlein, S. B., R. P. Schwarzenbach. 1995. "Environmental Processes Influencing the Rate of Abiotic Reduction of Nitroaromatic Compounds in the Subsurface." In J. C. Spain (Ed.), *Biodegradation of Nitroaromatic Compounds*, pp. 199-225. Plenum Press, New York, NY.

Klausen, J., S. P. Trober, S. B. Haderlein, R. P. Schwarzenbach. 1995. "Reduction of Substituted Nitrobenzenes by Fe(II) in Aqueous Mineral Suspensions." *Environ. Sci. Technol.* 29:2396-2404.

Kostka, J. E., G. W. Luther. 1994. "Partitioning and Speciation of Solid Phase Iron in Saltmarsh Sediments." *Geochim. Cosmochim. Acta* 58(7):1701-1710.

Kostka, J. E., K. H. Nealson. 1995. "Dissolution and Reduction of Magnetite by Bacteria." *Environ. Sci. Technol.* 29:2535-2540.

Liu, S. M., W. J. Jones. 1995. "Biotransformation of Dichloroaromatic Compounds in Nonadapted Freshwater Sediment Slurries." *Appl. Microbiol. Biotechnol.* 43:725-732.

BIOLOGICAL STRATEGIES FOR THE REMEDIATION OF CHLOROPHENOL-CONTAMINATED SOILS

Patrick Steinle[1], Philipp Thalmann[2], and Gerhard Stucki[3], [1]University of Zurich, Zurich, [2]Swiss Federal Institute of Technology, Zurich, [3]Ciba Specialty Chemicals, Pratteln, Switzerland

ABSTRACT: For the cost- and ecoefficient remediation of a site contaminated with chlorophenols (CP), biological clean-up strategies were developed. Soil contaminated with moderate concentrations (0.6 to 20 mmol/kg) of several CP congeners can be remediated by a combination of alkaline extraction and mineralization of the extracted CP in a bioreactor. Alkaline extraction was optimized, using 2,6-dichlorophenol (2,6-DCP) as model compound. It yielded 74% removal efficiency in one and 97% in three subsequent steps. The procedure is applicable to different types of soil and a wide range of CP concentrations. The resulting aqueous extract (6.8 mM 2,6-DCP) was treated in an aerobic fixed-bed bioreactor. 2,6-DCP was degraded to below the quantification limit (1.8 μM) and significant detoxification was reached at volumetric loading rates up to 2.1 g/(L$\times$d). The effect of environmental factors on the *in situ*-degradation of low concentrations of CP (< 0.6 mmol/kg) in unsaturated soil was studied to predict *in situ* degradation rates. The presence of sufficient oxygen and CP-concentrations below the toxicity level were crucial for rapid degradation of the CP. Bioaugmentation and elevated temperature did significantly increase the degradation rate, whereas soil moisture and the presence of a secondary substrate had no significant effect.

INTRODUCTION

CP are found as soil contaminants in many places, due to their direct release into the environment as pesticides and due to improper handling of intermediary chemical products and wastes (Häggblom and Valo, 1995). Owing to their relatively high water solubility, CP often spread from the original core of the contamination into the surroundings, creating zones of progressively lower contaminated soils. Considering cost- and ecoefficiency, it is advisable to treat different CP-concentrations differently, according to the degree of contamination. For highly contaminated soil, removal and incineration seems to be the only feasible solution (Acharya, 1998). For moderately CP-contaminated soils, composting in bio-piles (Häggblom and Valo, 1995), thermal desorption (Koustas and Fischer, 1998) and deposition have been proposed and applied as remediation technologies. Low concentrations may be treated *in situ* (Feidiecker et al., 1995).

The methods mentioned for the remediation of moderately CP-contaminated soils can be applied only if important conditions are met: Composting is very space-requiring and complete CP-degradation will take months to years; thermal desorption needs appropriate facilities and consumes large amounts of energy; deposition does not really eliminate the contamination and requires both appropriate landfills and legal permission. In lab-scale experiments, we have developed a very attractive alternative for the remediation of moderately CP-contaminated soil, combining alkaline soil-washing with subsequent mineralization of the dissolved CP in a fixed-bed bioreactor.

Many factors may influence the *in situ* degradation of CP in unsaturated soil. We have conducted statistically designed experiments which allow to assess the relative importance of the individual parameters and of interactions between them. The results allow better predictions of natural attenuation on one hand, and the design of well-focused *in situ*-bioremediation measures on the other hand.

MATERIALS AND METHODS

Soil Properties. Landfill-samples were taken at 7 m depth at a site highly contaminated with CP from waste-dumping. Soil A is a loamy sand that was chosen for its similarity to the landfill samples, but has had no history of previous CP contamination. Soil B is a forest soil with a sandy-clayish loam texture, which was collected from an industrial area. Soil properties are listed in table 1.

TABLE 1. Soil properties.

soil type and designation in text	% coarse sand > 250 µm	% fine sand > 45 µm	% silt	% clay	max. WHC[1] (ml/kg)	organic matter (% w/w)	pH	K_D[2]
sand	61.5	33.5	5.0	0	210	0	7.4	0.8
loamy sand, soil A	57.6	28.1	7.8	6.5	245	2.1	7.5	2.8
forest soil, soil B	29.8	28.8	21.1	20.4	492	6.0	7.3	3.5
landfill	59.3	27.9	8.0	4.8	263	2.25	7.6	ND[3]

[1] Water holding capacity, [2] linear adsorption constant for 2,6-DCP at pH 7, [3] not determined

Alkaline Soil Extraction. Two kg of soil A, containing 6.14 mmol 2,6-DCP/kg, were placed in a dough mixer and 2 L of NaOH (10 mM, pH 12) were added and the slurry was stirred for 30 min. The soil was then settled for 4 h and the supernatant was removed. To obtain a high extraction yield, three extraction steps were done.

For the extraction of larger amounts of soil, the washing was performed in a semi-continuous countercurrent mode as shown in figure 1 (left side), using fresh NaOH only for the third extraction step. The supernatant of step 3 was used for the previous extraction step 2, and the supernatant of step 2 for extraction of the crude soil in step 1.

Fixed-bed bioreactor. A schematic drawing of the reactor set-up is shown in figure 1 (right). The aerated fixed-bed reactor consisted of a glass column with an inner diameter of 2 cm and a total volume of 220 mL. It was filled with 60 mL of sintered glass beads of 2 - 4 mm diameter. The flow rate was held at 3 L/d. Increasing amounts (38 - 200 mL) of the elute from alkaline soil extraction step 1 were fed to the mixing chamber.

When the biological degradation of 2,6-DCP started, a pH gradient developed in the column. The pH-drop of 0.1 to 0.4 over the column was counterbalanced by dosing NaOH (0.5 M) to the mixing chamber, (pH 7.2 ± 0.1).

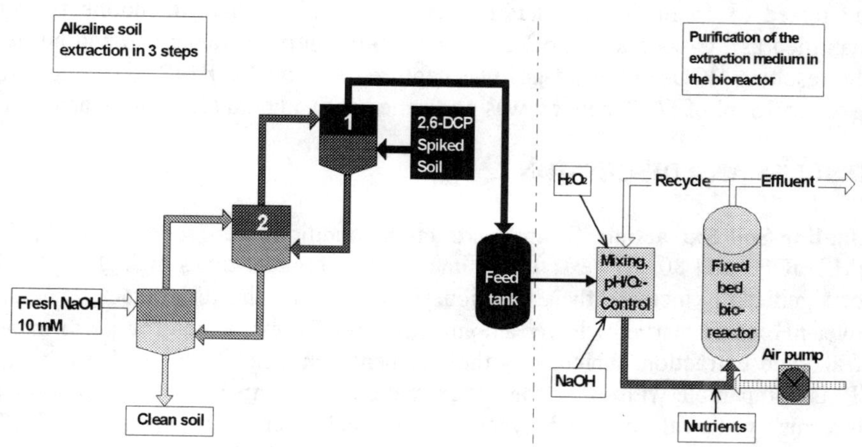

Figure 1. Experimental set-up for the alkaline soil-extraction in combination with mineralization of the dissolved 2,6-DCP in a fixed-bed bioreactor.

Microcosms with Unsaturated Soil. The impact of several environmental parameters on 2,6-DCP degradation in soil was assessed with a fractional factorial experimental design. Table 2 lists the chosen parameters examined at two levels each (high and low). To weighten the importance of the individual parameters and to assess possible interactions, all parameters were combined with each other (2^n combinations, where n is the number of parameters examined). The statistical experimental design allowed to reduce the number of combinations to 2^{n-1} without loosing information on the level of two-factor interactions. Soil microcosms were established and incubated as described in detail by Steinle et al. (1999). The results were analyzed using the half-live period of 2,6-DCP as response variable.

Table 2. High and low values of the examined environmental parameters for the fractional factorial experimental design.

Combination n°	Temp. [°C]	Inoculation [CFU/g dw]	Soil moisture [% max. WHC]	Phenol conc. [µmol/kg dw]	2,6-DCP conc. [µmol/kg dw]
high value	20	10^6	90	106	617
low value	8	10^3	30	0	62

Analytical Methods. 2,6-DCP was measured by reversed-phase HPLC equipped with a nucleosil C-18 column and a UV-detector.

The fate of the disappearing 2,6-DCP in soil was examined using U-^{14}C-labeled 2,6-DCP. 50 g of soil with 61 µmol 2,6-DCP/kg and an activity of 155 Bq/g were filled in 250 ml Schott-flasks, equipped with a Falcon tube containing 5 ml of 2 M NaOH. Mineralization was determined by trapping CO_2 into NaOH: The soil was acidified by addition of 10 ml 5 M H_3PO_4, then the NaOH was removed and mixed to 15 ml of scintillation cocktail (SC). Remaining components were quantified as ^{14}C-counts: 1 g of soil was weighted into a ceramic cup and burned. The resulting $^{14}CO_2$ in the off-gas was captured in 5 ml NaOH (2 M), which was mixed to 15 ml of SC. Counting was performed with a liquid scintillation analyzer.

RESULTS AND DISCUSSION

Alkaline Soil Extraction. Optimal extraction conditions were found at pH 12 (10 mM NaOH) and 30 min. extraction time. Longer contact times or higher pH did not significantly increase the extraction yield, whereas shorter contact times and lower pH led to markedly lower amounts of 2,6-DCP extracted. The performance of alkaline extraction, which uses the excellent water solubility of deprotonated CP, is comparable with other solvent extractions and with Soxhlet extraction and performs well with different types of soil and over a broad range of CP-concentrations (extraction efficiency between 72 and 88%). Alkaline extraction is solvent-free and enables subsequent aerobic microbial degradation of the dissolved 2,6-DCP. With the semicontinuous-counterflow extraction as shown in figure 1, the overall extraction efficiency was as high as 97%.

Fixed-bed bioreactor. Figure 2 shows the time-course of feeding of the soil extract to the bioreactor. On day 74, the oxygen source was changed from (cell-toxic) H_2O_2 to air. The volumetric loading rate could be increased up to 2.1 g 2,6-DCP/(L×d) at a 100% conversion efficiency.

Mass Balance. The treatment of 16 kg of soil contaminated with 6.17 mmol 2,6-DCP/kg consumed 16 L of water, 9 g NaOH, 1.3 L of a nutrient solution and some energy for stirring and pumping. About 97% of the soil contaminant were removed and mineralized completely. Both the soil and the extraction medium were detoxified to a great extend, as measured with a luminescent-bacteria test kit.

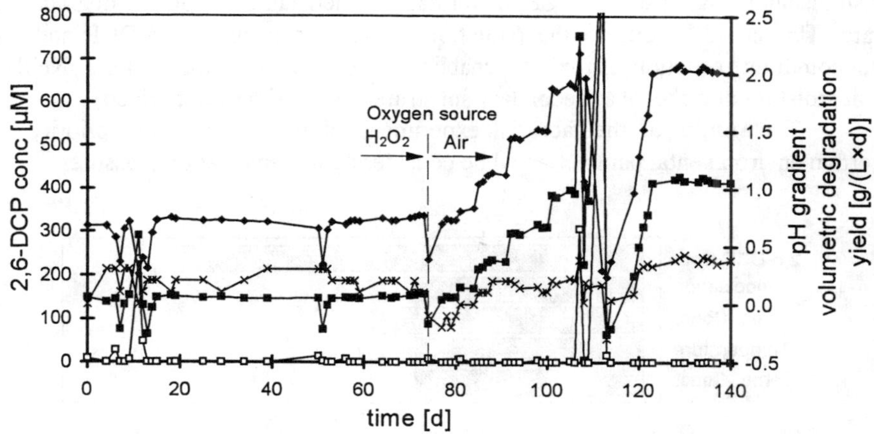

FIGURE 2. Mineralization of 2,6-DCP in a fixed-bed bioreactor. Left scale: ■ feed conc.; □ effluent conc.; right scale: ◆ volumetric degradation yield (2,6-DCP degraded per reactor volume and time); × pH decrease over column.

2,6-DCP Degradation in Unsaturated Soil. Using microcosm-experiments with radiolabled 2,6-DCP in soil we were able to prove its complete mineralization to CO_2 (Figure 3). The activity remaining in the soil reflects ^{14}C incorporated into biomass or humic substances, as no remaining 2,6-DCP could be detected.

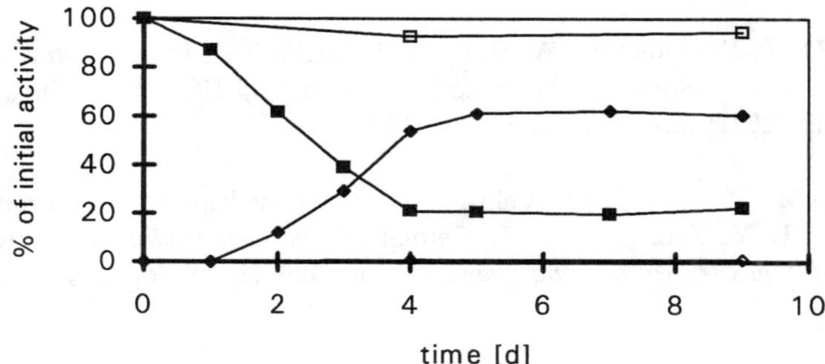

FIGURE 3. Mineralization of U-^{14}C labled 2,6-DCP in soil. ■: remaining activity in soil, after acidification; □: remaining activity in soil of the sterile control; ◆: [14]C trapped as CO_2; ◊: [14]CO_2 evolving from sterile control.

With the fractional factorial design, three factors and two interactions between factors were identified as crucial for the efficient 2,6-DCP degradation in soil (Figure 4). The most important factor was the contaminant's concentration itself. Concentrations above 0.43 mmol 2,6-DCP/kg strongly delayed the degradation rate. Factors contributing to fast conversion were the inoculation with 2,6-DCP degrading bacteria and increased temperature. Strong interactions were observed between 2,6-DCP concentration and temperature and inoculation.

Astonishingly, aging of 2,6-DCP in soil had a beneficial effect on the degradation rate. This could be due to the reduction of directly available 2,6-DCP and the concomittant recuction of toxicity, enabling faster growth of the bacteria. Neither soil moisture nor phenol as secondary substrate made a significant effect.

The results of the factorial experiment allow to focus with priority on certain environmental parameters while considering bioremediation measures.

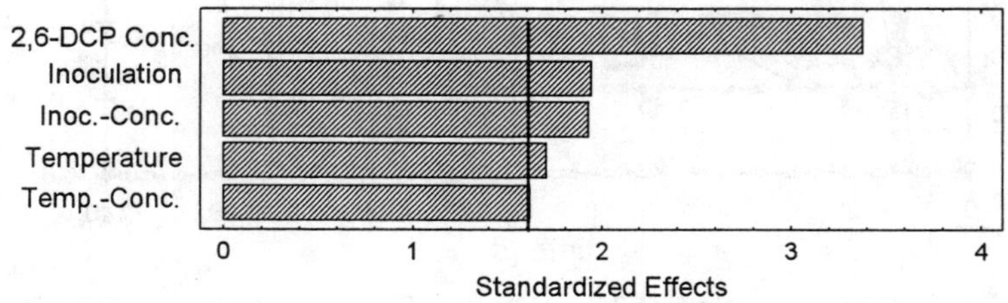

FIGURE 4. Standardized effects of environmental parameters and their interactions on the biodegradation of 2,6-DCP in unsaturated soil. The vertical line represents 90% confidence interval.

REFERENCES

Acharya, P. 1998. "Twenty Years of Site Remediation via Incineration in the United States." *Environ. Prog. 17*: 31-37.

Feidieker, D., P. Kämpfer, and W. Dott. 1995. "Field-Scale Investigations on the Biodegradation of Chlorinated Aromatic Compounds and HCH in the Subsurface Environment." *J. Contam. Hydrol. 19*: 145-169.

Häggblom, M. M., and R. J. Valo. 1995. "Bioremediation of Chlorophenol Wastes." In: L. Y. Young, and C. E. Cerniglia (Eds.), *Microbial transformation and degradation of toxic organic chemicals*, pp. 389-434. Wiley-Liss Inc., New York, NY.

Koustas, R. N., and D. Fischer. 1998. "Review of Separation Technologies for Treating Pesticide-Contaminated Soil." *J. Air and Waste Manage. Assoc. 48*: 434-440.

Steinle, P., G. Stucki, R. Stettler, and K. W. Hanselmann. 1998. "Aerobic Mineralization of 2,6-Dichlorophenol by *Ralstonia sp.* Strain RK1." *Appl. Environ. Microbiol. 64*: 2566-2571.

Steinle, P., P. Thalmann, P. Höhener, K. W. Hanselmann, and G. Stucki. 1999. Effect of Environmental Factors on 2,6-Dichlorophenol Degradation in Soil. In press.

PERFORMANCE OF *DESULFITOBACTERIUM FRAPPIERI* PCP-1 IN A CONTINUOUS ANAEROBIC REACTOR TREATING PENTACHLOROPHENOL

B. Tartakovsky, M.-J. Levesque, R. Dumortier, *S. R. Guiot*
(Biotechnology Research Institute, NRC, Quebec, Canada)
R. Beaudet (Institute Armand-Frappier, Quebec, Canada)

ABSTRACT: In this work, a mixed bacterial community of an anaerobic upflow sludge bed reactor degrading pentachlorophenol (PCP) was augmented in a strain of anaerobic PCP degraders *Desulfitobacterium frappieri* (strain PCP-1). To estimate the efficiency of augmentation, the population of PCP-1 in the reactor was enumerated using a competitive polymerase chain reaction (cPCR) technique. Strain PCP-1 appeared to compete well with other microorganisms of the mixed bacterial community with its population increasing from 10^6 to 10^{10} cell/g volatile suspended solids over a period of 70 days. Proliferation of the strain PCP-1 allowed for a substantial increase of the volumetric PCP load from 5 to 80 mg/L/day.

INTRODUCTION

Augmentation of natural bacterial populations with highly efficient laboratory strains is attractive in maximizing bioprocess performance. Anaerobic degradation of pentachlorophenol (PCP) is an example of a process that may benefit from the addition of a laboratory strain. PCP can be completely mineralized to methane and CO_2 under anaerobic conditions in a sequential process of dechlorination and mineralization by a mixed bacterial consortium (Christiansen et al., 1995; Mikesell et al., 1986). In this sequential biotransformation the dechlorination step is rate limiting due to high toxicity of PCP (Duff et al., 1995).

In this work, a strain of anaerobic PCP degraders *Desulfitobacterium frappieri* (Bouchard et al., 1996) was used to augment a mixed bacterial community of anaerobic sludge. This organism is a strictly anaerobic gram-positive bacterium isolated from a methanogenic consortium degrading PCP. It is the first known anaerobic microorganism that can degrade PCP to 3-Cl-phenol via the formation of 2,3,4,5- and 3,4,5-Cl-phenols. To estimate the efficiency of the augmentation, the population of PCP-1 in the reactor was enumerated using a competitive polymerase chain reaction (cPCR) technique.

MATERIALS AND METHODS

Analytical methods. Reactor off-gas composition (CH_4 and CO_2) was determined by gas chromatography. Details of the method are provided elsewhere (Shen et al., 1996). The inorganic chloride content was determined using the colorimetric Mercuric Thiocyonate method (APHA, 1995). Concentration of PCP and its metabolites was determined by high performance liquid chromatography (HPLC) as described in Arcand et al. (1995).

Microorganisms. *D. frappieri* strain PCP-1 was grown under anaerobic conditions at 37°C in a mineral salts medium supplemented with 55 mM pyruvate and 0.1% yeast extract. Anaerobic sludge (62 g/L VSS) was obtained from a UASB reactor treating wastewater from a food industry (Champlain Industries, Cornwall, On, Canada).

DNA extraction and competitive Polymerase Chain Reaction (cPCR). 1-2 mL sludge samples were centrifuged for 5 min at 10,000 g. The pellet was re-suspended in 700 µL of TEN (10 mM TRIS/HCl pH 8.0, 1mM EDTA, 100mM NaCl) and DNAs were extracted with a glass mill homogenizer according to a protocol given in Levesque et al. (1997).

The enumeration of PCP-1 cells by competitive PCR was based on the procedure of Zachar et al. (1993) and described by Levesque et al. (1998). Briefly, PCR amplifications were performed on serial dilutions of an internal standard each mixed with a constant amount of sludge DNA in the presence of specific primers PCP1G and PCP4D (5'AGGTACCGTCATGTAAGTAC3'). The internal standard was composed of the same primers-binding sites, size and sequence as the target DNA (16S rRNA gene) except that an EcoRV restriction site was introduced by PCR-mediated site directed mutagenesis. The resulting PCR products were digested by EcoRV, fractionated by 2% agarose gel electrophoresis and colored with Vistra Green (Molecular Dynamics, CA, USA). For each PCR amplification, the ratio between the fluorescent intensities of the internal standard (408 bp DNA fragment) and those of the target sludge DNA (544 bp) were plotted against the amount of added internal standard. The point at which the ratio was equal to 1 was taken as a measure of the amount of target sludge DNA molecules (PCP-1 rDNA gene copies) initially present in the PCR mixture.

Reactor setup and operation. Experiments were performed in a 5 L upflow anaerobic sludge bed (UASB) reactor (Duff, et al., 1995). To minimize adsorption of PCP and its metabolites, the reactor was made of glass and recirculation and feeding lines were made of glass and viton. The reactor was inoculated with 1.5 L (39.2 g volatile suspended solids, VSS) of anaerobic sludge and 10 mL of pure culture (strain PCP-1). The reactor was operated at a temperature of 35°C and a residence time of 28 hours. The pH controller maintained the pH at 7.3 ± 0.1 using a 20 g/L NaOH solution.

All influent solutions were pumped into the recirculating reactor stream in different sidestreams. The dilution stream contained a bicarbonate buffer ($NaHCO_3$ 1.36 g/L, $KHCO_3$ 1.74 g/L). The PCP solution contained 2 g/L PCP dissolved in 20 g/L NaOH solution. The solution of nutrients contained (in g/L): sucrose, 304; butyric acid, 96; yeast extract, 7; EtOH (95%), 70; KH_2PO_4, 6;

K$_2$HPO$_4$, 7; NH$_4$HCO$_3$, 68. The feeding rate of the nutrient solution was 28 mL/day.

The PCP loading rate was related to methane production, i.e. toxicity of PCP towards the anaerobic consortium expressed as a normalized methane yield determined the PCP load. An increase in the load was imposed only if no decline in the methane production was observed. The load was unchanged or even decreased if a decline in the methane production was noted. At each time, this feeding strategy insured a maximization of the feeding rate while avoiding reactor overload.

RESULTS AND DISCUSSION

While bioaugmentation of natural bacterial communities in target strains is an obvious way of bioprocess design, the retention of a laboratory strain is not easy to achieve. In fact, disappearance of introduced strains was observed as often as retention. Not only do indigenous species outcompete target strains, they often demonstrate similar or even better performance making the use of laboratory strains unnecessary. However, the bioaugmentation approach could be advantageous in certain situations, i.e. if a fast start-up period is required or if the target compound is extremely toxic. In the later case, a lengthy adaptation period is required to achieve high biodegradation rates. Moreover, throughout the adaptation system the microbial population is extremely sensitive to the contaminant toxicity. An accidental spike of the toxicant may require a restart of the system.

Fig. 1a shows the volumetric PCP load and normalized methane production throughout the experiment. Due to a control strategy which related the PCP load to the methane production rate, the normalized methane production remained steady over the course of the experiment. We observed a substantial increase in the reactor dechlorination capacity. A steady state was reached at a volumetric PCP load of 80 mg/L/day. At this load no further increase was allowed by the control algorithm indicating that a maximal dechlorination capacity had been reached.

Analysis of the effluent composition showed a PCP concentration below 0.2 mg/L throughout the experiment, i.e. the removal efficiency was at about 99% (Fig 2a). Phenol and 3-CP were the only observed intermediates. Traces of other chlorophenols were noted but their concentrations were too low to be quantified. While phenol concentration remained below 0.5 mg/L during the entire run, a distinct increase in the 3-CP concentration was observed with increasing PCP load. At the highest PCP load, 27 mg/L of 3-CP was observed in the effluent (Fig. 2b).

Chloride concentration in the effluent corresponded well with the observed disappearance of PCP and production of 3-CP. At steady state (a PCP load of 80 mg/L/day) the dechlorination efficiency was 90.5 %. Also, 3-CP accounted for 13.1% of chloride fed in the reactor as PCP. Thus, the PCP material balance was completed with an acceptable accuracy. The dechlorination rate, estimated using the chloride release data, was 45.4 mg Cl/L/day or 70 mg PCP/L/day.

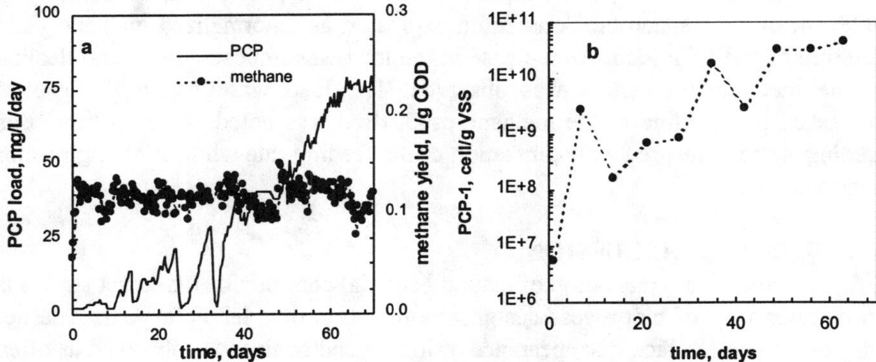

Figure 1. PCP load, normalized methane production (a) and enumeration of strain PCP-1 in the reactor using cPCR (b).

Strain PCP-1 enumeration using competitive PCR of the sludge samples suggested an increase of at least two orders of magnitude during the first week of reactor operation (Fig. 1b). After the initial fast proliferation, the growth rate slowed down. Nevertheless, the cell number approached 10^{10} to 10^{11} cell/g-VSS by the end of the run. Also, a PCR signal characteristic of strain PCP-1 was detected in the reactor effluent at day 60, suggesting intensive growth of this microorganism in the reactor.

Retention and proliferation of strain PCP-1 in the reactor could be explained by its complementary role within the anaerobic consortium exposed to PCP. Literature review (Christiansen, et al., 1995) suggests the existence of at least two dechlorination pathways. If PCP is dechlorinated by initial *meta* cleavage, the transformation occurs via a 2,3,6-TCP → 2,6-DCP → 4-CP route. Initial *para* or *ortho* cleavage results in the formation of 3-CP via either 2,3,5- or 3,4,5 TCP. In a mixed bacterial system, different bacteria are responsible for sequential dechlorination steps. These microorganisms have different growth rates and as a consequence, accumulation of partially dechlorinated products is to be expected throughout the adaptation period. Since strain PCP-1 is capable of dechlorination at the *ortho*, *meta*, and *para* positions, its presence in the reactor reduced the number of strains required for successful dechlorination. Fewer members of the consortium were required, thus eliminating rate-limiting steps in the dechlorination process. Indeed, 3-CP which is a known product of PCP dechlorination by strain PCP-1 was found to be the main chlorinated phenol in the reactor effluent while higher chlorinated phenols were not detected. Thus, the presence of strain PCP-1 minimized the PCP degradation pathway as well as the number of toxic intermediates in the reactor. Consequently, methanogens and other members of anaerobic consortium were protected and allowed to thrive in this less toxic environment.

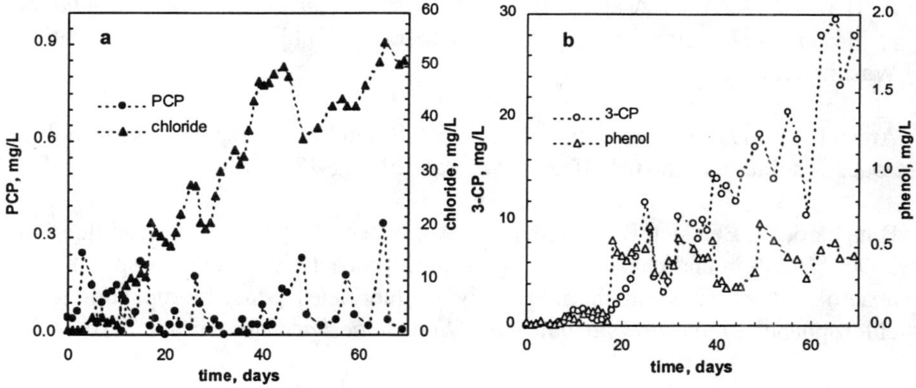

Figure 2. Effluent concentrations of PCP and dechlorination products.

In fact, a mutualistic consortium of PCP-1 and other anaerobic bacteria was established in the reactor. While strain PCP-1 dechlorinated PCP, it required pyruvate as a carbon source (Bouchard, et al., 1996). This carbon source was provided by the metabolism of other members of the anaerobic consortium. Thus, mutualistic ties were established between the introduced strain and indigenous bacteria. It could be hypothesized that upon attachment to anaerobic granules, strain PCP-1 proliferated on the granular surface, thus protecting methanogenic bacteria in the granular core from PCP toxicity. This hypothesis was confirmed by the presence of strain PCP-1 in the reactor effluent. According to the model of a layered anaerobic granular biofilm (Tartakovsky et al., 1997), faster growing species are located on the granule surface. These species will be present in the detached biofilm particles as well.

A further improvement of the system performance can be achieved by augmentation of the anaerobic population in an effective strain of 3-CP dechlorinators.

Alternatively, an enrichment in 3-CP degraders could be achieved not only by the addition of a laboratory strain but by a proper reactor control strategy aimed at the enrichment of a bacterial community in target bacteria. This control strategy, which we call "a selective stress control strategy", involves the presence of an elevated concentration of a toxicant under consideration. The concentration of such a compound, however, should be below a certain level to avoid reactor failure. Automated reactor control is a valuable approach to obtain fast enrichment in the desired population.

Acknowledgement. The technical support of M. Manuel and S. Deschamps is greatly appreciated.

REFERENCES

APHA, APHA, AWWA and WEF. 1995. *Standard methods for the examination of water and wastewater*, 19 ed. American Public Health Association, Washington, DC.

Arcand, Y., Hawari, J., Guiot, S. 1995. "Solubility of pentachlorophenol in aqueous solutions: The pH effect". *Wat. Res. 29*: 131-136.

Bouchard, B., Beaudet, R., Villemur, R., McSween, G., Lepine, F., Bisailllon J.G. 1996. "Isolation and characterization of Desulfitobacterium frappieri sp. nov., an anaerobic bacterium which reductively dechlorinates pentachlorophenol to 3-chlorophenol". *Internartional Journal of Systematic Bacteriology. 46*: 1010-1015.

Christiansen, N., Hendriksen, H.V., Jarvinen, T., Ahring, B.K. 1995. "Degradation of chlorinated aromatic compounds in UASB reactors". *Wat.Sci. Tech. 31*, 249-259.

Duff, S.J.B., Kennedy, K.J., Brady, A.J. 1995. "Treatment of dilute phenol/PCP wastewaters using the upflow anaerobic sludge blanket (UASB) reactor". *Wat. Res. 29*: 645-651.

Levesque, M.-J., La Boissiere, S., Thomas, J.-C., Beaudet, R., Villemur, R. 1997. "Rapid method for detecting Desulfitobacterium frappieri strain PCP-1 in soil by the polymerase chain reaction". *Appl. Microbiol. Biotechnol. 47*, 719-725.

Levesque, M-J., Beaudet, R., Bisaillon, J-G., Villemur, R. 1998. "Quantification of *Desulfitobacterium frappieri* strain PCP-1 and Clostridium like strain 6 in mixed bacterial populations by competitive polymerase chain reaction". *J. Microbiol. Methods. 32*: 263-271.

Mikesell, M.D., Boyd, S.A. 1986. "Complete reductive dechlorination and mineralization of pentachlorophenol by anaerobic microorganisms". *Appl. Environ. Microbiol. 52*: 861-865.

Shen, C.F., Guiot, S.R. 1996. "Long-term impact of dissolved O2 on the activity of anaerobic granules". *Biotechnol. Bioeng. 49*:611-620.

Tartakovsky, B., Guiot, S.R. 1997. "Modeling and analysis of layered stationary anaerobic granular biofilms". *Biotechnol. Bioeng. 54*: 122-130.

Zachar, V., Thomas, R.A., Goutin, A.S. 1993. "Absolute quantification of target DNA: a simple competitive PCR for efficient analysis of multiple samples". *Nucleic Acids Res. 21*: 2017-2018.

SIMULATION OF IN SITU CHLOROPHENOL BIOREMEDIATION FROM GROUNDWATER IN PILOT-SCALE

Jörg. H. Langwaldt and *Jaakko. A. Puhakka*
(Tampere University of Technology, Tampere, Finland)

Abstract: *In situ* chlorophenol-bioremediation of contaminated groundwater in pilot-scale soil reactor (2.44 m^3) was simulated to show the ability of indigenous microorganisms to mineralize chlorophenols (CP) at ambient groundwater temperature. CP-removal upto 99% and effluent concentrations for trichlorophenol (TCP), tetrachlorophenol (TeCP) and pentachlorophenol (PCP) below 0.02 mg l^{-1} were achieved. Biodegradation of CPs remained stable for over 16 weeks under aerobic conditions at hydraulic retention·times (HRT) of 0.05-2.0 d and CP-loads of 2.5-219 g CP (m^3 saturated soil d)$^{-1}$. Dissolved organic carbon (DOC) of 7.9 mg l^{-1} was reduced by 26% to a background value of 5.7 mg l^{-1}. The groundwater contained upto 8.5 mg l^{-1} of ferrous-iron. During reactor operation iron precipitated and caused plugging of the system.

INTRODUCTION

The use of chlorophenols for timber protection has led to groundwater contamination around sawmills (Valo et al. 1985). Groundwater contamination is of public health and environmental concern. In southern Finland a large-scale CP-contamination of an aquifer close by a sawmill was observed in 1987. CP-concentrations upto 190 mg l^{-1} were found in the groundwater (Lampi et al., 1990). According to the initial estimation 100,000 m^3 of groundwater had been contaminated with approximately 3,000 kg of CPs. The CP-plume migrated contaminating municipal drinking water well 800m away from the sawmill.

Various on-site pump-and-treat and off-site treatment systems to clean-up the site were studied (Ettala et al., 1992, Järvinen et al., 1994, Järvinen, 1995, Melin et al., 1998, Valo and Hakulinen, 1990). A full-scale fluidized-bed bioreactor was installed on-site, which has operated successfully at ambient groundwater temperature (Puhakka and Melin, 1998). Due to tailing of CP-concentration in groundwater during operation of the pump-and-treat system (Puhakka and Langwaldt, submitted), a long treatment time is expected. The ability of the indigenous bacteria to degrade the CPs was shown in simulation experiments in laboratory-scale (Langwaldt et al., 1998). The aquifer harbours a diverse microbial community with a widely spread ability to degrade CPs (Männistö et al., in press).

Aim of this study was to demonstrate the potential of the indigenous microbial population to mineralize CPs under site conditions in pilot-scale.

MATERIALS AND METHODS

In situ bioremediation of CPs at pilot-scale was simulated with a horizontal flow soil reactor with the following dimensions: length 3.0 m, width 0.76 m and height 1.5 m (Figure 1). The reactor contained 2.44 m^3 of CP-contaminated sandy saturated subsurface soil from the site, which had been excavated from depths between 7 and 12 m. The pilot system was operated at ambient groundwater temperature, (6-7 °C). Groundwater containing (mg l^{-1}): TCP, 1.0-2.0; TeCP, 4.0-7.6 and PCP, 0.4-0.6, was pumped with a rate of approximately. 0.1 to 3.6 m^3 d^{-1} into the horizontal flow soil reactor. In the groundwater concentrations of following parameters were (mg l^{-1}): DOC, 7.9; ferrous-iron, 8.5; total-nitrogen, 0.6 and total-phosphorus, 0.015. The pH (6.3-6.7) of the groundwater was not buffered. Aerobic conditions were maintained by aeration of the inlet groundwater in the reactor with a rate of 2-20 m^3 d^{-1}. CP-degradation was confirmed by CP analyses, reduction of DOC and release of inorganic chloride.

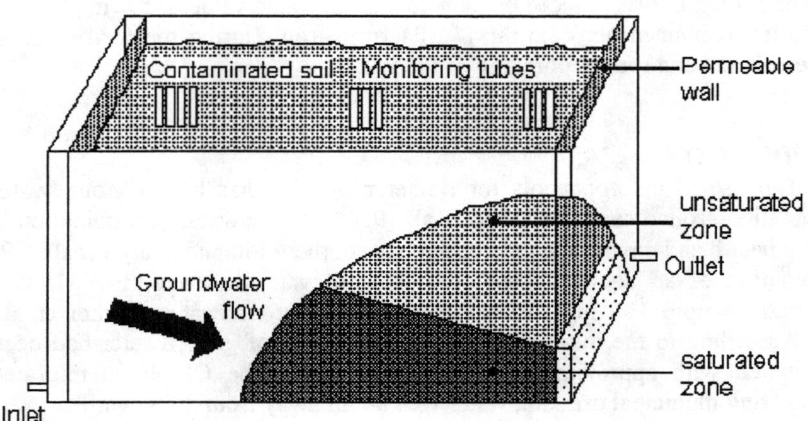

FIGURE 1. Horizontal flow soil reactor design

RESULTS AND DISCUSSION

During 16 weeks, 205 m^3 of contaminated groundwater were treated at CP-loads of 2 to 219 g CP (m^3 saturated soil d)$^{-1}$ and HRTs of 0.05 to 2.0 days. CP-concentration in the groundwater was between 5.0-10.6 mg l^{-1}. Mineralization of CPs started with the degradation of TCP and TeCP 4 and 10 days, respectiveley, after start-up of continous groundwater feed (Figure 2). Degradation of PCP started after 23 days, CP removal of 99 % and effluent concentrations of 0.02 mg l^{-1} for TCP, TeCP and PCP were achieved. Biodegradation started without addition of nutrients.

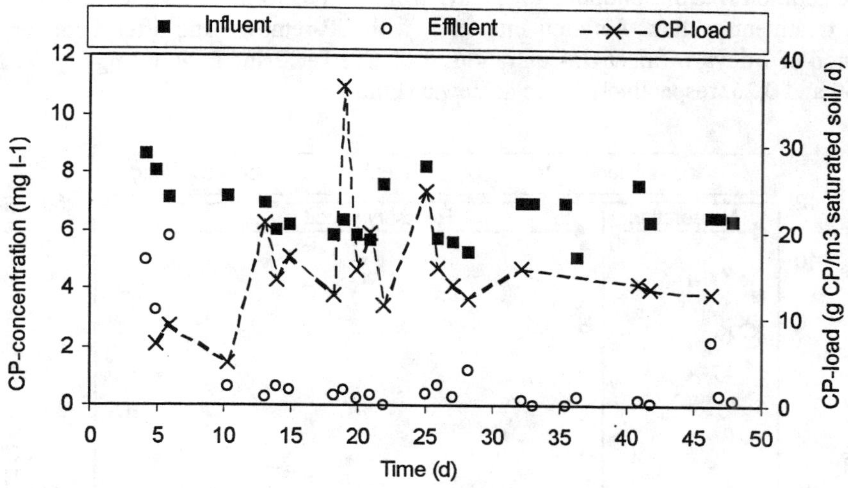

FIGURE 2. Start-up and maintenance of CP-biodegradation

During high CP-load due to elevated CP-concentration in the groundwater CP-removal decreased and the CP-concentration in the effluent increased close to the influent CP-concentration (Figure 3). After 4 days of CP-loads of over 200 g CP $(m^3$ saturated soil d)$^{-1}$ the CP-concentration in the effluent had decreased to 1.15 mg l^{-1}.

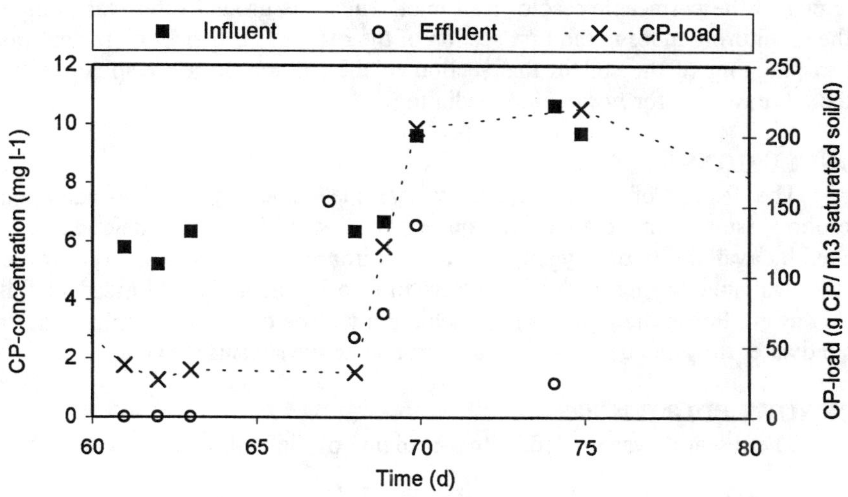

FIGURE 3. Process performance during high CP-load

The aeration was discontinued for 9 days from the reactor to study the recovery of the treatment system. Aeration interfered with CP-removal and after a recovery period of 7 days, effluent concentrations for TCP, TeCP and PCP of (mg l^{-1}): 0.02, 0.69 and 0.26, respectively, were achieved (Figure 4).

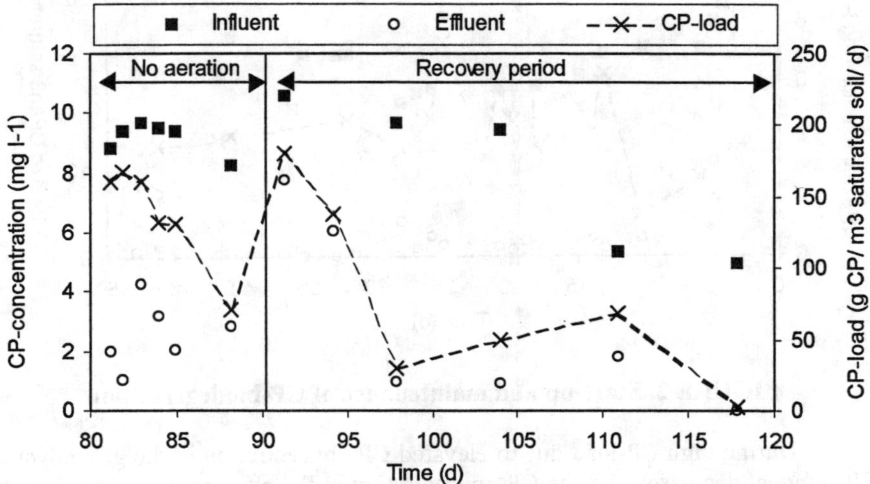

FIGURE 4. Process performance during discontinuation of aeration and during the recovery period

Iron was completely oxidized in the system. Iron precipitation caused clogging of the permeable reactor wall inlet. Then, the groundwater was pumped to the monitoring tubes in the first section of the reactor. Further iron precipitation led to plugging of the soil in this section of the reactor. Iron precipitation is a potential drawback for *in situ* bioremediation.

CONCLUSIONS

The results of this study show that the capability of the indigenous microorganisms to mineralize chlorophenols *in situ*. *In situ* biodegradation is limited by availability of oxygen, only. *In situ* iron precipitation is also controlled by the available oxygen and needs, therefore, to be carefully addressed in full-scale design. Full-scale *in situ* site remediation will be designed to utilize natural groundwater flow and generating of a microbial active subsurface zone.

ACKNOWLEDGMENTS
The research was funded by the Academy of Finland.

REFERENCES

Ettala, M., J. Koskela and A. Kiesilä. 1992. "Removal of chlorophenols in a municipal sewage treatment plant using activated sludge." *Water Research*, 26(6): 797-804.

Järvinen, K. T., E. S. Melin, and J. A. Puhakka. 1994 "High-rate bioremediation of chlorophenol-contaminated groundwater at low temperatures." *Environmental Science and Technology*. 28: 2397-2392

Järvinen, K. 1995. "Activated carbon treatment of chlorophenol contaminated groundwater." *Uusimaa Regional Environment Centre*. (in Finnish).

Langwaldt, J. H., M. K. Männistö, R. Wichmann and J. A. Puhakka. 1998. "Simulation of in situ subsurface biodegradation of polychlorophenols in air-lift percolators." *Applied Microbiology and Biotechnology*. 49: 663-668

Männistö, M. K., M. A. Tiirola, M. S. Salkinoja-Salonen, M. S. Kulomaa, and J. A. Puhakka. "Diversity of chlorophenol-degrading bacteria isolated from contaminated boreal groundwater." *Archives of Microbiology*. (in press).

Melin, E., K. Järvinen and J. Puhakka. 1998. "Effects of temperature on chlorophenol biodegradation kinetics in fluidized-bed reactors with different biomass carriers." *Water Research*. 32(1): 81-90

Puhakka, J. A., and J. A. Langwaldt. "On-site biological remediation of contaminated groundwater: a review." *Environmental Pollution*. (submitted).

Puhakka, J. A., and E. S. Melin. 1998. "Chlorophenol-contaminated groundwater bioremediation at low temperatures." In R. E. Meyers (Eds.), *Encyclopedia of environmental analysis and remediation*. pp. 1111-1120. John Wiley and Sons, NY

Valo, R., and R. Hakulinen. 1990. "Bioremediation of chlorophenol-contaminated ground water." *Kemia-Kemi*. 17(10): 811-813. (in Finnish).

BIODEGRATION OF 2,4 -DICHLOROPHENOXY ACETIC ACID UNDER FIELD CONDITIONS IN WHEAT CROP

Ajit S. Nehra and I.S. Hooda
CCS HAU Hisar - 125 004
Haryana (India)

ABSTRACT : Experiment was conducted to study the degradation of 2,4 - dichlorophenoxy acetic acid (2,4 -D) by *Azotobacter chroococcum* strain 33 and its effects on wheat (cv. WH - 283, sensitive to 2,4 - D) growth and yield. Wheat seeds were inoculated with culture of *A. chroococcum* Strain 33. Crop was sprayed with two concentrations of 2,4 -D (ester) viz., d_1 @ 570 and d_2 @ of 1140 g a. i. per hectare at 35 days after sowing (DAS). Degradation of 2,4- D and microbial count were followed in the soil collected from the rhizosphere on different days (0,7,14,21,28) after spraying. It was found that 2,4 -D degraded at a faster rate in case of inoculation in comparison to that in which inoculation was not carried out. Application of nitrogen @ of 90 kg ha^{-1} also enhanced the degradation rate of 2,4 - D. Microbial count in soil samples increased with inoculation. Spraying of 2,4 -D and application of nitrogen increased soil microbial count. The degradation rate of 2,4 -D was found to be at a faster rate under higher dose of it and where inoculation was carried out along with nitrogen application. Seed inoculation with *A. chroococcum* strain 33, with or without nitrogen application increased growth and yield of wheat. It was also recorded that seed inoculation lessened the adverse effect of both the concentrations of 2,4-D on wheat growth and yield in comparison to non-inoculation.

INTRODUCTION

A number of pesticides - insecticides and herbicides are presently used in agriculture to control the insects, weeds and other pests and increase the crop yields. Their increasing use is creating serious pollution problems making it necessary to look for the ways to lower this level in the environment. Phenoxy herbicides are known to be degraded in soil by some microorganisms. In tropical conditions such as in India, biological transformations of pesticides is expected to be relatively higher and rapid because of high levels of heat, humidity, light and microbial activities. 2,4 -D is well known herbicide used for control of broad leaf weeds in cereals. The fate and behaviour of 2,4-D in soil is critically affected by microbial degradation (Kaufman and Kearney, 1976). Studies concerning the relationship between soil population of 2,4 -D degrading micro-organisms and the rates of 2,4 -D break down (Cullimore, 1981; Ou, 1984) are very important for degradation studies.

The ability of *Azotobacter chroococcum* in degrading pesticide like 2,4 - D has not been studied in detail under Indian field conditions. *A. chroococcum* is a highly useful biofertiliser for cereals, oil seeds and other non - legume crops complementing the nitrogen requirement of field grown crops (Lakshminarayana *et al.*, 1992).

Besides nitrogen fixation *A. chroococcum* possesses a wide range of metabolic capabilities (Robson *et al.*, 1984). Balajee and Mahadevan (1990) reported the degradation of 2,4 ,6-T and lindane. Gahlot and Narula (1996) studied the degradation of 2,4 -D by a number of *A. chroococcum* strains and found that one strain number 33 possesses a very high ability to degrade 2,4 -D. So we conducted a field experiment to study the biodegradation of 2,4 -D by *A. chroococcum* strain 33.

Site Description. The field experiment was conducted with wheat (*Triticum aestivum*) at the Research Farm (Agronomy) of CCS Haryana Agricultural University, Hisar during the winter season (*rabi*) of 1995 - 96. The soil of the exprimental field was sandy loam, low in organic carbon, medium in available phosphorus and high in potash and sligtly alkaline in reaction (soil pH 7.8) The expriment was laid in randomized block design with three replications. The wheat variety " WH - 283 " was sown on 8 th December, 1995.

MATERIALS AND METHODS :

For the determination of the growth of the wheat, three rows of one metre each representing whole plot were selected at random in each plot. Effective tiller number per running metre row length, number of fertile spikelet per earhead were determined from these selected rows. Test weight, straw and grain yield were determined after sun drying.

Preparation of Azotobacter Culture. *A. chroococcum* strain 33 was procured from culture collection by the Department of Microbiology and maintained on Jensen's agar slants (Jensen, 1951) and when required it was grown in N-free liquid medium (Jensen broth) and grown for 3-4 days at 30 ± 2 °C on a rotary shaker. The fully grown broth was used to prepare culture packets by mixing with charcoal, cured and packed in polyethylene bags as per standard procedure. Wheat seeds were treated with this inoculant before sowing.

Application of 2,4 -D. Wheat cv. WH - 283, susceptible to 2,4 -D, was sown in three replications on 8th Dec. 1995 in randomised block design. 2,4 -D was sprayed 35 days after sowing uniformly with two concentrations of d_1 =570 and d_2 = 1140 g a.i. of ester formulation per hectare with and without inoculation of *A. chroococcum* strain 33.

Soil Sample. Soil sample from the field (0.29 per cent total carbon) were collected at 0-15 cm depth from rhizosphere at 0,7,14,21 and 28 days after application of 2,4 - D.Each sample was mixed thoroughly and used for various determinations including 2,4 -D degradation and microbial counts.

Measurement of 2,4 -D Degradation. The degradation of 2,4 -D was determined using the improved gas chromatographic method as described for analysis of 2,4 - D free acid in soil by Donald *et al.* (1970).

Viable Cell Counts. Ten grams soil sample was taken in 90 ml sterile distilled water and mixed thoroughly. Serial dilutions were made four times by taking one ml of suspension and mixing it in 9 ml sterilised distilled water each time. Each of the dilution series was plated on Jensen's agar medium by taking 0.1 ml of suspension. The plates were incubated at $30 \pm 2\,^{\circ}C$ for 3 days in a BOD incubator and colonies for *Azotobacter* were counted.

RESULTS AND DISCUSSION

Degradation of 2,4 -D in Soil. The inoculation of wheat seeds with *A. chroococcum* strain 33 showed 2,4 -D degradation at a faster rate at both the concentrations tested as compared to non-inoculation conditions (Table1 & Fig1). The fact that *A. chroococcum* is a rhizospheric bacterium and in addition it possess high 2,4-D degrading ability could have played the role in the observed higher degradation rate of the pesticide. In the soil availability of metabolisable carbon substrate is normally limited under conditions when sources of carbon and energy (such as farm yard manure) is not applied. Therefore, there is less likelihood of 2,4-D being co-metabolised under such conditions (Rozenberg and Alexander, 1980). At both the concentrations of 2,4-D, it was observed that up to 95 % of the weedicide degraded at 14th day after spraying in case of inoculation whereas in case of non-inoculation it was 75%. In case of non - inoculation on 28 th day after spraying, 2,4-D was present in detectable amount in soil whereas in case of inoculation it was negligible. A perusal of data in Table1 indicate that addition of nitrogen @ 90kg per hectare enchanced degradation of 2,4-D. This is because application of nitrogen increased the population of degrading bacteria as nitrogen served as source of energy for the bacteria. At 0 days of spraying 8-10 % dissipation of 2,4-D was due to drifting by air and adsorption by soil.

TABLE 1. Rate 2,4-dichlorophenoxy acetic acid degradation (per cent) as influenced by various treatments

Treatments	Days after spraying				
	0	7	14	21	28
Crop + d$_1$	10	50	80	95	99
	(3.1X10^3)	(1.5 X 10^4)	(2.0X10^5)	(3.0X10^5)	(3.93 X10^5)
Crop + d$_2$	9	45	75	96	100
	(3.0X10^3)	(1.8 X 10^4)	(2.4X10^5)	(3.8X10^5)	(4.1 X10^5)
Crop + Culture + d$_1$	8	68.33	95	100	100
	(5.3X10^3)	(7.0 X 10^5)	(6.8X10^5)	(6.8X10^5)	(6.0 X10^5)
Crop + Culture +d$_2$	9.5	75	95	100	100
	(5.1X10^3)	(8.0 X 10^5)	(9.2X10^5)	(7.1X10^5)	(6.3 X10^5)
Crop +90 N +d$_1$	10.5	65	85	96	99
	(5.0 X10^3)	(1.0 X 10^5)	(5.0X10^5)	(4.1X10^5)	(4.9 X10^5)
Crop + 90 N +d$_2$	9	63	82	98	99
	(5.0X10^3)	(1.4 X 10^5)	(5.5X10^5)	(6.0X10^5)	(5.3 X10^5)
Crop + Culture +90 N+d$_1$	8	80	95	100	100
	(8.1X10^3)	(8.0 X 10^5)	(8.1X10^5)	(8.0X10^5)	(7.0 X10^5)
Crop + Culture +90 N+d$_2$	10	82	95	100	100
	(7.9X10^3)	(9.3 X 10^5)	(8.9X10^5)	(8.4X10^5)	(7.3 X10^5)

Where d$_1$ = 570 and d$_2$ = 1140 g a. l. of 2,4 -D (ester) per hectare.
Values in parenthesis indicate viable count per gram of soil.

Viable Count of *Azotobacter* in Soil.   Due to inoculation increased number of cells per gram of soil were observed in the rhizosphere of wheat (Table 1). With both the concentrations of 2,4 -D, increase in count was observed. In case of no inoculation, the count increased uptill 28 days after spraying while in case of inoculation, the maximum count was obtained on 14th day after spraying, thereafter, the cell count decreased. This might be because of fast degradation of 2,4 -D by *A. chroococcum* strain 33 and its higher multiplication in presence of 2,4 -D in comparision to native *Azotobacter* present in soil. These results are in agreement with findings of Cullimore (1981) who showed that persistence and kinetics of degradation of 2,4 -D in soil are

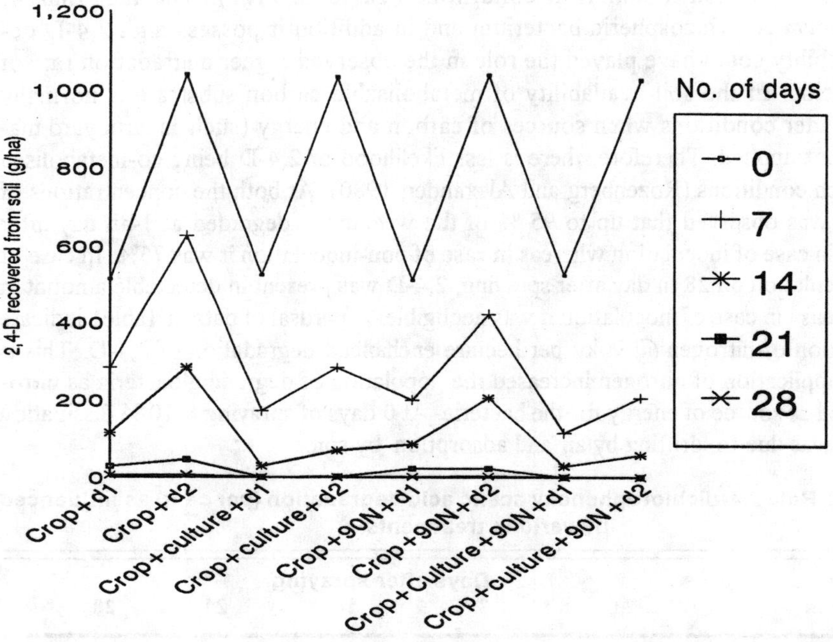

FIGURE 1. 2,4-D degradation as influenced by various treatments at different days after spraying.

associated with the population dynamics of 2,4 -D degrading soil microorganisms. Degradation rate increased with the application of nitrogen as it supported increased population of degrader.

Yield and Yield Attributes.   Inoculation of crop with *Azotobacter* produced 10.85, 6.64, 8.30, 3.81 % higher effective tiller number, spikelet number per earhead, number of grains per ear and test weight respectively over control (Table 2). Which is in the same line of study by Lakshminarayanan *et al.* (1992). An increase of 16.93 % in grain yield was obtained by inoculation as compared to control plot. At the same time, corresponding increase in straw yield observed was 7.62 % (Table 2). Spraying of 2,4 -D resulted in phytotoxic effects and had adverse effects on yields and yield attributes. Significantly lowest grain yield (1501 kg ha^{-1}) and straw yield (2530 kg

TABLE 2. Yield and yield attributes as affected by various treatments

Treatments	Effective tiller/m	Spikelet number/ earhead	Grain Number/ earhead	Test wt. (g)	Grain Yield (kg ha^{-1})	Straw Yield (kg ha^{-1})
Control	58.03	12.19	34.82	38.00	2286	3478
Crop + Culture	64.33	13.00	37.86	39.21	2673	3743
Crop + d$_1$	56.73	11.01	33.54	37.58	2037	3139
Crop + d$_2$	47.42	10.10	30.40	35.74	1501	2530
Crop + Culture + d$_1$	63.16	12.31	36.93	38.80	2443	3576
Crop + Culture +d$_2$	60.12	11.98	34.68	37.95	2207	3395
Crop + 90 N +d$_1$	100.61	14.17	43.94	41.12	3700	5718
Crop + 90 N + d$_2$	92.45	12.88	40.89	38.06	3239	5251
Crop + Culture+90 N+ d$_1$	108.43	15.29	46.90	42.66	4175	6263
Crop + Culture+ 90N + d$_2$	102.50	15.00	44.46	41.39	3913	5997
SEm ±	2.08	0.30	0.99	0.37	113	174
C.D.	6.19	0.90	2.94	1.10	337	516

Where d$_1$ = 570 and d$_2$ = 1140 g a. i. 2, 4 -D (ester) per hectare.

ha^{-1}) were observed when sprayed with higher dose of 2,4 -D. This can be attributed to significantly lesser number of effective tiller, spikelet per earhead, number of grains per earhead and abnormal morphological changes in ears by deforming their shape and shrivelled grain (low test weight) due to application of 2,4-D (Gregg and Allan, 1965; Singh and Sharma, 1984). Beneficial effects of inoculation was more apparent with higher dose of 2,4 -D and when crop was fertilized with 90 kg N/ha. It is because inoculation reduced injurious effects of weedicide at both the concentrations of 2,4 -D as it degraded the weedicide rapidly (Gahlot and Narula , 1996) in soil because inoculation increased the population of degrader in soil. In case of inoculation the *A. chroococcum* strain 33 present in the rhizosphere of plant also trasformed toxic herbicide material in root exudates of plants.

ACKNOWLEDGEMENT

The author are highly thankful to Dr. K. Laximinarayanan, Prof., Department of Microbiology, HAU, Hisar for providing necessary facilities.

REFERENCES

Balajee, S. and A. Mahadevan. 1990. " Utilisation of Chloroaromatic Substance by *Azotobacter chroococcum*." *Sys. Appl. Microbiol.* 13 : 194 - 198

Cullimore, D.R. 1981. "The Enumeration 2,4 -D Degraders in Saskatchwan Soils". *Weed Sci.* 29 : 440 - 443.

Donald W. Woodham, G. Mitchell William, D. Loftis Camella, and W. Collier Carroli. 1970. " An Improved Gas Chromatographic Method for Analysis of 2,4 -D free acid in soil." *J. Agric. Fd. Chem.* 19 (1) : 186 - 188.

Gahlot, R., and N. Narula. 1996. "2,4 -D Degradation by Resistant Strains of *Azotobacter chroococcum.*" *Indian J. Microbiol.* 36 : 141-143.

Gregg, B.R., and A.E. Allen. 1965. "2,4 -D Effects on Wheat Selections of Different Genetic Height Levels." *Crop Sci.* 5 : 93 - 95.

Jensen, V. 1951. " Notes of the biology of *Azotobacter.*" *Proc. Soc. Appl. Bacteriol.* 74 : 89 - 93.

Kaufman, D.D., and P.C. Keaney. 1976. " Microbial Transformations in the Soil." In : L.J. Audus (Ed) *Herbicides : Physiology, Biochemistry, Ecology,* Vol 2. Academic press, Landon, pp 29 - 64.

Lakshminarayanan, K., N. Narula, I.S. Hooda, and A.S. Faroda. 1992. " Nitrogen Economy in the Wheat Through the Use of *Azotobacter chroococcum.*" *Indian J. Agric. Sci.* 62 (1) : 75 - 76.

Ou, L.T. 1984. "2,4 -D Degradation and 2,4 -D Degrading Microorganisms in Soils." *Soil Sci.* 137 : 100-107.

Robson, R.L., J.A. Chesshyre, C. Wheeler, R. Jones, P.R. Woodley, and J.R. Postage. 1984. " Genome Size and Complexity in *Azotobacter chroococcum.*" *J. Gen. Microbiol.* 130 : 1603 - 1612.

Rosenberg, A., and M. Alexander. 1980. "2,4, 5 - Trichlorophenoxy acetic acid (2,4,5-T) Decomposition in Tropical Soil and its Co-metabolism *in vitro.*" *J. Agric. Fd. Chem.* 28 : 705 - 709.

Singh, B.P., and H.C. Sharma. 1984. " Effect of 2,4 -D and Hand Weeding on Weed Population and Growth and Yield of Wheat cv HD - 2009." *Haryana Agric. Univ. J. Res.* 14 : 346 -349.

FULL-SCALE BIOREMEDIATION OF CHLORINATED PESTICIDES

N.C.C. Gray, P.R. Cline, G.P. Moser, L.E. Moser, H.A. Guiler, A.L. Gray and D.J. Gannon (Zeneca Corp., Mississauga, ON, Canada)

ABSTRACT: A commercial-scale demonstration has been carried out using the Xenorem[SM] process of the Stauffer Management Company (SMC), a subsidiary of Zeneca Group PLC, to treat 1000 yd^3 of chlorinated pesticide contaminated soil. The goal of the demonstration was to show that the laboratory and pilot-scale results could be scaled-up, and that the process could remediate the soil to acceptable levels. The initial concentrations were 779 ppm toxaphene, 243 ppm DDD, 88 ppm DDT, 48 ppm chlordane, 11 ppm DDE, and 3 ppm dieldrin. The final target levels were achieved for all compounds. The U.S. EPA has recommended in the site Record of Decision that bioremediation be used as the preferred technology to treat the pesticide contaminated soil.

INTRODUCTION

SMC has been evaluating and developing remedial technologies for the treatment of pesticide contaminated soil for a number of years to remediate its own properties. In the past, the company has used low temperature thermal desorption and/or capping technologies to remediate sites. Neither of these technologies/approaches destroys the contaminants, but rather partitions them into a smaller matrix or reduces the overall risk by separating the soil from potential receptors.

For the past six years SMC has explored bioremediation as a cost-effective technology to treat pesticide contaminated soils. In this case, bioremediation is the utilization of the indigenous microflora associated with the historically contaminated soil, under controlled environmental conditions. The initial target of this research was DDT and its major metabolites, DDD and DDE (the DDX family), and then the program extended to other chlorinated pesticides, such as toxaphene and chlordane. Innovative technology and approaches have been developed and at this time three U.S. patents have been granted.

Laboratory and pilot-scale research demonstrated that the DDX family could be degraded under appropriate environmental conditions. Early model research, using uniformly labelled C^{14}-U-DDT, indicated that 100% of the DDT could be degraded without the build-up of any major metabolites, as indicated by autoradiography and thin layer chromatography.

Laboratory work using contaminated soil from the SMC, Tampa, Florida site, also indicated that the DDT could be degraded, but that greater control of the environmental parameters was required. Screening work was carried out to identify various triggers that promoted or hindered the DDT degradation. Such parameters as macro- and micronutrients, redox potential, aeration flow rates, anaerobic/anoxic/aerobic cycling patterns, temperature, moisture, pH and the addition of surfactants were evaluated in the different screens.

From these results, three technologies were identified as potential modes of application: aerobic/anoxic soil treatment, biopile and enhanced anaerobic/ aerobic composting. This preliminary field trial evaluated each technology of a 14 yd^3 scale, in enclosed cells, to obtain mass balance data. All three technologies were able to degrade DDT. Under controlled, but non-optimized conditions, excellent DDT degradation was achieved, even starting at a fairly high concentration (> 700 ppm), using the enhanced composting system.

Objective. The objective of this field trial was to demonstrate that the earlier results obtained in the field pilot-trials could be obtained under full-scale conditions, and that the technology could be used to degrade other chlorinated pesticides found as co-contaminants in the soil.

Site Description. The SMC site in Tampa, Florida was used for this trial. This facility was operated as a pesticide formulation and packaging plant from 1951 to 1986.

MATERIALS AND METHODS

Initially five "hot" zones were identified from a field-sampling program at the Tampa site. Random samples were collected from each zone and submitted to the laboratory for microcosm evaluations. Based on the sampling verification program and the results from the microcosm studies, two "hot" zones were selected for the full-scale trial.

The soil was excavated, screened, mixed and used in the full-scale trial. The soil was amended with organics, giving rise to a final volume of ca. 1000 yd^3. The amended soil matrix was engineered into a compost windrow, built inside a warehouse. Due to potential odor issues from some of the pesticides, and the high groundwater table in the area, it was prudent to carry out the trial inside the warehouse.

Six soil samples were collected from the windrow, on a biweekly schedule. Each soil sample consisted of a mix of individual composite (based on depth) samples, which were mixed together to form a single sample. The soil samples were air dried and ground prior to analysis (EPA 8080/1). The geomean of the data was plotted on a temporal basis for each chemical.

The windrow was monitored continuously using a data acquisition/ controller system for changes in temperature, redox, oxygen and moisture levels. This system supported multi-tasking in the Windows '95 environment, and permitted real-time access and modifications of the system's configuration parameters. Measurement confirmation was also carried out using hand-held probes.

Aerobic and anaerobic periods in the windrow were controlled by the use of organic amendments, and mechanical mixing.

RESULTS AND DISCUSSION

Preliminary Laboratory Study. As a preliminary assessment of the soil to be used in the field-scale trial, microcosm experiments were carried out using grab samples from the different "hot" zones of the site. The microcosms were carried out over a 6-week period, 4 weeks under anaerobic conditions, followed by 2 weeks under aerobic conditions to represent one anaerobic/aerobic cycle.

As illustrated in Figures 1 and 2, the microcosms were very successful in showing the potential of the enhanced anaerobic/aerobic composting technology to degrade all of the chlorinated pesticides. For quite a few of the pesticides, such as aldrin, BHC, dieldrin, heptachlor epoxide and methoxychlor, the minimum detection level was reached (0.5 ppm). No inhibition was seen by higher levels of pesticides than previously evaluated in the soil. These favourable results supported the decision to carry out the field trial at a larger scale.

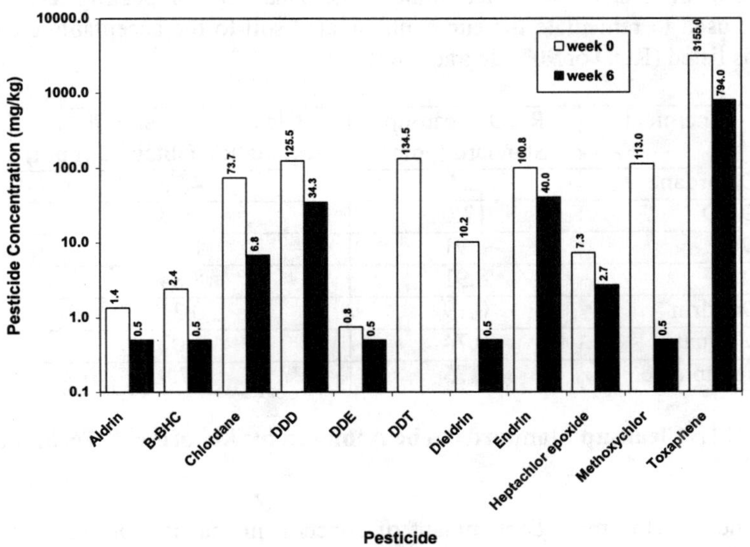

FIGURE 1. Results of Microcosm 7132 After 6 Weeks

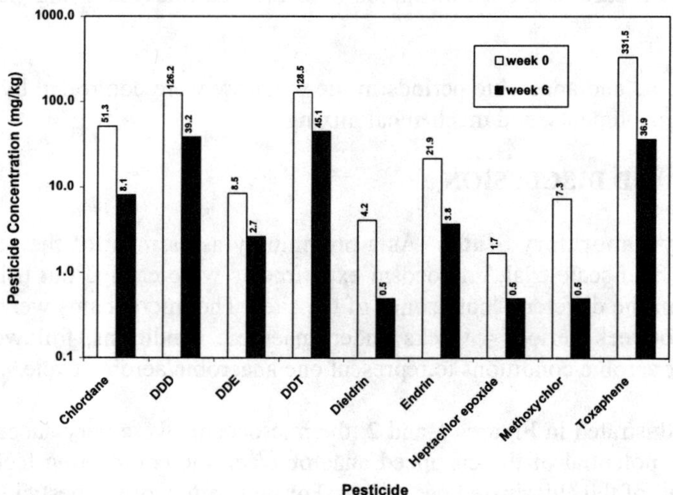

FIGURE 2. Results of Microcosm 7140 After 6 Weeks

Field Study. The chemicals of primary concern in this study are those listed on the Record of Decision (ROD) for the Tampa site (Table 1). The overall objective of the study was to determine if the enhanced composting technology could be used to remediate the site contaminated soil to the acceptable clean-up standards listed (ROD or 90% degradation).

Chemical	ROD Clean-up Standard (ppm)	Clean-up Level if 90% Degradation Obtained (ppm)
Chlordane	2.3	4.75
DDD	12.6	16.3
DDE	8.91	1.13
DDT	8.91	8.84
Dieldrin	0.19	0.31
Molinate	0.74	1.02
Toxaphene	2.75	77.9

TABLE 1. Clean-up Standards to be Achieved by Xenorem[SM] Technology

Toxaphene was the main contaminant of concern, in that it represented nearly 70% of the total contaminants, followed by the DDX family. Toxaphene was analyzed using two different methods in order to follow (a) the degradation of the parent congeners and the corresponding breakdown products, and (b) to follow only the parent congener degradation using EPA Method 8081. Figure 3 illustrates the degradation curve for toxaphene and related by-products using the two methods. Each data point on the figure represents the geomean value of at least 6 samples. Under the optimal environmental conditions, rapid losses of toxaphene and metabolic byproducts are seen, giving rise to an overall loss greater than 90% (Table 2).

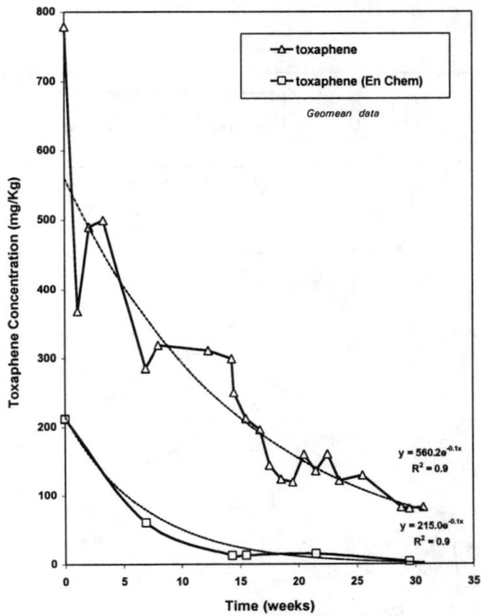

FIGURE 3. Toxaphene Degradation Curves

Pesticide	ROD Level (ppm)	T_o Value (ppm)	T_f Value (ppm)	% Degraded	Comments
Chlordane	2.3	47.5	5.2	89	Ca. 90%
DDD	12.6	242	23.1	90.5	> 90%
DDE	8.91	11.3	6.8	40	ROD
DDT	8.91	88.4	1.2	98	ROD
Dieldrin	0.19	3.1	< MDL	< MDL	< MDL
Molinate	0.74	10.2	< MDL	< MDL	< MDL
Toxaphene	2.75	469	29	94	> 90%
Tox +metab	NA	779	68	91	> 90%

TABLE 2. Summary of End-Points Achieved for the Pesticides

DDT was rapidly biodegraded in the first stage of the field trial, dropping from 88.4 ppm to ROD level by week 14 (Figure 4). DDD was measured in the original soil as the second largest contaminant (T_0 = 163 ppm), which increased very quickly with the degradation of DDT to 242 ppm. DDD was then degraded throughout the field trial, achieving more than 90% degradation. DDE was the slowest of the DDX pesticides to degrade, with only 40% being degraded, but still to a level that was below the ROD for the site.

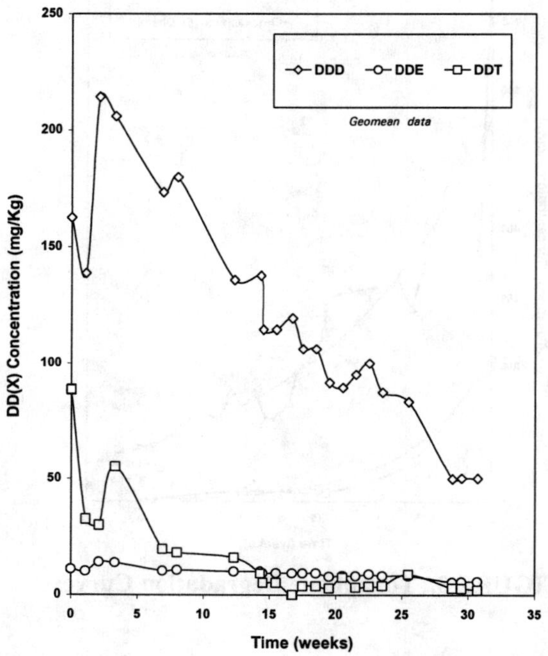

FIGURE 4. DDX Degradation

CONCLUSIONS

Data collected in this field trial and the previous laboratory and pilot-scale work shows that the ROD chemicals of concern are biodegradable. The Xenorem[SM] technology, when operating within the optimal environmental/process windows, is effective in reducing the target pesticides to at least 90% for the SMC Tampa site.

It has been demonstrated that the field conditions can be manipulated/ controlled to provide/create the necessary conditions for the technology to work successfully.

BIODEGRADATION OF AROMATIC ETHERS
BY NITRIFYING BACTERIA

Soon W. Chang, Kyonggi Univiersity, Korea
Michael R. Hyman, North Carolina State University, Raleigh
Kenneth J. Williamson, Oregon State University, Oregon

ABSTRACT: In this study we examined the oxidation of a variety of aromatic ethers by the soil nitrifying bacterium *Nitrosomonas europaea*. Our experiments demonstrate that simple methoxylated aromatics such as anisole (methoxybenzene) are rapidly oxidized by this bacterium. Our evidence also suggests that all of the observed products are compatible with reactions catalyzed by the non-specific ammonia-oxidizing enzyme; ammonia monooxygenase (AMO). The results presented describe our studies into the effect of aromatic ether structure on the rates of ether degradation and the relative ratios of aromatic hydroxylation and O-dealkylation reactions catalyzed by AMO in *N. europaea*.

INTRODUCTION

Aromatic ethers are widespread in nature and are most abundant as the polymeric components of lignin. The ether linkage is also recognized as an important structural feature that limits the biodegradability of many anthropogenic compounds. A variety of methoxylated aromatic compounds are known to be degraded by aerobic bacteria and many of these reactions are initiated by non-specific monooxygenase enzymes. For example, the methane-oxidizing bacterium *Methylosinus trichosporium* OB3b is known to oxidize simple aromatic ethers such as anisole (methoxybenzene). However, O-dealkylation reactions directed at the methoxy group of anisole are minor reactions relative to the more prominent ring-directed hydroxylation reactions (2, 3).

Nitrosomonas europaea is a lithoautotrophic bacterium that derives energy for growth primarily from the oxidation of ammonia to nitrite. The initial conversion of ammonia catalyzed by this bacterium results in the product of hydroxylamine (NH_2OH), a reaction that is catalyzed by the membrane-bound enzyme, ammonia monooxygenase (AMO). The reductant required to sustain all AMO-catalyzed reactions, including ammonia oxidation itself, is provided by the further oxidation of hydroxylamine to nitrite. This reaction is catalyzed by the complex soluble enzyme hydroxylamine oxidoreductase (HAO). Previous studies have demonstrated that AMO in *N. europaea* is a highly non-specific catalyst that is capable of transforming a wide range of non-growth supporting hydrocarbons, including simple alkyl ethers (1). In this study we have examined the ability of the soil nitrifying bacterium *N. europaea* to oxidize simple aromatic ethers.

MATERIALS AND METHODS

Growth Conditions. Cells of *N. europaea* were grown in batch cultures (1.5 liters) and harvested by centrifugation and finally resuspended in buffer (50 mM sodium phosphate [pH 7.8] with 2 mM $MgSO_4$), as described previously (1)

Degradation Studies of Aromatic Ethers. Experiments following aromatic ethers degradation by *N. europaea* were conducted in serum vials (37 ml) sealed with Teflon-lined silicone septa (Sun Brokers™, Wilmington, NC). The incubation medium (5 - 10 ml) consisted of phosphate buffer (50 mM potassium phosphate [pH 7.8], 2 mg $MgSO_4$) and $(NH_4)_2SO_4$ (5 mM initial concentration) The reactions were initiated by the addition of cells (0.25 - 0.5 ml; *ca* 2.5 - 5 mg of protein) to the reaction vials. The vials were then placed in a heated shaking water bath (30 °C, 200 rpm). At the indicated times samples (200 µl) of the liquid phase were removed from the reaction using microsyringes. The cells were removed by centrifugation in a microfuge and a sample (100 µl) of the supernatant was injected into an HPLC apparatus for analysis, as described below.

Chemical Analysis. Most analyses to quantify ethers and ether oxidation products were conducting using an HPLC equipped with a reversed-phase Ultramex C_{18} column (150 mm * 4.60 mm). Individual aromatic ethers were eluted under isocratic conditions using a solvent system of acetonitrile (30—50% [v/v]) in deionized water at the flow rate of 0.8 - 1.0 ml/min. All compounds other than 4-nitrophenol were detected by UV absorbance at 254 nm. Acetaldehyde production was monitored by gas chromatography and formaldehyde was measured by a coupled assay with formaldehyde dehydrogenase from *Pseudomonas putida*. Nitrite was determined colorimetrically and protein concentrations were determined by the Biuret assay, as described previously (1).

RESULTS AND DISCUSSION

Anisole Degradation by *N. europaea*. Our initial investigations focused on degradation of anisole (methoxybenzene) as a model aromatic ether. The results of an experiment following the time course of anisole degradation (Fig. 1) demonstrates that anisole was rapidly degraded by *N. europaea* and that phenol, 4-methoxyphenol, catechol and hydroquinone were detected as aromatic oxidation products. Formaldehyde was also detected as a minor and transient product during the early stages of the reaction. The mass balance for all detected oxidation products was greater than 85% at all points in the reaction. The product profile observed in this experiment suggested that both O-dealkylation and hydroxylation reactions were involved in anisole degradation although ratio of the products compatible with O-dealkylation (phenol, catechol and hydroquinone) exceeded products compatible with hydroxylation reactions (4-methoxy-phenol) by a ratio of 2.3:1 (at 30 min). In the presence of acetylene (1% v/v) no oxidation products were observed in an identical experiment. This result indicates the AMO is responsible for initiating the oxidation of anisole.

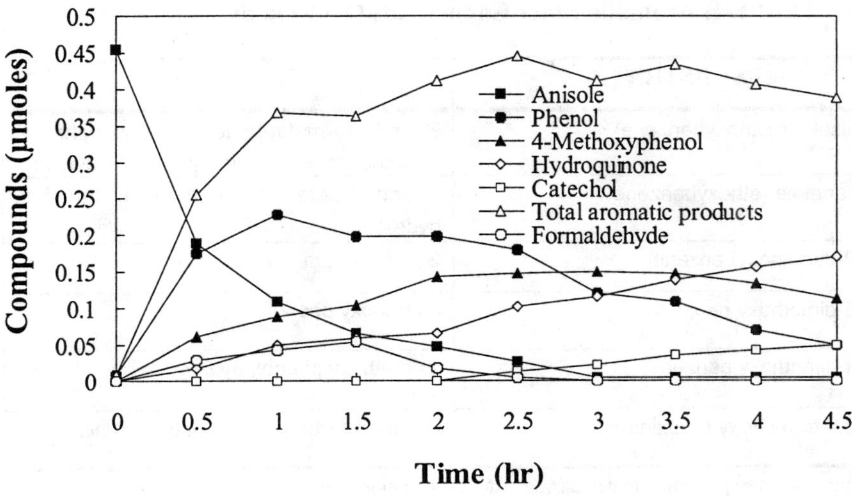

FIGURE 1: Time course of anisole degradation by *N. europaea*

Subsequent studies using intermediates identified in Fig. 1 and selective inhibitors of AMO activity (C_2H_2) allowed us to verify the following overall pathway for anisole degradation by *N. europaea* (Fig. 2).

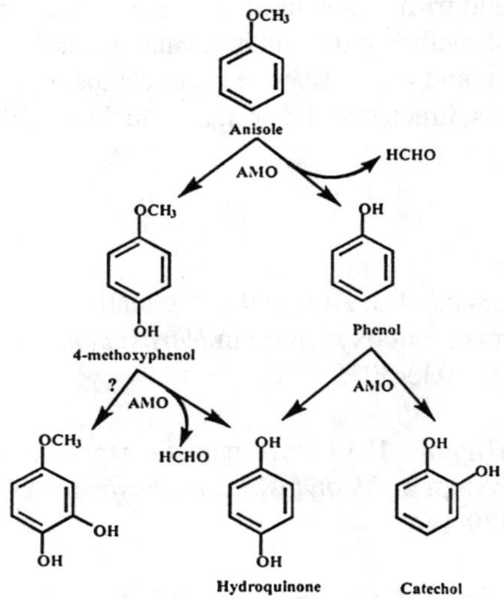

FIGURE 2. Overall Pathway of Anisole Degradation by *N. europaea*.

We subsequently investigated the reactivity of *N. euroapea* towards several methoxylated aromatic ether compounds listed in Table 1.

TABLE 1. Aromatic Ether Compounds Oxidized by *N. euroapea*.

SUBSTRATE	PRODUCTS
Anisole (methoxybenzene)	Phenol, formaldehyde, 4-methoxyphenol, 2-methoxyphenol, hydroquinone, catechol
Phenetole (ethoxybenzene)	Phenol, acetaldehyde, ethanol, catechol, hydroquinone
1,2-dimethoxybenzene	2-methoxy phenol, catechol
1,3-dimethoxy benzene	3-methoxy phenol
1,4 dimethoxy benzene	4 –methoxy phenol, hydroquinone
1,3,5-trimethoxy benzene	Oxidized (product was not determined)

The substrates are presented in decreasing order of oxidation rate

Our results indicate that most of methoxylated aromatic ether compounds were oxidized through both O-dealkylation and/or hydroxylation reactions. Since all of these transformation reactions were inhibited by the presence of acetylene it also appears that all of these reactions are catalyzed by AMO. The oxidation of phenetole (ethoxybenzene) generated a mixture of phenol, catechol, hydroquinone, acetaldehyde and ethanol. This product distribution suggests that anisole and phenetole are degraded through substantially similar pathways. In the case of the series of di- and tri-methoxylated benzene isomers we observed that an increase in the number of methoxylated substituents generally decreased both the rate of substrate oxidation and the number of products that could be accounted for by hydroxylation reactions directed at the aromatic nucleus of the compounds.

REFERENCES:

Hyman, M. R., C. L. Page and D. J. Arp. 1994. "Oxidation of methyl fluoride and dimethyl ether by ammonia monooxygenase in *Nitrosomonas europaea*". *Appl. Environ. Microbiol.* 60 (8) 3033-3035.

Jezequel, S. G. and I. J. Higgins. 1983. "Mechanistic aspects of biotransformation by the monooxygenase system of *Methylosinus trichosprium* OB3b". *J. Chem. Tech. Biotechnol.* 33B: 139-144.

Jezequel, S. G., B. Kaye, and I. J. Higgins. 1984. "O-dealkylation: a newly discovered class of reactions catalyzed by the soluble mono-oxygenase of the methanotroph *Methylosinus trichosporium* OB3b". *Biotechnol. Lett.* 6: 567-570.

Chang, S., M. Hyman, and K. Williamson 1998. "Nitrifying bacteria as "priming catalysts" for the biodegradation of xenobiotics", In G. B. Wickramanayake and R. E. Hinchee (Eds.), *Natural Attenuation: Remediation of Chlorinated and Recalcitrant Compounds*, pp.51-55 Battelle Press, Columbus, OH,

BIODEGRADATION OF 2,4-D AND DDT AT HIGH CONCENTRATIONS IN LOW-COST PACKAGING BIOFILTERS

Luis G. Torres, Germán Santacruz (Instituto Mexicano del Petróleo. México, D.F., MEXICO) and
Erick R. Bandala (Instituto Mexicano de Tecnología del Agua (Jiutepec, Morelos. MEXICO

ABSTRACT: Degradation of two pesticides (2,4-D and DDT) using a 54-ml glass column packed with tezontle (a low cost packaging consisting of a red basaltic scoria) was tested. Bacteria were cultured in a yeast, peptone and glucose (YPG) liquid medium at 32°C. To inoculate the rich medium, it was pumped over 24 hours through the column. Later, the wasted medium was discharged and the pesticide added. Optical densities, total organic carbon and pesticide concentrations were determined. Pesticide removals for 2,4-D (with initial concentrations between 100 and 500 mg/L) were around 99%. DDT removals (at initial concentrations up to 150 mg/L) were as big as 58-99%. TOC removals for 2,4-D were in the 36-87 % interval while for DDT were as high as 36-78%.

INTRODUCTION

Like many other countries, Mexico has big environmental problems regarding the presence of pesticides in water bodies and even in underground water. Dichlorodiphenyltrichloroethane (DDT) and 2,4-dichlorophenoxyacetic acid (2,4-D) are two of the most used pesticide in Mexico, despite the fact that DDT is a restricted use pesticide (CICOPLAFEST, 1993). The volume of DDT used in Mexico decreased from near 8,000 metric tons in 1988 to around 1,000 metric tons in 1995 (DGMP, 1997). Nevertheless consumption of both pesticides reached almost 7,000 metric tons in 1995 (ANIQ, 1996).

Herbicide 2,4-D has been used in Mexico since 1993 for aquatic weed control in many reservoirs that also are used as supply sources for drinking water (Martinez and Bandala, 1995). DDT has been detected at high concentrations in underground and surface water from many places of the country (Audelo, 1990; Bandala et al., 1998). These data and the knowledge of the high persistence and toxicity of these compounds, has resulted in awareness and in the inclusion of both pesticides in Mexican legislation for drinking water (DOF, 1996).

Many papers dealing with the aerobic and anaerobic biodegradation of DDT and 2,4-D have been reported in the past specifically for soils with moderate concentrations of them (Degher et al., 1997; Dooley et al., 1997). Nevertheless, relatively few works have been published about the removal of high concentrations of these pesticides in water.

Objective. The main objective of this work is to demonstrate that a low-cost biofilter with immobilized *Pseudomonas fluorescens* can effectively carry out the degradation of streams containing 2,4-D and DDT at high concentrations.

MATERIALS AND METHODS

Pesticide solution preparation. The pesticides (2,4-D and DDT) were selected for this study because of their presence in Mexican water bodies and their chemical and toxicological properties. Pesticide standard solutions were prepared as stock solutions in methanol and spiked in the column to obtain the desired concentrations. Figure 1 shows the chemical structures of the tested pesticides.

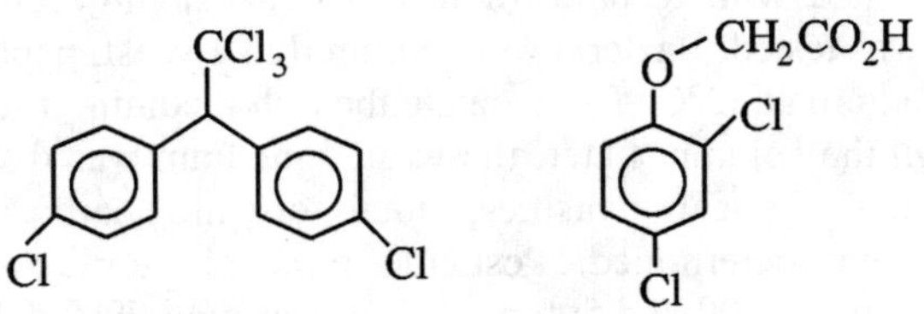

FIGURE 1. Chemical Structures of DDT and 2,4-D.

Bacteria culture. The bacteria *Pseudomonas fluorescens* were grown in liquid YPG media (in mg/L: yeast extract, 10; peptone of casein, 10 and glucose, 10) for 24 hours at 32°C. The minimum media employed was the one reported by Dapaah and Hill (1992). Media were adjusted at pH = 7.0, and cell growth was registered as the increment on optical density (O.D.) at 610 nm (1:10 dilution). This strain was previously used in the biodegradation of phenol; chloro, nitro and methylphenols; aldrin; dieldrin; heptachlor; and heptachlor epoxide (Bandala et al., 1998).

Biodegradation assessments Preliminary tests were developed, culturing the bacteria at the conditions described above and adding the pesticide after 24 hours. Optical densities were measured to investigate if bacterial growth occurred. A 54-ml glass column packed with *tezontle* (a low-cost packaging, consisting of a red basaltic scoria), in which a pure strain was immobilized, was used for the biotransformation of two pesticides. Pesticide concentrations were measured by means of a gas chromatography/mass spectrometry (GC/MS) system. Total organic carbon (TOC) was determined in a Beckman Carbon Analyzer. $FCU/g_{support}$ were measured as detailed in previous works.

Analytical methods. Pesticide concentration were monitored as follows. The sample extractions were carried out using methylene chloride nanograde. The organic phases were dried with anhydrous sodium sulfate. Once dried, the extracts were concentrated until reaching 0.5 ml volume at reduced pressure and a temperature not higher than 40°C. Extracts for 2,4-D determination were derivatizated before injection by methylation with diazomethane. Approximately one microliter of the concentrate sample was injected to a Hewlett-Packard 5890 Series II GC coupled with a Hewlett Packard 5971 Series MS. A 25 m × 0.2 mm × 0.33 µm Ultra 2 capillary column was employed. The opeational conditions were as follows: carrier flow, 80 ml/min; split/splitless ratio, 1:25; initial temperature, 70°C; final temperature 250°C; increment, 15°C/min.

RESULTS AND DISCUSSION

Free cells assessments. Maximum bacterial growth, measured as O.D./O.D.o (optical densities at a given time/optical density at the beginning of the experiment) ratios are shown in Table 1 for the blank, DDT, and 2,4-D assessments. Values reported are the average of a duplicate. It is noticeable that in most of the cases the O.D./O.D.o ratios are equal or higher for the assessments with pesticide than those without it, reaching values up to 1.72. These values were reached at 24-48 hours. On the other hand, the relationship between O.D./O.D.o and compound concentration seems to be a direct function in the case of 2,4-D, but not for DDT. This assessments are a preliminary demonstration of the feasibility of the pesticide degradation by means of *Pseudomonas fluorescens*.

TABLE 1. Maximum bacterial growth (O.D./O.D.o) in presence of the pesticides DDT and 2,4-D. Averages of duplicates.

Concentration	Blank	DDT	2,4-D
10	1.16	1.56	1.22
25	1.47	1.45	1.57
50	1.75	1.72	1.72
100	1.66	1.63	1.70

Immobilized cells. As noticed before, the biofilter designed was packed with tezontle and inoculated with a 24 hours YPG culture of *Pseudomonas fluorescens*. The average FCU/$g_{support}$ was around 3.9×10^6. The wasted media was discharged and the runs with the Dapaah media + pesticides were developed. Regarding the TOC degradation (DDT initial concentration = 25 mg/L), the typical profile of TOC as a function of time is shown in Figure 2. From this graph, it is clear that most of the TOC disappears during the first 48 hours (almost 59%) and reaches a maximum value of 78.7% at the 168 hours, time where the process was stopped

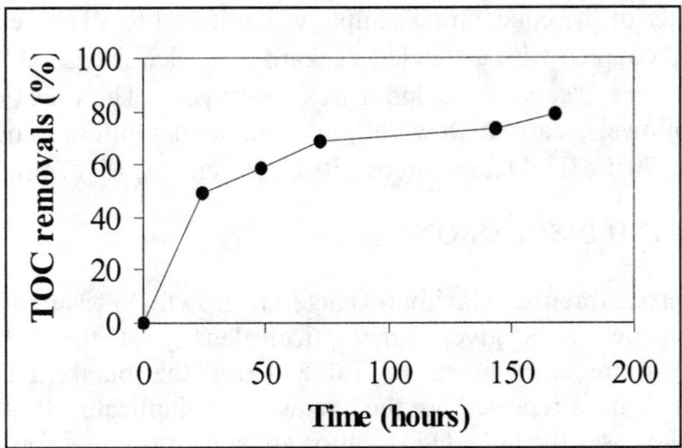

FIGURE 2. Typical Profile for the TOC (for an assessment with DDT) Degradation.

Table 2 shows the DDT-TOC removals as a function of the initial concentration of the assessed pesticide. In the case of the TOC removals, these values were in the range of 58.4-78.67%. As a general observation, the higher the DDT concentration, the lower the TOC removal (in %). In the case of the DDT (with initial concentrations between 25 and 150 mg/L) removals were as high as 58.39-99.99% (data not shown).

TABLE 2. DDT -TOC Removals as a Function of the Pesticides Initial Concentration Average of Duplicates.

Initial concentration (mg/l)	Maximum TOC removal (%)
25	78.67
50	59.49
100	58.39
150	62.11

On the other hand, 2,4-D-TOC removals as a function of the initial concentration, are shown in Table 3. The TOC removals were in the range of 36.3 and 87.4%.. It seems that the higher the 2,4-D concentration, the higher the TOC removal (in %).

TABLE 3. 2,4-D-TOC Removals as a Function of the Pesticide´s Initial Concentration Average of Duplicates.

Initial concentration (mg/L)	Maximum TOC removal (%)
100	36.28
300	80.26
400	87.38
500	78.02

In summary, the pesticide removals for 2,4-D (with initial concentrations between 100 and 500 mg/L) were around 98.79 and 99.99%. In the case of the DDT (with initial concentrations between 25 and 150 mg/L) removals were as high as 58.39-99.99%. TOC removals for 2,4-D were in the 36.3-87.32% interval, while removals for DDT were as high as 36.12-79.67%.

The next steps of the project will be: (a) upgrading of the column performance, including the optimization of the saline medium, (b) scaling-up of the process, and (c) evaluation of the system performance with real environmental samples (obtained from contaminated underground waters).

ACKNOWLEDGEMENTS

This work was carried out at IMTA. Thanks to Manuel de la Torre for English style revision.

REFERENCES

Audelo, V.J. 1990. Identificacion y evaluacion de plaguicidas en aguas del distrito de riego numero 004 del valle de Culiacan Sinaloa, Mexico. Memorias del VII Congreso Nacional SMISAAC. Oaxaca, Oax. Mexico.

Asociacion Nacional de la Industria Quimica (ANIQ). 1996. Anuario estadistico de la industria quimica mexicana 1996. Mexico, D.F.

Bandala, E.R., J.A. Octaviano, V. Albiter and L.G.Torres. 1998. Degradation of pesticides by free *Pseudomonas flourescens* cell cultures. Designing and Applying Treatment Technologies 177-182. In Wickramanayake, G.B. and R.E. Hinchee (eds), The First International Conference on Remediation of Chlorinated and Recalcitrant Compounds. Battelle Press. Columbus, USA.

CICOPLAFEST. 1993. Catalogo oficial de plaguicidas. Secretaria de Agricultura y Recurso Hidraulicos. Mexico, D.F.

Dapaah S.Y and G. A. Hill. 1992. Biodegradation of chlorophenol mixtures by *Pseudomonas putida*. *Biotechnology and Bioengineering*, 40:1353-1358.

Degher, A.B., K.G. Wang, K.J. Williamson, S.L.Woods and R.Strauss. 1997. Bioremediation of soils contaminated with 2,4-D, 2,4,5-T, Dichlorprop, and Silvex. Proceedings of the Fourth International *In Situ* and On-Site Bioremediation Symposium: Volume 2. New Orleans, USA. Battelle Press.

Direccion General de Medicina Preventiva (DGMP). 1995. Informe bienial 1994-1995. Secretaria de Salud. Mexico, D.F.

Diario Oficial de la Federacion. 1996. Norma Oficial Mexicana NOM-127-SSA1-1994, Saludo ambiental, agua para uso y consumo humano. Limites permisibles de calidad y tratamientos a que debe someterse el agua para su potabilizacion.

Dooley M., F. Yakusijin and P. Thorn. 1997. Biodegradation of pesticide contaminated soil. Proceedings of the Fourth International *In situ* and On-site Bioremediation Symposium: Volume 2. New Orleans, USA. Battelle Press.

A BIOTECHNOLOGICAL APPROACH OF DETOXIFYING HERBICIDE-CONTAMINATED BUILDING RUBBLE

Roland H. Müller, Roland A. Müller, Yvonne Jahn and Wolfgang Babel
UFZ Centre for Environmental. Research Leipzig-Halle,
D-04318 Leipzig, Germany

ABSTRACT: Investigations were performed into bioremediating the concrete rubble of a demolished herbicide production plant contaminated by a mixture of various chlorinated/methylated phenols and phenoxyalkanoic acids. Aqueous eluates exhibited pH values of up to 12.5. Bacterial strains with herbicide-degrading activities were isolated from this material, grown *ex situ* and applied as an inoculum for bioremediation. Decontamination was studied with the heaping and fixed-bed reactor techniques, the latter proving more successful. Degradation rates amounted to 20–30 g herbicides/m^3*h. The desorption characteristics showed that the herbicides were almost quantitatively removed from debris < 16 mm but persisted to a large extent in debris > 30 mm. Under appropriate conditions concrete material can be bioremediated within 2–4 weeks.

INTRODUCTION

Sites used by the chemical industry are often polluted including soil, (ground)water and the facilities themselves. In a specific case, phenoxyalkanoic acid herbicides were produced for several decades in a chemical plant in Schwarzheide/Brandenburg, Germany. Production ceased in 1990 and the plant was demolished in 1992. The building material underwent coarse crushing (producing some 24,000 m^3 of rubble) and was temporarily deposited on site. The solid matrix was heavily polluted by chloroorganics. It was covered up to avoid polluting the surroundings. Contamination was heterogenous in various samples, amounting to > 1000 mg/kg; the spectrum of compounds is shown in Table 1. Recent analytical screening indicated a mean contaminant concentration of 50–200 mg chloroorganics/kg concrete material. Owing to its toxicity, either this material must be safely disposed of or preferably the pollutants should be removed.

Objective. For both ecological and economic reasons we favor the decontamination of the building rubble by microbially mediated degradation. Applying this philosophy ought to enable the material to be reused. When we started work, virtually nothing was known about the bioremediation of concrete material (Müller and Babel 1994). The metabolism and the degradation of chlorinated phenols and phenoxyalkanoic acids has been well described for various neutrophilic bacterial strains (cf. Häggblom 1992), but aqueous eluates from concrete debris are very alkaline, exhibiting pH values of 12.5 at a debris:water ratio of 10:1. Thus success hinged on the finding bacteria which tolerate alkaline conditions. Such strains have been isolated (Hoffmann et al. 1996; Ehrig et al. 1997; Mertingk et al. 1998; Müller et al. 1998) and proven to function under *in situ* conditions. This paper discusses various aspects of tackling this problem, particularly its technical solution.

MATERIALS AND METHODS

Bioremediation experiments were performed by applying fixed-bed reactor techniques and heaping techniques. In the former, eluates derived from a dump were treated in a fixed-bed reactor, whereas in the latter, heaps were treated by permanent elution and re-circulating eluates. The principles are illustrated in Fig. 1.

TABLE 1. Chloroorganic contaminants of the building rubble

Compound (Abbreviation)	Concentration (mg/kg)	
	Sample 1	Sample 2
2,4-dichlorophenol (DCP)	1102	6.4
4-chloro-2-methylphenol (MCP)	105	5.2
4-chloro-2-methylphenoxyacetate (MCPA)	228	23.9
2,4-dichlorophenoxyacetate (2,4-D)	118	4.9
2,4-dichlorophenoxypropionate (2,4-DP)	60	29.8
4-chloro-2-methylphenoxypropionate (MCPP)	53	12.7
4-(2,4-dichlorophenoxy)-butyrate (2,4-DB)	60	3.9
4-(4-chloro-2-methylphenoxy)-butyrate (MCPB)	210	5.2

Sample 1: individual sample taken in 1992 during the crushing phase
Sample 2: average sample from Dump VI taken in 2/1998

Fixed beds and heaps were inoculated with *ex situ* grown microorganisms in different combinations depending on the state of the art at the time of investigation. The heaps were 0.05 to 5 m^3 in size while the fixed-bed reactors measured 0.005 to 0.05 m^3. The desorption characteristics of the chloroorganics were determined depending on the particle size in separate experiments without inoculation using solid:water ratios of (5-10):1.

Chloroorganic contaminants were determined individually by HPLC analysis and as a sum parameter by AOX (adsorbable organically bound halogenides) measurements via standard procedures. Eluates from solid samples were derived from 24 h extraction of particles ≤10 mm.

RESULTS AND DISCUSSION

Microbiological Terms. Percolation of the concrete rubble with a mineral salt solution using the mixed contamination as a carbon and energy source for growth resulted in the enrichment of a bacterial consortium which could be cultivated and shown to degrade the mixture of chloroorganics *in situ* (Müller and Babel 1994). Consecutive selective enrichment procedures led to the isolation of *Comamonas acidovorans* P4a (Hoffmann et al. 1996), *Rhodoferax* sp. P230 (Ehrig et al. 1997), *Aureobacterium* sp. K2-17 and *Rhodococcus erythropolis* K2-12 (Mertingk et al.

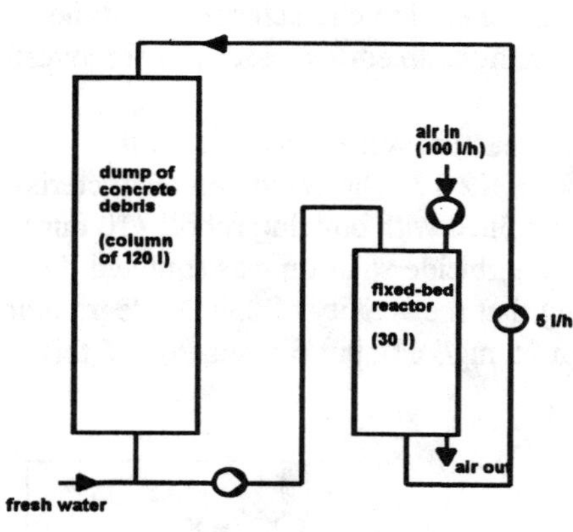

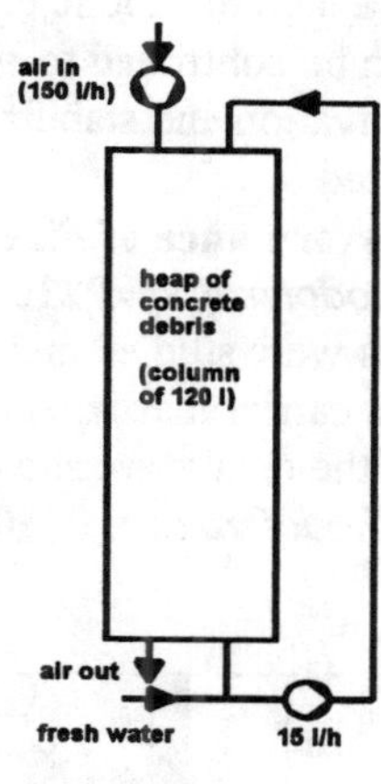

FIGURE 1. Diagram of treatment of building rubble with fixed-bed (a) and heaping technique (b)

1998) and *Ochrobactrum anthropi* K2-14 (Müller et al. 1998). Essential physiological properties concerning the degradation profile and the pH pattern are summed up in Table 2. It can be seen that activities required for the degradation of the whole spectrum of contaminants were available. Moreover, the strains were alkalitolerant/alkaliphilic.

TABLE 2: Degradative properties and pH pattern of strains isolated from building material

Strain	DCP	MCP	2,4-D	MCPA	2,4-DP	MCPP	2,4-DB	MCPB	pH
K2-14	+	+	-	-	-	-	-	-	9.5 (10.2)[1]
P4a	+	+	+	+	-	-	-	-	8.5 (10.5)
P230	+	+	+	+	+	+	-	-	7.5 (10.5)
K2-17	-	-	-	-	-	-	+[2]	+[2]	8.5 (11.5)[3]
K2-12	-	-	-	-	-	-	+[2]	+[2]	8.5 (11.5)[3]

[1] Optimum and maximum (in brackets) pH of growth or [3]degradation
[2] Only cleavage of the ether bond of the substrates by forming DCP or MCP

Degradation Procedures. As no significant degradation resulted from the activation of autochthonous microorganisms, inoculation was necessary. The feasibility of bioremediation was tested on a semi-technical scale using about 200 kg of concrete material with either the fixed-bed or heaping technique. In the former case a reactor containing building material was inoculated and supplied with eluates obtained from a model dump of concrete debris. The heaping technique involved inoculating a heap and circulating water from bottom to top (Fig. 1). Degradation was found to be effective on this scale in both cases (not shown; cf. Müller et al. 1999). However, the fixed-bed reactor technique was favored for practical reasons as a high metabolic capacity can be established in a small volume and the process can be controlled to some extent. Therefore, the elution characteristics and the activation and stabilization of the consortium in the fixed-bed reactor were investigated.

Performance of Fixed-bed Reactors. The reactors were inoculated with *Rhodoferax* sp. P230 and *Aureobacterium* sp. K2-17. The activation characteristics were studied in 5-liter fixed-bed reactors filled with building rubble (10 mm) as the carrier matrix. A two-component model herbicide solution was supplied. Some of the results are shown in Fig. 2, indicating that the activity of 2,4-DP degradation (*Rhodoferax* sp. P230) was raised to about 25 mg/kg*h but did not exceed this

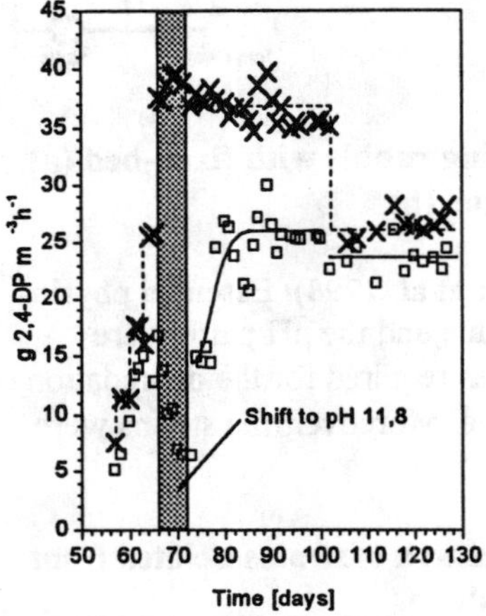

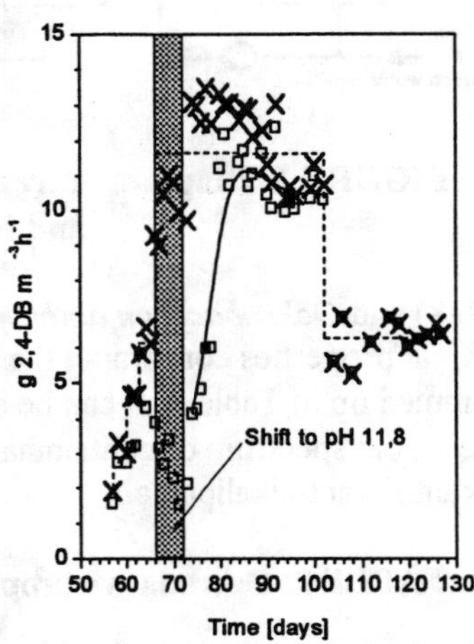

FIGURE 2. Characteristics of bacterial activity in a fixed-bed reactor supplying 2,4-DP (160 mg/l) and 2,4-DB (40 mg/l), pH 11.2 (×, feed rate; □, degradation rate)

value even with excess substrate. The degradation of 2,4-DB (*Aureobacterium* sp. K2-17) was quantitatively related to the feeding profile. The initial pH value was 11.2; it can be seen that a shift up to 11.8 resulted in a significant decrease in the degradation rate. This loss of activity was negligible in experiments in which the feeding rate was adapted to the actual conversion rate (cf. Müller et al. 1999). Similar results were obtained in a 50-liter reactor; when applying initial pH values of 8.5–10 and feeding the whole spectrum of contaminants, activation of the reactor proceeded within about 10 days (Fig. 3). The pollutants were not completely degraded but the residual concentration followed the feeding profile; this presumably results from inhomogeneities in the reactor. Taking these results into account, control of the pH to values not exceeding about 11 would be favorable in stabilizing the performance of the consortium in the fixed-bed reactor.

Elution of the Contaminants from Building Material. Size analysis of the crushed material showed that some 80% of the particles were smaller than 16 mm in diameter (cf. Fig. 4). The unfractionated material was eluted by water, either by changing the eluent only after reaching saturating concentrations (3 to 7 days) or by changing the water daily. Although the former mode was initially more effective, only after 4 cycles was a degree of desorption similar to that of the second

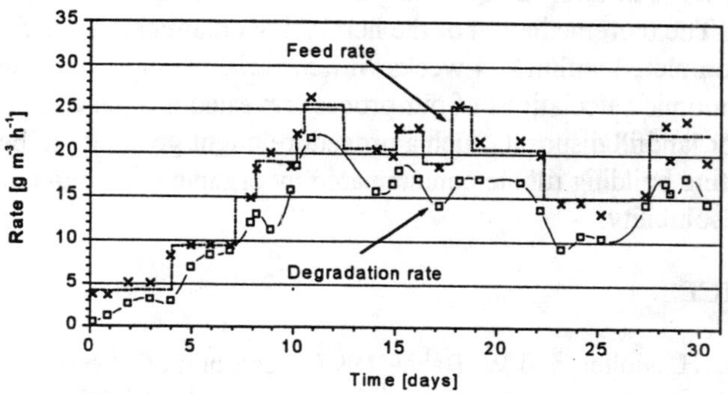

FIGURE 3. Activity profile of a fixed-bed reactor fed with the whole spectrum of contaminants in a sum of 50 mg/l (100 mg/l after 8 days)

mode achieved. About 60–80% of the total pollution was removed after 7 cycles in the latter case which is shown in more detail in Fig. 4. In the smaller fractions (< 16 mm) an average AOX value of $\leq$ 5 mg/kg was even obtained after 4 cycles; in the 16-30 mm fraction the pollutants were desorbed by about 70% within 7 cycles. The desorption efficiency in the fractions of larger particles was significantly lower, with about 60% of the pollutants remaining in the solid matrix after 80 days of treatment. Overall, an AOX value $\leq$ 5 mg/kg, the remediation target, can easily be reached in a continuous aqueous treatment procedure with a high content of particles < 20 mm.

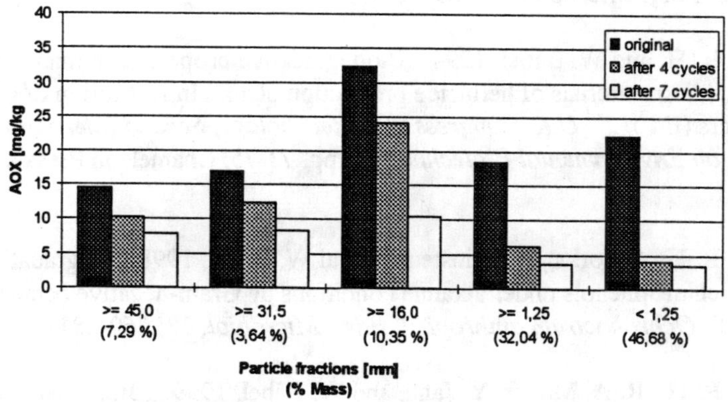

FIGURE 4. Size-dependent desorption of contaminants from building rubble (daily change of water)

Conclusion. The microbial decontamination of building rubble has been shown to be feasible. However, inoculation with specific bacteria is necessary to bring about significant, almost complete degradation. When using the fixed-bed reactor technique to decontaminate aqueous eluates from building rubble with a pH as high as 12 or more, some titration is useful to stabilize the microbial activity in the fixed-bed reactor. The bioremediation of the herbicide-contaminated building rubble should be completed within 2–4 weeks with material mainly comprising particles < 20 mm. Economic calculations of our process revealed it to be competitive with the prices for landfill disposal. Such a procedure ought generally to be applicable for remediating building rubble contaminated by organic compounds with at least finite water solubility.

REFERENCES

Ehrig, A., R. H. Müller, and W. Babel, 1997. „Isolation of phenoxy herbicide degrading bacteria from contaminated building material, broad substrate specificity with *Rhodoferax* sp. strains." *Acta Biotechnol*. 17: 351-356.

Häggblom, M. M., 1992. „Microbial breakdown of halogenated aromatic pesticides and related compounds." *FEMS Microbial. Rev*. 103: 29-72.

Hoffmann, D., R. H. Müller, B. Kiesel, and W. Babel, 1996. „ Isolation and characterization of an alkaliphilic bacterium capable of growing on 2,4-dichlorophenoxyacetic acid and 4-chloro-2-methylphenoxyacetic acid." *Acta Biotechnol*. 16: 121-131.

Mertingk, H., R. H. Müller, and W. Babel, 1998. „Etherolytic cleavage of 4-(2,4-dichlorophenoxy)butyric acid and 4-(4-chloro-2-methylphenoxy)butyric acid by species of *Rhodococcus* and *Aureobacterium* isolated from an alkaline environment." *J. Basic Microbiol*. 38: 257-267.

Müller, R. H. and W. Babel, 1994. „Biodegradative properties of microorganisms from building materials of herbicide production plant." In Institution of Chemical Engineers (Ed.), *2nd U.K. Congress of Biotechnology, Second International Symposium on Environmental Biotechnology*, pp. 73-75. Chameleon Press Ltd., London, U.K.

Müller, R. H., S. Jorks, S. Kleinsteuber, and W. Babel, 1998. „Degradation of various chlorophenols under alkaline conditions by Gram-negative bacteria closely related to *Ochrobactrum anthropi*. *J. Basic Microbiol*. 38: 269-281.

Müller, R. H., R. A. Müller, Y. Jahn, and W. Babel, 1999. „Bioremediation of building material contaminated by herbicides." In D. L. Wise, D. J. Trantolo, and E. J. Cichon (Eds.), *Remediation of hazardous waste contaminated soils*, in press. Marcel Dekker Inc., New York, N.Y.

ISOLATION AND IDENTIFICATION OF MICROORGANISMS FROM A SOIL MIXED CULTURE AEROBICALLY DEGRADING POLYCHLORINATED BIPHENYLS

N.G. Rojas-Avelizapa; R. Rodríguez-Vázquez; J. Martínez-Cruz; and H.M. *Poggi-Varaldo*. (CINVESTAV del IPN, Dept. Biotechnology, Mexico)

ABSTRACT: The objective of this study was to isolate and characterize the native microflora in a Mexican soil contaminated with PCBs and evaluate its potential to remove PCBs contained in a transformer oil. PCBs-degrading microorganisms were isolated from a soil contaminated with transformer oil using enrichment procedures with media including 1% v/v transformer oil and later they were characterized using standard microbiological procedures and the API identification kit. Batch, aerobic degradation tests were carried out using inocula obtained from the last enrichment and 1% v/v transformer oil as the only carbon source. Eight morphologically distinct bacterial strains, three of them Gram negative and five Gram positive were observed. They were identified as *Bacillus lentus, Bacillus thuringiensis, Bacillus mascerans, Pseudomonas, Achromobacter, Comamonas acidovorans Flavobacterium devorans* and *Acinetobacter calcoaceticus*. An *Actinomyces* strain and a fungus *Paecilomyces* were also isolated but not completely characterized. The mixed culture could remove up to 75% of PCBs present in the transformer oil at the 9th day of incubation.

INTRODUCTION

Transformer oils are of environmental concern due to their high content in polychlorinated biphenyls (PCBs). Toxic, carcinogenic and reproductive effects of PCBs on animal and humans have been demonstrated (Crine *et al.*, 1986). In Mexico there are large areas contaminated with PCBs and important amounts of stored, spent transformer oil waiting for appropriate remediation and/or treatment and disposal (INE, 1995). Reports on aerobic degradation of PCBs have been focused on the role and performance of pure cultures or well-defined mixed cultures (Abramowicz, 1990). Only few works using aerobic consortia have been published (Tucker *et al.*, 1975). On the other hand, information on PCBs degradation via anaerobic pathway by mixed cultures has become more abundant (Bedard *et al.*, 1997; Campos-Velarde *et al.*, 1995). This work aimed at isolating and characterizing the native microflora in a Mexican soil contaminated with PCBs and evaluate its potential to aerobically remove the PCBs contained in a transformer oil.

MATERIALS AND METHODS

Culture media. Medium A contained (in g/L) KNO_3, 1; $FeCl_3$ 0.02; $MgSO_4$, 0.2; NaCl, 0.1; $CaCl_2$, 0.1; K_2HPO_4 1; yeast extract, 0.05; and 1%v/v transformer oil was added at 1% v/v previously emulsified with an aqueous solution of Triton X 100 (1:1) at 1% v/v. Medium B contained the components of medium A plus noble agar at 1.5% w/v. The pH of the media was adjusted to 6.3 with HCl 0.1 N.

Enrichment and isolation of microorganisms. A soil contaminated with PCBs from transformer oils (7000 mg/Kg of soil) and other hydrocarbons was sampled. Transformer oil-degrading microorganisms were isolated using an enrichment procedure described elsewhere (Rojas-Avelizapa, 1999). Colonies isolated were then recultured on transformer oil medium (Medium B) to show their growth capacity.

Bacterial identification. Colony morphological characteristics were determined. Also, staining (Gram and spores), cell motility and biochemical tests such as catalase, cytochrome oxidase, *etc.* were done. The isolates were identified using an API identification kit (BioMérieux, France). The identity of an isolate was obtained by comparing its profile to the profiles in the database by the Apilab Plus computer program and through the keys and description in Bergey's Manual. The fungus isolated from mixed culture was grown on potato dextrose agar (PDA) during 48 h and after that examined under the microscope.

Mixed culture growth and transformer oil degradation tests. Inoculum was obtained from 5 mL of the third transfer, transferred to 250 mL Erlenmeyer flasks with 50 mL of medium A and incubated in an orbital shaker at 180 rpm and 28°C during 9 days. Abiotic controls were also included: medium A without inoculum (N) and medium A with only yeast extract as the sole carbon source (AYI).

Extraction and analysis of transformer oil from liquid culture. The PCBs content in the transformer oil corresponded to 88% w/v; the most abundant PCBs were the penta to heptachlorinated isomers. All the solvents were HPLC quality. For PCBs purification Florisil (Sigma, mesh 60-100) was used. Cultures were filtered through a 0.2-μm Millipore membrane and the filtrate was post-treatead as reported by Rojas-Avelizapa (1999). The purified transformer oil was analyzed by gas chromatography according to 8080 USEPA method (1982) using a Perkin Elmer Autosystem equipped with an electron capture detector (Rojas-Avelizapa, 1999). The PCBs removal was determined after normalization of nondegradable peaks areas in experimental chromatograms with those from control samples (Mondello, 1989).

RESULTS AND DISCUSSION

Isolation and characterization of the mixed culture. All the isolated colonies were able to grow in medium B, where the transformer oil was the sole carbon source. Eight morphologically distinct bacterial strains, three of them Gram negative and five Gram positive were observed on nutrient agar plates. The isolates were identified as *Bacillus lentus, Bacillus thuringiensis, Bacillus* *mascerans*, *Pseudomonas*, *Achromobacter*, *Comamonas acidovorans*, *Flavobacterium devorans* and *Acinetobacter calcoaceticus* using API traits and Bergey's Manual (Tables 1 and 2). Another bacterial strain was also isolated and it was partially identified as an *Actinomyces*. A fungus isolated was identified as *Paecilomyces* in PDA medium. Abramowickz (1990) and Barriault *et al.* (1997) reported that bacteria from the genera *Pseudomonas*, *Comamonas*, *Achromobacter* and *Acinetobacter* were involved in PCBs degradation/transformation. On the other hand, there is scarce information about *Bacillus* species able to degrade aromatic compounds such as PCBs and results of this work suggest an interesting area for future research. Regarding the fungus *Paecilomyces*, so far it has not been reported as a microorganism present in PCB's-contaminated matrices. Whether the fungus is able to degrade PCBs and/or their metabolites, or just plays a casual role, is unclear so far. There are published reports of *Actinomyces* present in and isolated from oil wells, oil storage tanks and hydrocarbon-polluted soil (Lacey, 1988) which were able to degrade a variety of organic compounds, including chlorinated organics such as PCBs.

TABLE 1. Morphological characteristics of the bacterial strains from mixed culture

	Colony morphology							
	1	**2**	**3**	**4**	**5**	**6**	**7**	**8**
Shape	Circular	Circular	ND	Circular	ND	ND	Circular	ND
Elevation	Plain	Plain	Plain	Raised	ND	ND	Plain	ND
Surface	Smoth	Smoth	Smoth	Smoth	ND	ND	Smoth	ND
Borders	Even	Even	Even	Wave	ND	ND	Even	ND
Color	White	White	Colorless	Yellow	ND	Colorless	Yellow	ND
Pigment secretion	None	None	None	None	None	None	None	None
Consistence	Soft	Sticky	Sticky	Soft	ND	Sticky	Soft	ND
Gram	-	+	+	+	+	+	-	-
Spores*	-	+ [a]	+ [a]	+ [b]	+	+ [c]	-	-

Notes:

ND= Not determined

*Type of spores: [a]Oval or cylindrical, central or subterminal, rarely swelling the sporangia; [b] Ellipsoidal, subterminal or terminal, swelling the sporangia; [c] Ellipsoidal, central or paracentral, and not swelling the sporangia appreciably.

Mixed culture growth and transformer oil degradation tests. The kinetics of biomass production (growth of the mixed culture) in medium A with transformer oil as the sole carbon source is illustrated in Fig. 1. A net growth of 2.16 g dry weight/L was observed for 7 days incubation, corresponding to a high level of PCBs removal (approximately 60%) during exponential phase of mixed culture growth. Transformer oil depletion continued over the full incubation period and removal increased up to 75% after 9 days of treatment. Peaks below 15 min retention time were dramaticaly reduced (corresponding to PCBs with 2 to 5 chlorine atoms per molecule) and peaks above 15 min retention time showed a reasonable reduction (corresponding to hexa-heptachlorobiphenyls). Control cultures did not show any biomass increase. However, it was observed a slight decrease of transformer oil concentration probably due to abiotic removal mechanisms (up to 8.5%). Our results compared favourably with those from

Table 2. Biochemical characteristics of the isolated bacteria.

Biochemical assays	1	2	3	4	5	6	7	8
Catalase	+	+	+	+	+	+	+	+
Cytochrome oxidase	-	+	-	-	-	+	+	-
Motility	+	+	+	+	+	+	+	+
H$_2$S	-	-	-	-	-	-	-	-
Glycerol	ND	-	-	+	-	ND	ND	+
Erythritol	ND	-	-	-	-	ND	ND	-
D-arabinose	+	-	-	+	-	ND	-	-
L-arabinose	+	-	-	+	+	ND	-	-
Ribose	ND	+	+	+	+	ND	ND	+
D-xylose	ND	+	+	+	-	ND	ND	-
L-xylose	ND	-	-	-	-	ND	ND	-
Adonitol	ND	-	-	-	-	ND	ND	-
B-methyl-D-xyloside	ND	-	-	-	-	ND	ND	-
Galactose	ND	-	-	+	+	ND	ND	+
Glucose	+	+	+	+	+	+	+	+
Fructuose	ND	+	+	+	-	ND	ND	-
Mannose	-	-	-	+	-	+	-	-
Sorbose	ND	-	-	-	-	ND	ND	-
Rhamnose	ND	-	-	+	-	ND	ND	-
Dulcitol	ND	-	-	-	-	ND	ND	-
Inositol	ND	-	-	-	-	ND	ND	-
Manitol	ND	-	-	+	-	ND	ND	-
Sorbitol	ND	-	-	-	-	ND	ND	-
Methyl-D-manoside	ND	-	-	+	-	ND	ND	-
Methyl-D-glycoside	ND	-	-	+	-	ND	ND	-
N-acetyl-glucosamine	-	+	-	+	+	+	-	+
Amigdaline	ND	-	-	+	-	ND	ND	+
Arbutine	ND	-	-	+	-	ND	ND	+
Esculine	ND	+	+	+	+	ND	ND	+
Salicine	ND	-	-	+	+	ND	ND	+
Cellobiose	ND	-	-	+	-	ND	ND	+
Maltose	-	+	+	+	+	+	-	+
Lactose	ND	-	-	+	-	ND	ND	-
Melibiose	+	-	-	-	-	-	-	-
Saccharose	ND	-	-	+	-	ND	ND	+
Trehalose	ND	+	+	+	+	ND	ND	+
Inuline	ND	-	--	-	-	ND	ND	-
Melezitose	ND	-	-	-	-	ND	ND	-
D-rafinose	ND	-	-	-	-	ND	ND	-
Starch	ND	-	-	-	-	ND	ND	-
Glycogene	ND	-	-	-	-	ND	ND	-
Xylitol	ND	-	-	-	-	ND	ND	-
Gentibiose	ND	-	-	-	-	ND	ND	-
D-turanose	ND	-	-	-	-	ND	ND	-
D-lyxose	ND	-	-	-	-	ND	ND	-
D-tagatose	ND	-	--	-	-	ND	ND	-
D-fucose	ND	-	-	-	-	ND	ND	-
L-fucose	ND	-	-	-	-	ND	ND	-
D-arabitol	ND	-	-	-	-	ND	ND	-
L-arabitol	ND	-	-	-	-	ND	ND	-
Gluconate	-	-	-	-	-	ND	+	-
2-keto-gluconate	ND	-	-	-	-	ND	ND	-
5-keto-gluconate	ND	-	-	-	-	ND	ND	-
Orto nitro phenyl galactoside	+	ND	ND	ND	ND	-	-	ND
Arginine	+	ND	ND	ND	ND	-	-	ND
Lysine	-	ND	ND	ND	ND	-	-	ND
Ornithine	-	ND	ND	ND	ND	-	-	ND
Sodium citrate	+	ND	ND	ND	ND	-	+	ND
Urea	-	ND	ND	ND	ND	-	+	ND
Tryptophane	-	ND	ND	ND	ND	+	+	ND

Table 2 (Continued)

Sodium piruvate	+	ND	ND	ND	ND	+	-	ND
Gelatin	+	ND	ND	ND	ND	-	-	ND
Nitrates	-	ND	ND	ND	ND	-	+	ND

Notes:

ND: not determined; strains 1 to 8 were identified as *Acinetobacter calcoaceticus var. anitratus*, *Bacillus lentus*, *Bacillus mascerans*, *Flavobacterium devorans*, *Bacillus thuringiensis*, *Achromobacter sp.*, and *Pseudomonas sp.*, *Comamonas acidovorans* respectively.

Tucker *et al.* (1975) who found removals from 18 to 80% of low-chlorinated PCBs with aerobic mixed cultures.

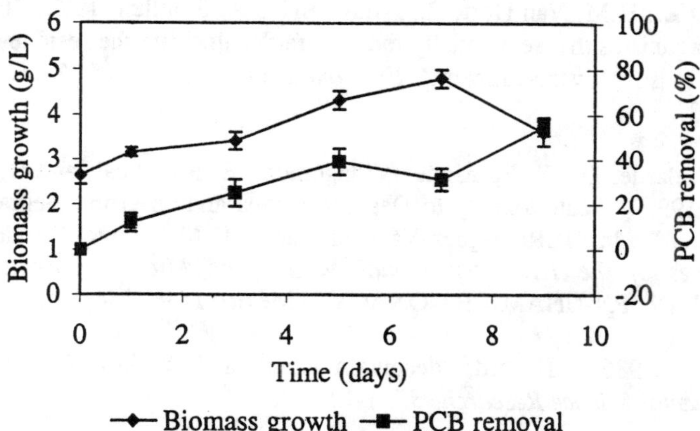

FIGURE 1. PCB removal and biomass growth in aerobic, batch mixed culture

CONCLUSIONS

An aerobic microbial consortium was isolated from contaminated soil and characterized.

1.- Eight morphologically distinct bacterial strains, three of them Gram negative and five Gram positive were observed on nutrient agar plates. The isolates were identified as *Bacillus lentus, Bacillus thuringiensis, Bacillus mascerans, Pseudomonas, Achrromobacter, Comamonas acidovorans Flavobacterium devorans* and *Acinetobacter calcoaceticus* using API traits and Bergey's Manual.

2.- Another bacterial strain and a fungus were also isolated and characterized as belonging to genera *Actinomyces* and *Paecilomyces* respectively.

3.- The mixed culture could remove up to 75% of PCBs present in the transformer oil at the 9th day of incubation in aerobic, batch tests using the transformer oil as the sole carbon source.

Results of this work suggest the possibility for using cultures from contaminated soils as sources for bioaugmentation in the treatment of PCBs-contaminated matrices and wastes.

REFERENCES

Abramowickz, D.A.1990. "Aerobic and anaerobic biodegradation of PCBs: a review." *Critical Reviews in Biotechnoogy. 10*: 241-251.

Barriault, D., C. Pelletier, Y. Hurtubise, and M. Sylvester. 1997. "Substrate selectivity pattern of *Comamonas testosteroni strain B-356* towards dichlorobiphenyls." *International Biodeterioration and Biodegradation. 39*: 311-316.

Bedard, D.L.; H.M. Van Dort; R.J. May, and L.A. Smullen. 1997. "Enrichment of microorganisms that sequentially meta, para-dechlorinate the residue of Aroclor 1260 in Hudson River sediment." *Environmental Science and Technology. 31*: 3308-3313.

Campos-Velarde, D.; G. Fernández-Villagómez.; E. Ríos-Leal, and H.M. Poggi-Varaldo. 1995. "Acute toxicity to Daphnia magna and anaerobic degradability of PCB 1260." In R.Rodríguez-Vázquez and H.M. Poggi-Varaldo (Eds.) *Proceedings of the 1st. International Meeting on Microbial Ecology* p.160. CINVESTAV-IPN-UNAM-CP-CONACYT. México D.F.; México.

Crine, H. 1986. "Hazard, decontamination and replacement of PCBs." *Environmental Science Research. 37*: 1-12.

INE. 1995. *Bifenilos policlorados*, National Institute of Ecology, Secretary of the Environment and Natural Resources, México, D.F. In Spanish.

Lacey, J. 1988. In Goodfellow, M., S.T. Williams and M. Mordansky (Eds.). *Actinomycetes as biodeteriogens and pollutants of the environment, Actinomyces in Biotechnology*. pp. 359-432. Academic Press, New York.

Mondello, F. J. 1989. "Cloning and expression in Escherichia coli of *Pseudomonas strain LB400* genes encoding polychlorinated biphenyl degradation." *Journal Bacteriology. 171*: 1723-1732.

Rojas-Avelizapa, N. 1999. "Biodegradation of PCBs in contaminated soils." Sc.D. Thesis. Dept. Biotechnology. CINVESTAV-IPN. México D.F., México.

Tucker, E.S., V.W. Saegan, and O. Hicks. 1975. "Activated sludge primary degradation of polychlorinated biphenyls." *Bulletin of Environmental Contamination and Toxicology 14*: 705-712.

U.S. Environmental Protection Agency. 1982. *The determination of polychlorinated biphenyls in transformer fluid and waste oils*. 8080 Method.

BIOAVAILABILITY, ESTROGENICITY, AND TOXICITY OF SURFACTANT-PCB MIXTURES AFTER BIODEGRADATION AND PHOTOLYSIS

John Sanseverino, Fu-Min Menn, Betsy Gregory, Alice Layton, Eric Sampsel, Zhou Shi, Mriganka Ghosh, T. Wayne Schultz, and Gary Sayler, University of Tennessee, Knoxville, Tennessee, USA.

Abstract: Bioavailability and measuring associated toxicity is fundamental to risk assessment and defining environmentally acceptable endpoints for biologically treating polychlorinated biphenyl (PCB)-contaminated soils and debris. Two molecular-based assays were used to measure bioavailability, estrogenicity, and toxicity of surfactant/Aroclor 1248/1260 mixtures before and after photolysis and biodegradation. The 2 assays, a biphenyl-bioluminescent reporter strain for measuring inducers of the *bph* operon and general toxicity, and a yeast-based reporter that contains the human estrogen receptor for measuring environmental estrogens was applied to surfactant/PCB solutions generated in this study. A third assay still under development is a yeast-based assay incorporating the human aryl hydrocarbon receptor (hAHR) for toxicity measurements. Preliminary results indicated that PCBs and their bacterial metabolites are not estrogenic. The bioluminescent bioreporter demonstrated that photolysis produced inducers of the *bph* operon (biphenyl and mono-chlorinated biphenyls) that were not present in nonphotolyzed solutions. Estrogenic activity was present in nonphotolyzed solutions but not photolyzed solutions. However, a control experiment utilizing Aroclor 1248 in surfactant gave no estrogenic responses. This information combined, with other recent data, implies that PCBs and their metabolites were not responsible for the observed estrogenic activity.

INTRODUCTION

Biological degradation of PCBs was first reported in 1973 with the work of Ahmed and Focht. Since then, the biochemistry and genetics of biphenyl metabolism and PCB metabolism has been characterized. However, the use of bioremediation for PCB-contaminated sites has not been accepted for several reasons. First, bioremediation cannot reach mandated clean-up levels due to the high hydrophobicity, low aqueous solubility of these compounds. Second, bacteria cannot aerobically degrade highly chlorinated PCB congeners; generally congeners with greater than 5-6 chlorines are considered recalcitrant. Third, reductive dechlorination can remove chlorines from PCBs but it is a slow process.

A major factor in developing PCB bioremediation technology and gauging endpoints of effective PCB treatment is the relative insolubility of PCB and its sorptive properties, hence low bioavailability in many soils and sediments. The problem of poor bioavailability (defined as availability of substrate to the microorganism) has been overcome by soil washing with nonionic surfactants and feeding the surfactant/PCB mixture to 2 previously developed genetically-engineered microorganisms, *Pseudomonas putida* IPL5::TnPCB and *Ralstonia eutrophus* B30P4::TnPCB (Lajoie et al., 1997; Layton et al., 1998a). These organisms degrade surfactant and constitutively express a *bph* operon for degradation of degradable PCB congeners.

Bioavailability (defined now as availability to human and ecological receptors) and measuring associated toxicity is fundamental to risk assessment and defining environmentally acceptable endpoints for surfactant washing and biologically treating PCB-contaminated soils and debris. PCBs removed and remaining on contaminated soil should not pose an unacceptable risk to human health and the environment. Likewise, PCB-degradation products generated as the result of any treatment process should pose no threat.

This study is part of a larger effort to further develop a treatment train approach to remediate PCB-contaminated soil. An accompanying paper describes soil washing of a PCB-contaminated soil and photolysis of the surfactant/PCB mixture (Sampsel et al. 1999). This report presents preliminary data on the application of 2 molecular-based assays to measure bioavailability, estrogenicity, and toxicity of the surfactant/PCB mixtures before and after photolysis and biodegradation. The long-term goal is to further develop these assays to monitor bioavailability and toxicity of residues remaining on the surfactant-washed soil.

MATERIALS AND METHODS

Soil. A PCB-contaminated soil (designated PR3-1) was obtained from a site in the northeastern United States. The soil had an organic carbon content (f_{oc}) of 0.0187 and contained 1100 mg/kg and 80 mg/kg of Aroclor 1248 and 1260, respectively, and 8.5 g/kg TPH. This soil also contained low levels of polynuclear aromatic hydrocarbons (data not available).

Soil Washing and Photolysis. Soil washing using a nonionic surfactant (polyoxyethylene 10 lauryl ether [POL 10]) and photolysis of surfactant/PCB solutions is described in a companion paper (Sampsel et al., 1999).

Biodegradation of Surfactant and PCBs. Bioreactors consisted of 2 L flasks with 500 mL of PAS minimal salts medium (pH 7.0) (Bopp, 1986). Surfactant/PCB solutions were added to a final concentration of 10,000 mg surfactant/L. Aeration was by agitation and an air flow rate into the reactors of 60 mL/min. Initial experiments included 4 treatments. Later experiments added a fifth treatment. The treatments were:
- Reactor 1 – Unseeded control
- Reactor 2 – *P. putida* IPL5::TnTc + *R. eutrophus* B30P4::TnTc
- Reactor 3 - *P. putida* IPL5::TnPCB + *R. eutrophus* B30P4::TnPCB
- Reactor 4 - *P. putida* IPL5::TnPCB + *R. eutrophus* B30P4::TnPCB + *R. eutropha* ENV307
- Reactor 5 – acidified control (pH 2.0)

R. *eutropha* ENV307 is a PCB-degrading bacterium with a broad congener specificity (Lajoie et al., 1994). Surfactant/PCB solutions from washing of soil PR3-1 were used in these experiments. In addition, a control experiment using approximately 50 ppm Aroclor 1248 in 10,000 ppm POL 10 was also conducted. Optical density (OD_{600}), pH (adjusted to 7.0 with 1 N NaOH), surfactant concentration and surfactant measurements were made daily. Surfactant concentration was determined with the CTAS (cobalt thiocyanate active substances) assay (Standard Methods, 1985). PCBs were analyzed by a modification of EPA Method 8081.

Bioluminescent Reporter Assay. Construction and development of the bioluminescent reporter has been reported previously (Layton et al., 1998). Briefly, the reporter strain, R. *eutropha* ENV307 (pUTK60) was exposed to surfactant/PCB solutions generated from each phase of the treatment process. Bioluminescence measurements were assayed as photon counts per minute in each sample. The level of induction or repression was based on the comparison of bioluminescence of bacteria with test substrate versus the bioluminescence of bacteria with no substrate (Layton et al., 1998).

Human Estrogen Receptor Assay. The estrogenicity of test samples was assayed as described previously (Routledge and Sumpter, 1996; Schultz et al., 1998). Briefly, a *Saccharomyces cerevisiae* strain contains the human estrogen receptor integrated into its chromosome. The estrogen responsive sequences are upstream of the B-galactosidase (*lac*Z) reporter gene. An estrogen-like compound binds to the receptor protein, induces the *lac*Z gene producing B-galactosidase. B-galactosidase transform the chromogenic substrate chlorophenol red-B-D-galactopyranoside (CPRG) to a red product measured by absorbance (540 nm).

RESULTS AND DISCUSSION

Biodegradation of Surfactant/PCB Solutions. In all seeded reactors, approximately 95% of the surfactant was degraded in 2-3 days. PCB removal data is summarized in Table 1. The combination of photolysis and biodegradation significantly enhanced the total removal of PCBs from the pseudo-aqueous phase than biodegradation alone. The purpose of photolysis was to dechlorinate the heavily chlorinated congeners economically and shift the congener profile towards the more biodegradable congeners (Sampsel et al., 1999). Either method used alone is not practical for 2 reasons: (1) Photolysis as a means to completely dechlorinate PCBs is too expensive, and (2) biodegradation is not effective with the highly chlorinated congeners. The combination of both methods may provide an efficient, economical means to treat PCB-contaminated soil.

Bioluminescent Response to Surfactant/PCB Solutions. Figure 1 shows the bioluminescent response of R. *eutropha* ENV307 (pUTK60) after exposure to surfactant/PCB solutions from each stage of the treatment process.

Table 1. PCB removal in nonphotolyzed and photolyzed experiments using surfactant/PCB solutions from soil PR3-1.

Reactor	T_0[1] (mg)	Nonphotolyzed Exp. % PCB Removed by	Photolyzed Exp. % PCB Removed by		
		Bioreactor	Photolysis	Bioreactor	Total[3]
R1	21	-[2]	61.6	1.6	62.0
R2	21	41.9	60.8	9.7	65.0
R3	21	58.8	61.1	51.8	80.9
R4	21	59.9	61.7	44.0	78.6

[1]Initial mass of PCB in the surfactant solution before biodegradation and photolysis. [2]na - not available. [3]Total PCB removed by the combined treatment was calculated by the removal of PCB (mg) from each treatment relative to the initial mass of PCB in the surfactant solution (21 mg).

A bioluminescent response was observed with pure Aroclor 1242 and with the photolyzed surfactant/PCB solution before biodegradation. The nonphotolyzed solution, which was composed primarily of Aroclors 1248 and 1260, did not elicit a response implying that no inducers of the *bph* operon were present. The photolysis treatment generated inducers (biphenyl, mono-chlorinated biphenyls) which are capable of inducing the *bph* operon. No response was observed in any sample after biodegradation.

Presence of Environmental Estrogens in Surfactant/PCB Solutions. Surfactant/PCB solutions generated from soil washing, photolysis, and biodegradation treatments were screened for estrogenic activity using the yeast-based hER system. The nonphotolyzed surfactant/PCB solution from soil PR3-1 before and after biodegradation elicited a significant estrogenic response. Pure solutions of Aroclors 1242 and 1248 and the surfactant POL 10 (before and after biodegradation) did not elicit an estrogenic response implying that unmodified PCBs are not estrogenic. This is supported by other reports which indicated that 4'-hydroxylated PCBs rather than non-hydroxylated PCBs are estrogenic in mammalian systems (Fielden et al., 1997; Korach et al., 1998; Schultz et al., 1998). A more detailed analysis is being conducted to determine which compound(s) were estrogenic. Photolysis of the surfactant/PCB solution appeared to remove estrogenic activity. In addition to dechlorination of PCBs, photolysis also destroys polynuclear aromatic hydrocarbons and other organic contaminants.

In addition to the hER assay, a yeast-based human aryl hydrocarbon receptor (hAHR) test strain has been developed (Miller, 1997) and will be applied to our work. The hAHR binds aromatic compounds without activation by the mammalian systems. This will provide a faster measure of compound toxicity. Further, future work will examine bioavailability and toxicity of PCBs bound to soil after surfactant washing. The objective is to quantify toxicity and/or estrogenicity reduction directly in the soil and determined environmentally acceptable endpoints.

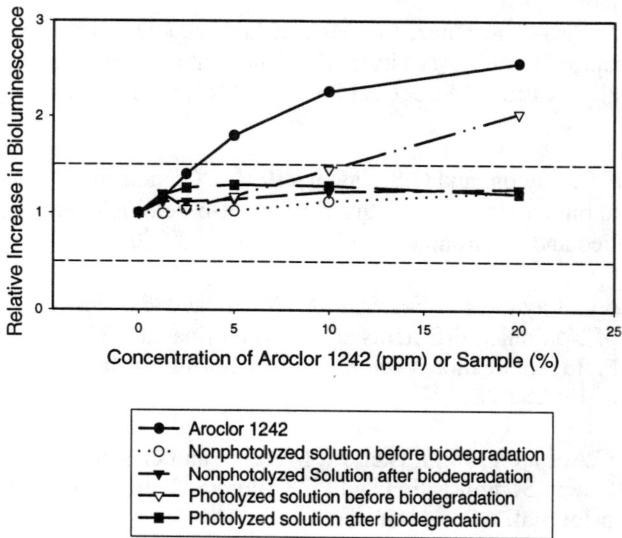

FIGURE 1. The bioluminescent response of *R. eutropha* ENV307 (pUTK60) to surfactant/PCB solutions before and after photolysis and biodegradation. A relative increase >1.5 or <0.5 indicates induction and repression, respectively.

ACKNOWLEDGMENTS

This work was supported by the Department of Energy (DOE Contract No. DE-FG02-97ER62350) and the Waste Management Research and Education Institute at the University of Tennessee, Knoxville. We also thank Envirogen, Inc., for permission to use *R. eutrophus* ENV307.

REFERENCES

Ahmed, M. and D.D. Focht. 1973. "Degradation of Polychlorinated Biphenyls by Two Species of *Achromobacter*." Canadian Journal of Microbiology. 19:47-52.

Bopp, L.H. 1986. "Degradation of highly chlorinated PCBs by a *Pseudomonas* strain LB400." Journal of Industrial Microbiology. 1: 23-29.

Fielden, M.R., I. Chen, B. Chitteim, S.H. Safe, and T.R. Zacharewski. 1997. "Examination of the Estrogenicity of 2,4,6,2',6'-Pentachlorobiphenyl (PCB 104),

Its Hydroxylated Metabolite 2,4,6,2',6'-Pentachloro-4-biphenylol (HO-PCB 104), and a Further Chlorinated Derivative, 2,4,6,2',4',6'-Hexachlorobiphenyl (PCB 155)." Environmental Health Perspectives. 105(11): 1238-1248.

Korach, K.S., P. Sarver, K. Chae, J.A. McLachlan, and J.D. McKinney. 1998. "Estrogen Receptor-Binding Activity of Polychlorinated Hydroxybiphenyls: Conformationally Restricted Structural Probes." Molecular Pharmacology. 33: 120-126.

Lajoie, C.A., A.C. Layton, and G.S. Sayler. 1994. "Cometabolic Oxidation of Polychlorinated biphenyls in Soil with a Surfactant-based Field Application Vector." Applied and Environmental Microbiology. 60: 2826-2833.

Lajoie, C.A., A.C. Layton, J.P. Easter, F.-M. Menn, and G.S. Sayler. 1997. "Degradation of Nonionic Surfactants and Polychlorinated Biphenyls by Recombinant Field Application Vectors." Journal of Industrial Microbiology & Biotechnology. 19: 252-262.

Layton, A.C., C.A. Lajoie, J.P. Easter, M. Muccini, and G.S. Sayler. 1998a. "An integrated Surfactant Solubilization and PCB Bioremediation Process for Soils." Bioremediation Journal. 2(1): 43-56.

Layton, A.C., M. Muccini, M.M. Ghosh, and G.S. Sayler. 1998b. "Construction of a Bioluminescent Reporter Strain to Detect Polychlorinated Biphenyls." Applied and Environmental Microbiology. 64(12): 5023-5026.

Miller, C.A., III. 1997. "Expression of the Human Aryl Hydrocarbon Receptor Complex in Yeast." The Journal of Biological Chemistry. 272: 32824-32829.

Routledge, E.J. and J.P. Sumpter. 1996. "Estrogenic Activity of Surfactants and Some of Their Degradation Products Assessed using a Recombinant Yeast Assay." Environmental Toxicology and Chemistry. 15: 241-248.

Sampsel, E, M. Ghosh, J. Sanseverino, F.-M. Menn, Z. Shi, M.M. Wong, B. Gregory, T.W. Schultz, and G.S. Sayler. 1999. "Desorption and Destruction of Polychlorinated Biphenyls (PCBs) in Micellar Media." In *In Situ and On-site Bioremediation Symposium Proceedings*. Battelle Press, Columbus, Ohio.

Schultz, T.W., D.H. Kraut, G.S. Sayler, and A.C. Layton. 1998. "Estrogenicity of Selected Biphenyls Evaluated Using A Recombinant Yeast Assay." Environmental Toxicology and Chemistry. 17: 1727-1729.

Standard Methods for the Examination of Water and Wastewater, 16th ed. Section 512C. Nonionic Surfactants as CTAS. pp. 585-588. American Public Health Association, Publishers.

DESORPTION AND DESTRUCTION OF PCBs IN MICELLAR MEDIA

Eric Sampsel (University of Tennessee, Knoxville)
Mriganka Ghosh, Zhou Shi, John Sanseverino, Fu-Min Menn, Meng Meng Wong, Betsy Gregory, and Gary Sayler (University of Tennessee, Knoxville)

ABSTRACT: Laboratory-scale studies have been conducted to determine the feasibility of using a surfactant washing system for the removal of polychlorinated biphenyls (PCBs) from a contaminated soil. Further, photodechlorination was used to destroy the PCBs that were transferred into surfactant micelles. Eighty-five percent of the Soxhlet extractable PCBs could be removed from the soil with three successive column wash cycles using polyoxyethylene 10 lauryl ether (POL(10)). In excess of 50% of the PCBs in the surfactant wash water were photolyzed in 2 hours resulting in a quantum yield of 0.0104.

INTRODUCTION

A primary source of PCBs in the environment is the transformer oils that were produced until 1979. PCBs were added to make the oils fire-resistant and to impart excellent dielectric properties. However, the same properties that give PCBs their stability also cause them to be highly recalcitrant in the environment. Their strong hydrophobicity and low water solubility cause them to bind to soil particles very tightly with almost no noticeable migration. Currently, large amounts of time and energy are being spent on the disposal of PCB contaminated wastes. Previous studies have shown that surfactants are capable of removing PCBs from soil via solubilization (U.S. EPA, 1985).

Once PCBs are removed from the soil there are several options available for their destruction. Photolysis has been shown to successfully dechlorinate PCBs in surfactant micelles with minimum undesirable side reactions (Epling et al., 1988).

Objective. This study was conducted to determine the feasibility of treating aged PCB contaminated soils with an *in-situ* soil washing treatment scheme and to determine the effectiveness of photolysis as a means of toxicity reduction via PCB dechlorination. A nonionic surfactant, POL(10), was used to desorb PCBs from the soil matrix via solubilization.

Although the project involved assessing the toxicity of the soil wash solutions, the effects of photolysis on reduction of such toxicities, and on the enhancement of PCB biodegradability, photolysis is discussed in the current paper only in the context of PCB dechlorination. Experiments were performed to determine the most efficient surfactant:soil mass loading ratio in an *in-situ* treatment scheme. Further, the effect of other co-contaminants, such as transformer oils, on the treatability of soil and the quantum yield was studied.

MATERIALS AND METHODS

Soil samples. The soils used in this study were taken from a PCB contaminated Superfund site in the northeastern United States. Their characteristics are shown in the table below. Aroclor 1248 and 1260 were the major commercial PCB mixtures found in the soils.

Table 1: Characteristics of test soils

Sample	1248 mg/kg	1260 mg/kg	f_{oc}	% gravel	% sand	% silt+clay	TPH g/kg
PR1-2	26.1	400	>0.05	15	63	22	0.23
PR3-1	1100	88	0.0187	6	71	23	8.5

Batch tests. All batch tests were conducted using 25 mL Corex glass centrifuge tubes with Teflon screw caps. 3 g of soil and 15 mL of surfactant solution were added to each tube. The tubes are mixed in an end-over-end rotary mixer for pre-selected time periods. The tubes were then centrifuged at 2000 rpm for 1 hour. The supernate was then carefully removed with a pipette and extracted for PCBs.

Column studies. The soils were washed in an up-flow column with continuous recycling of surfactant solution until PCB concentration in the effluent became time-invariant. A schematic of the experimental setup is shown in Figure 1.

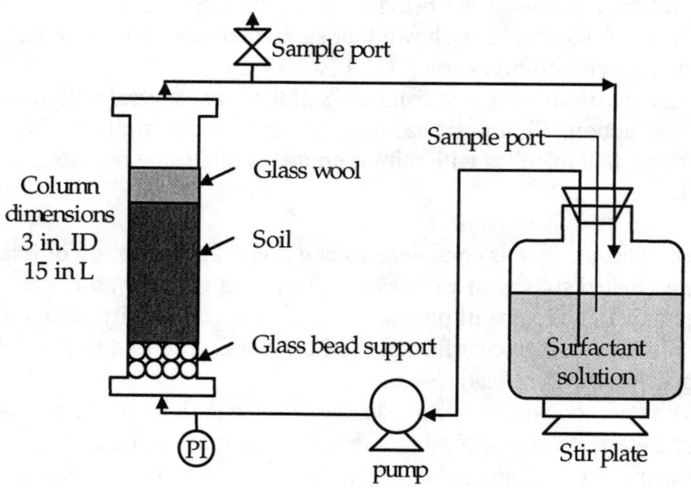

FIGURE 1: Schematic of soil washing column

One kilogram of soil was packed in the column to a bulk density of about 1.5 g/cm^3. A stainless steel screen support was placed at the bottom of the column along with a 1.5-in layer of 5 mm glass beads to allow the solution to fully disperse across the cross section. To minimize PCB adsorption, only stainless steel, glass, or Teflon was used in column construction. Five liters of a 20 g/L-POL(10) solution was pumped upwards through the soil at a rate of 2.5 mL/min.

Photolysis. A Rayonet Model 200 photochemical reactor was used for all photolysis experiments. The photoreactor was equipped with 254 nm low-pressure mercury lamps. A 4.25-cm diameter quartz cuvette was used as the sample holder in all photolysis experiments.

PCB analysis. PCBs were extracted for GC analysis using 1 mL of sample and 4 mL of hexane. To prevent emulsification, 0.3 g of tC$_{18}$ powder (Waters, Inc.) was added (Shi, 1998). Each sample was shaken horizontally for 30 min and thereafter centrifuged at 1000 rpm for 5 min. The hexane phase was then analyzed for PCBs according to EPA method 8081.

RESULTS AND DISCUSSION

Batch studies. Batch tests were performed on the test soils shown in Table 1. A comparison of overall PCB removal of the two soils is shown in Figure 1. Both soils attained equilibrium in 48 hours with an equilibrium solution-phase PCB concentration of 80 – 90 mg/L. Percent PCB removal values reported are based on Soxhlet extractable amounts. From the results of the batch tests, which simulate an *ex-situ* treatment scheme, a mass loading of 0.1 g POL(10)/g soil was chosen for use in all studies with soils PR3-1 and PR1-2. This mass loading provided over 20 times the amount of surfactant needed to solubilize all of the PCBs that were in the soil, thereby preventing any limitations on PCB solubilization.

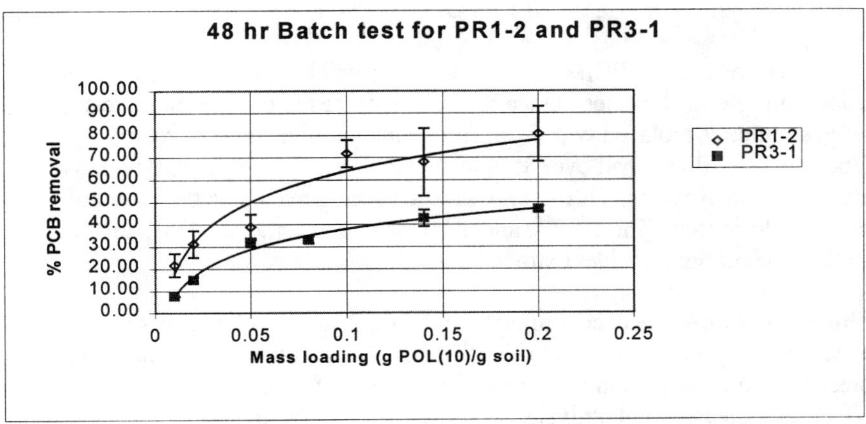

FIGURE 2. PCB removal at various surfactant:soil mass ratios

As seen in Figure 2, soil PR3-1 gave a much lower percentage removal than soil PR1-2. PR3-1 had a much higher TPH level than PR1-2, and therefore oil could be competing with the PCBs for the available surfactant micelles in solution. PR1-2 also had a lower initial concentration of PCBs than PR3-1. No noticeable change in the congener profile was seen as a result of soil washing.

Column studies. Once batch tests had been performed on the soils, the column system was started using a mass loading ratio of 0.1 g POL(10)/g soil. As was found in batch studies with the same mass loading of surfactant, the column system approached equilibrium in 48 hours with approximately 40% of the Soxhlet extractable PCBs removed from the soil after one week of washing (Figure 3). Thus, column washing, which mimicked *in-situ* treatment, produced removals comparable to those expected in *ex-situ* washing systems. Therefore, an *in situ* pump-and-treatment scheme can offer substantial cost savings over a comparable *ex situ* treatment scheme for treating excavated soils.

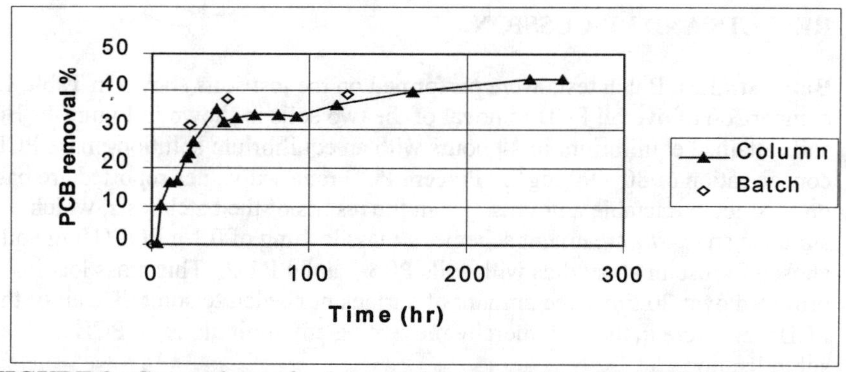

FIGURE 3. Comparison of column and batch washing of soil PR3-1 using
0.1 g POL(10)/g soil

The amount of PCBs removed from the soil can be greatly improved by using multiple wash cycles. Once the system comes to equilibrium, the spent wash solution is replaced with fresh surfactant solution and the process is repeated. Soil PR3-1 had over 85% of the Soxhlet extractable PCBs removed after three wash cycles. However, with each successive cycle the extent of removal decreased. The 1st, 2nd, and 3rd wash cycles removed 43%, 28%, and 15% of the original Soxhlet extractable PCBs, respectively.

Photolysis studies. In excess of 50% of the PCBs in PR3-1 soil wash solution were destroyed after two hours of photolysis. To quantify the effects of oil present in soil wash solutions, a micellar solution of only Aroclor 1248 in POL (10) was subjected to photolysis. As shown in Figure 4, the initial photodegradation of PCBs in the Aroclor 1248 solution was much faster than that of the column wash solution.

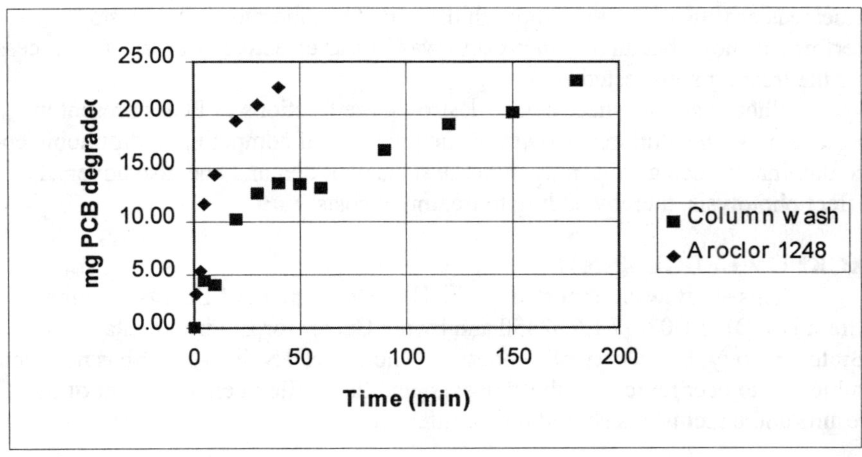

FIGURE 4. Comparison of Aroclor 1248 degradation in pure solution with PCB degradation in column wash solution

The following equation was used to calculate the quantum yield of photolysis (Zepp, 1978).

$$\Phi_P = \frac{k_0}{I_{\lambda,0}\,{A}\big/{V}}$$

Where Φ_P = quantum yield
  k_0 = zero order decay rate constant (mol m^{-3} s^{-1})
  $I_{\lambda,0}$ = light intensity (einstein m^{-2} s^{-1}) = 7.96 x 10^{-5} (this study)
  A/V = ratio of exposed area to volume (m^{-1}) = 94.1 (this study)

The calculated quantum yields of photolysis were 0.0479 and 0.0104 for the pure Aroclor 1248 solution and the column wash solution, respectively. The difference in the two values is most likely due to the interferences caused by oil and other contaminants present in the column wash solution. These additional components cause the column wash solution to turn dark yellow which conceivably absorbed available light. Therefore, a solution without any co-components would have had a markedly higher quantum yield.

CONCLUSIONS

Surfactant washing of PCB contaminated soil with POL(10) provided satisfactory, successful treatment of the two contaminated soils studied. By performing sequential washes of the soil, much larger amounts of PCBs could be removed. However, soils with large fractions of clay or silt may prove much more difficult to successfully wash because of their higher organic fraction. Both of the soils used in the study contained large amounts of sand, which allowed for much easier flow of solution through the soil. The laboratory-scale tests performed show that an *in-situ* process was just as effective as an *ex-situ* process for the treating soils tested.

Photolysis can successfully destroy large fractions of PCBs present in surfactant wash solutions. However, the presence of competing hydrophobic co-contaminants such as oils, may increase surfactant demand and also adversely affect photolysis, thereby adding to treatment costs.

ACKNOWLEDGEMENTS

This study was supported by the U.S. Department of Energy through Grant No. DE-FG02-97-ER62350 and by the Center for Environmental Biotechnology, University of Tennessee, Knoxville, TN. The paper has not been subjected to peer review by the sponsors, and thus official endorsement of the results and conclusions should not be inferred.

REFERENCES

Epling, G. A., Florio, E. M., Bourque, A. J., Qian, X., Stuart, J. D. 1988. "Borohydride, Micellar, and Exciplex-enhanced Dechlorination of Chlorobiphenyls." *Environ. Sci. Technol.* 22: 952-956.

Shi, Z. 1998. "Surfactant-Enhanced Dissolution and Photolysis of Chlorinated Aromatic Compounds (CACs)." Ph.D. Dissertation, University of Tennessee, Knoxville, TN.

U.S. Environmental Protection Agency. 1985. *Treatment of Contaminated Soils with Aqueous Surfactants*. Report EPA/600/2-85/129.

Zepp, R. G. 1978. "Quantum Yields for Reaction of Pollutant in Dilute Aqueous Solution." *Environ. Sci. Technol.* 12: 327-329.

MANAGING SEDIMENT CONTAINMENT FACILITIES TO PROMOTE PCB SEPARATION AND DEGRADATION

Jeffrey R. Chiarenzelli and Ronald J. Scrudato (Environmental Research Center, SUNY Oswego, New York)

ABSTRACT: On an annual basis millions of cubic yards of sediment will be dredged from our nation's waterways for remedial or navigation purposes. The fate of much of this material will be determined by the concentration of PCBs and other toxic compounds. Economics and policy decisions may result in placement of much of this material in containment facilities designed to entomb, rather than degrade, the associated contaminants, essentially transferring environmental liability to future generations. Understanding of the composition and properties of the polychlorinated biphenyl (PCB) fraction likely to be found in anaerobically dechlorinated sediments suggests the design, operation, and management of containment facilities to promote degradation is possible. This would include separation of the more mobile lower chlorinated congeners by enhancing physical desorption processes (solubilization and volatilization) and treatment of the derived aqueous and volatile fractions by existing technologies, with the ultimate goal of decreasing the mobility of the residual contaminant fraction remaining sorbed to the sediment.

INTRODUCTION

In New York State alone, dredging actions will be carried out on the Grasse, Raquette, and St. Lawrence rivers, is required for navigational purposes in New York Harbor, and is being considered to address PCB contamination of the Upper Hudson River. In each case PCBs, and to a lesser extent other contaminants, complicate disposal of the dredge spoils. With the prohibition of offshore disposal as an option to New York Harbor's dredge material, several million cubic yards of sediment per year will need to be managed and disposed of; some of it contaminated by a mixture of organochlorides and metals. Upland disposal of contaminated dredge material has been proposed; however, public opposition is strong.

Dredging to remove PCB contaminated sediments has begun or is planned at three industrial sites near the St. Lawrence River in upstate New York. The EPA's Superfund Post-Decision Proposed Plan (U.S. EPA, 1998a) at the Reynolds Metals site includes off-site treatment of PCB contaminated material (>500 ppm), off-site landfilling (50-500 ppm), and on-site disposal in an industrial landfill (<50 ppm). At the nearby General Motors facility adjacent to the Akwesasne Mohawk Nation, the amended Superfund Post-Decision Proposed Plan (U.S. EPA, 1998b) calls for off-site disposal of PCB contaminated materials (>500 ppm) and on-site containment (10-500 ppm). No plans exist for treatment within containment facilities or landfills.

The Upper Hudson River is the site of perhaps the largest PCB release on the planet and impacts the entire downstream length of the river (Sanders, 1989). Most experts concur that the river's hotspots, areas of elevated PCB contamination in bottom sediments, will continue to release PCBs to the water, air, and biota for generations unless removed by dredging. However, the question remains in each of these areas on what should be done with contaminated material once it is removed?

Here we argue that the PCB fraction of anaerobic microbially dechlorinated sediments is more mobile than currently recognized and can be separated and degraded by existing technologies. We therefore conclude that if containment facilities are to be utilized to address the problem of contaminated sediments, they should be designed, operated, and managed to degrade contaminants rather than as sterile holding cells.

DESTRUCTION OF PCBS SORBED TO SEDIMENT

While numerous technologies are capable of destroying PCBs in the aqueous phase (Carey et al., 1976; Pignatello and Chapa, 1994) the destruction of sorbed contaminants continues to be a vexing problem. Thus most treatment technologies combine an extraction or concentration step with subsequent treatment or disposal. Because the cost associated with extraction and treatment of organochlorides from sediment is generally prohibitive and the public is resistant to on-site incineration, landfill disposal is often the chosen alternative (U.S. EPA, 1998a; 1998b).

The relative inefficiency of treating organochloride compounds sorbed to sediment compared with those in the aqueous phase has a variety of reasons. Technologies, which involve the formation of oxidative species, such as free radicals or ozone, can be limited by scavenging by non-target inorganic or organic components within the sediment. Some organic-rich sediments may contain several orders of magnitude more natural organic matter than contaminants, requiring unrealistically large reagent applications for economic treatment of the target compounds. Sequestering, particularly in aged sediment, may also play a role in determining the residual levels of contaminants likely to remain after treatment (Hatzinger and Alexander, 1995). Problems associated with mass and energy transfer in sediment or slurry systems also play a role in limiting degradation reactions. In some instances getting a viable oxidant species to the site of a sorbed contaminant or contaminants sorbed to a catalyst is problematic. The physical properties of the sediment such as porosity and permeability may limit handling, reagent or energy introduction, and treatment efficiency. Finally, large contaminant concentrations sorbed on solids may necessitate large amounts of reagents, lowering destruction efficiencies as reactive species react with each other rather than the target compounds. In general, more soluble species, even within a specific chemical class such as PCBs, are more susceptible to degradation.

PCBS IN ANAEROBICALLY DECHLORINATED SEDIMENT

Evidence of microbially mediated anaerobic and aerobic dechlorination of PCBs is widely reported in the literature (Abramowicz, 1995; Spain, 1997) and has been achieved in numerous laboratory experiments. In general, anaerobic processes result in the removal of meta- and para-chlorines from PCB congeners resulting in lower average chlorine/biphenyl ratios and the production of totally ortho-chlorinated congeners with 1-3 chlorines. Dechlorination often reaches a plateau in about 8-12 weeks after which little further dechlorination occurs. Current research into anaerobic dechlorination of PCBs seeks to enhance this process. Aerobic microbes have been reported to destroy ortho-chlorinated congeners produced during anaeorbic processes and a combined anaerobic/aerobic bioremediation approach may be viable in the future.

DECHLORINATION LEADS TO ENHANCED PCB MOBILITY

In general, the solubility of PCBs is inversely related to the degree of chlorination of the commercial mixture or aroclor. Thus, one consequence of anaerobic dechlorination is that the congener mixture produced is more soluble than the original mixture released into the environment (Chiarenzelli et al., 1998a). Investigation of aqueous phase PCBs in contact with contaminated sediment reveals that microbial anaerobic dechlorination by-products (lower ortho-chlorinated congeners) dominate. For example, in Hudson River water, ~ 65% of the PCBs recovered consist of three completely ortho-chlorinated congeners (Bush et al., 1985).

One consequence of enhanced aqueous solubility is that PCBs are more likely to desorb from particulates, enter the aqueous phase, and be transported away from the site of the original release. Once in the aqueous phase they can also volatilize during evaporative processes and enter the air. The lower ortho-chlorinated PCBs, having both a relatively high solubility and volatility, are particularly likely to partition into the water and air. For example, in sediment from the St.Lawrence River near Massena, New York two ortho-chlorinated congeners make up >20% of the PCB fraction; however, when this sediment was dried, these same congeners made up >50% of the volatile fraction collected during 24 hour periods (Chiarenzelli et al., 1998a). Experimental data suggests that PCBs are carried by evaporated water even at ambient temperatures.

In one study, ~75% of the PCB fraction in St. Lawrence River sediment was transferred to the air during a five day drying period. Shorter term experiments (24 hour), involving less water, showed that repeat wetting and drying cycles resulted in incremental losses and that PCB volatile loss essentially ceased when evaporation did (Chiarenzelli et al., 1998a). A positive correlation between PCB volatile loss and sediment moisture content was found.

SEPARATION AND TREATMENT OF SORBED PCBS

Many studies have investigated the recovery of PCBs from sediment, soil, tissue, or liquids by steam distillation (Veith and Kiwus, 1977). In most instances recoveries approach complete extraction relative to aggressive organic solvent distillation. In one such study, 500 grams of sediment from the St. Lawrence River was boiled for a twelve hour period during which 2 liters of condensate was collected. The mass balance for the experiment was 112%; with 12% recovered from the aqueous phase, 92% recovered as a yellow oil, and 8% remaining in the sediment. The floating oily film, less than a milliliter in volume, was found to be 12% PCBs by weight (Scrudato and Chiarenzelli, 1998).

The water fraction from this experiment was further treated by electrochemical peroxidation, an advanced oxidation technology which utilizes electrochemically generated ferrous iron and small doses of hydrogen peroxide to produce hydroxyl radicals. After three peroxide doses totaling 130 ppm, the concentration of PCBs was reduced from 181 to less than 9 μg/L.

Thus, simple physical processes can readily remove PCBs from sediment resulting in large volume reductions. In addition, PCBs transferred to the aqueous phase can be readily destroyed by existing advanced oxidative technologies. Sediment which has been microbially dechlorinated or contaminated with lower chlorinated aroclors (1221, 1242, 1248) would be particularly amenable to this type of treatment.

CONTAINMENT FACILITIES ARE NOT A LONG-TERM SOLUTION

Without question because of navigational or remedial needs, large quantities of contaminated sediment will be dredged from the nation's waterways. There may be environmental benefits derived from temporarily isolating contaminated sediments by containment. However, existing and proposed containment facilities are essentially large entombment structures designed only to prevent their toxic contents from escape. In many instances this is achieved through the use of impermeable barriers preventing the inflow, as well as outflow, of water and air. Studies of municipal waste landfills suggest that in the absence of sufficient moisture, degradation and gas production will cease resulting in mummification of the waste. Similar results have been reported in soil profiles developed on sterile foundry waste contaminated with Aroclor 1248 decades after disposal (Chiarenzelli et al., 1998b). Thus containment is unacceptable as a long-term solution to contaminated sediments.

CONCEPTUAL APPROACH: CONTAINMENT CELL MANAGEMENT

How could a containment facility with sediment contaminated with PCBs and other organochlorides be designed, operated, and managed to promote separation and contaminant destruction? A simple design would include cells allowing various treatment processes to be conducted. Systems to regulate water, enhance permeability, add reagents, allow mixing or tilling, collect and recirculate leachate and gases, and treat contaminants transferred to the aqueous or vapor phase would be needed. The material in each cell would go through a series of sequential in-situ treatment phases and would not be moved.

Anaerobic Phase. As outlined above, anaerobic microbial dechlorination of sediment results in the production of a PCB fraction which is substantially more mobile than the original mixture. Anaerobic microbial dechlorination has been observed in sediment from numerous waterbodies, reproduced and enhanced in the laboratory, and accomplished at pilot-scale. Dredged sediment that had not reached its dechlorination plateau could be managed for further dechlorination within a containment facility. Enhancements including bacterial innocula, nutrients, moisture, substrates, surfactants, etc. could all be added through a delivery system. Further a recirculation system could be used to remove, collect, and degrade contaminants in the aqueous phase and gases.

Aerobic Phase. The aerobic biodegradation of PCBs has been reported and is complimentary to anaerobic processes, in that aerobic microbes destroy the lower ortho-chlorinated congeners which accumulate during, and remain after, anaerobic dechlorination. However, as argued above these same congeners are also susceptible to partitioning into the aqueous or vapor phase and could be effectively removed by simple aeration or evaporation associated with existing processes such as land farming, composting, bioventing, or others. Evaporation rates could be further enhanced by solar heating within a plastic enclosure or by exothermic reactions during composting. Biological, chemical, and physical processes could be combined to enhance separation and treatment. If benefits are to be derived from anaerobic/aerobic cycling, moisture levels could be managed to promote these reactions.

Residual Phase. The goals of the first two phases of treatment would be to separate and destroy as much of the sorbed PCBs as possible given efficiency and economic constraints leaving a residual fraction with greatly decreased mobility. Depending on the concentration of the residual fraction, additional in-situ treatment technologies could be employed utilizing chemical additives such as hydrogen peroxide. If further treatment was not required reclamation, phytoremediation, and closure could occur.

REFERENCES

Abramowicz, D. A. 1995. "Aerobic and anaerobic PCB biodegradation in the environment." *Environ. Health Perspect. 103*: 97-99.

Bush, B., K. W. Simpson, L. Shane, and R. Koblintz. 1985. "PCB congener analysis of water and Caddisfly Larvae in the Upper Hudson River by glass capillary column chromatography." *Bull. Environ. Contam. Toxicol. 34*: 96-105.

Carey, J. H., J. Lawrence, and H. M. Tosine. 1976. "Photodechlorination of PCBs in the presence of titanium dioxide in aqueous suspensions." *Bull. Environ. Contam. Toxicol. 16*: 697-701.

Chiarenzelli, J. R., R. Scrudato, B. Bush, D. Carpenter, and S. Bushart. 1998a. "Do large-scale remedial and dredging events have the potential to release significant amounts of semivolatile compounds to the atmosphere?". *Environ. Health Perspect. 106*(2): 47-49.

Chiarenzelli, J. R., R. Scrudato, K. Jensen,T. Maloney, M. Wunderlich, J. Pagano, and J. Schneider. 1998b. "Alteration of Aroclor 1248 in foundry waste by volatilization?" *Water, Air, Soil Pollution 104*: 113-124.

Hatzinger, P. B. and M. Alexander. 1995. "Effect of aging chemicals in soil on their biodegradability and extractability". *Environ. Sci. Technol. 29*(2): 537-545.

Pignatello, J. J. and G. Chapa. 1994. "Degradation of PCBs by ferric ion, hydrogen peroxide and UV light." *Environ. Toxicol. Chem. 13*(3): 423-427.

Sanders, J. E. 1989. "PCB-pollution problem in the Upper Hudson River: From environmental disaster to "environmental gridlock"." *Northeastern Environ. Sci. 8*(1): 1-86.

Scrudato, R. J. and J. R. Chiarenzelli. 1997. "Steam extraction combined with electrochemical peroxidation to degrade organically contaminated sediments." *4th International Conference on Advanced Oxidation Technologies for water and air remediation*. Orlando, Florida.

Spain, J. 1997. "Synthetic chemicals with potential for natural attenuation." *Bioremed. J. 1*(1): 1-9.

U.S. EPA. 1998a. "General Motors Superfund Site". *Superfund Post-Decision Proposed Plan 8/98*: 1-11.

U.S. EPA. 1998b. "Reynolds Metals Superfund Site". *Superfund Post-Decision Proposed Plan 7/98*: 1-15.

Veith, G. D. and L. M. Kiwus. 1977. "An exhaustive steam-distillation and solvent-extraction unit for pesticides and industrial chemicals." *Bull. Environ. Contam. Toxicol. 17*:631-636.

POLYCHLORINATED BIPHENYLS TRANSFORMATION BY BIOAUGMENTATION IN SOLID CULTURE

Fernández-Sánchez José Manuel,
Ruiz-Aguilar Graciela and Rodríguez-Vázquez Refugio
(CINVESTAV-IPN, México, D. F., México).

ABSTRACT. In this study a model system for the treatment of PCB's contaminated soil through bioaugmentation process, using a fungus grown on sugarcane bagasse pith established at laboratory scale. Different age spores and mycelia of the fungus were cultivated on sugarcane bagasse pith and added to a PCB's contaminated soil. In general, dehalogenation of highly chlorianted isomers to the less chlorinated was observed. The best percentage of biotransformation was obtained in the samples with inoculum of young mycelia (90 %).

INTRODUCTION

Polychlorinated biphenyls (PCB's) are known to persist in the environment. As a result of their employment in industrial processes and products (Jensen, 1969; Fishbein, 1974), the contamination with PCBs occurs in high concentrations. Soils are polluted by industrial activities like accidental spills, leaks during the transport of the chemicals or from leaks and fires in transformers, capacitors or others products containing PCB's.

PCB's are universal, environmental contaminants and are present in most environmental compartments, abiotic and biotic, throughout the world. The evolution of the health effects of PCB's mixtures is complicated by numerous factors, the most important of them is their congeneric composition since ultimately the toxicity of the mixture may depend on the toxicity of individual congeneres (ATSDR, 1996).

The microorganisms degrade mono-, di-, and trichlorinated biphenyls relatively rapidly and tetrachlorobiphenyls slowly, while higher chlorinated biphenyls are resistant to biodegradation. Chlorine substitution positions on the biphenyl ring appear to be important in determining the biodegradation rate. PCB's containing chlorine atoms in the *para* positions are preferentially biodegraded (WHO, 1993).

Biotreatment systems using microorganisms for the degradation of toxic recalcitrant organopollutants hold the promise for detoxification of vast quantities of contaminated soil (Bumpus et al., 1985). Fungi have remarkable biodegradative abilities. The white rot fungus *Phanerochaete chrysosporium*, is able to degrade a wide variety of environmental persistent organopollutants to CO_2 (Bumpus et al., 1987). The bioaugmentation process permits to use mixed cultures in order to obtain a better transformation of the compounds where each species may has its own niche for growth and substrate degradation. *P. chrysosporium* has the ability to produce hydrolytic enzymes which may act synergistically with other microorganisms (Gutiérrez-Correa and Tangerdy,

1997). This technology offers the advantage to degrade a wide range of recalcitrant pollutants and therefore a high number of sites amenable for use of bioremediation technologies.

Objective. The goal of this study was to establish a model system for the treatment of PCB's contaminated soil through bioaugmentation process using *Phanerochaete chrysosporium* grown on sugarcane bagasse pith. This model system was established at laboratory scale.

MATERIALS AND METHODS

Soil. Contaminated soil was collected from 10 and 30 cm depth, sieved by mesh 4 and 16 and stored at 4 °C. This soil was obtained from a Chemical Company site in Mexico City. Bi-, tri-, tetra-, penta-, hexa-, hepta-, octa-, and nonachloribiphenyls were found in the soil. Its composition is similar to Aroclor 1260. The soil was composed of 78 % sand, 15 % silt and 4 % clay and had a pH of 6.3. Total organic C content was 1.88 % and total P and N contents were 0.0007 mg/Kg soil and 0.098 mg/Kg soil respectively. Cation exchange capacity was 5.5 mEq/100 g.

Microorganisms and inoculum development. *Phanerochaete chrysosporium* CDBB h-298 (Microbial Collection of Department of Biotechnology and Bioengineering, CINVESTAV-IPN) was maintained on malt extract agar (MEA) plates (2 g of malt extract and 1.5 g of agar per 100 mL of distilled water) stored at 4 °C and subcultured every month. Two kinds of inoculum were prepared. For spores inoculum, the fungus was grown on MEA plates at 39 °C for 8 days and the spores scraped into Ringer's solution, 2×10^7 spores/mL inoculum was used. The fungus was grown on MEA plates at 39 °C for 1.5 days and small discs were used as mycelia inoculum. Both kinds of inoculum were scraped on sugarcane bagasse and harvested at several times (0, 7 and 15 days).

Support. Sugarcane bagasse (SCB) was used as a support, donated by the sugar Mill Casasano of Cuautla, Morelos, Mexico. It was washed 5 times with boiling water and once with cold water. The bagasse was sun dried for 4 days, and then milled with a ball mill and sieved by mesh 20 and 30.

pH and moisture content. Moisture content was determined gravimetrically in 1 g of homogeneous sample taken from the bottle. The pH was determined from a homogenous suspension of 1 g of sample in 0.9 mL of water previously shaked for 15 min.

Carbon dioxide and oxygen analysis. The most widely used measure of soil activity is soil respiration either as O_2 uptake or CO_2 evolution. The CO_2 production and O_2 consumption were determined by gas chromatography (Gow/Mac serie 580, Instrument) equipped with a concentric Altech CTR I, and electric conductivity detector. The CO_2 and O_2 evolution were calculated according to Saucedo-Castañeda et al. (1994).

PCB's extraction and analysis. 5 g of PCB's extracted in a Soxhlet apparatus for 9 h with 250 mL of *n*-hexane and acetone (1:1). The extract was evaporated to 50 mL, in a rotary evaporator, and then purified in a florisil column (60 - 100 mesh, 30 x 1.5 cm). Removal of organic contaminants was done with H_2SO_4 concentrated. The PCB analysis was performed by gas chromatography-mass spectroscopy (GC-MS) and high pressure liquid chromatography (HPLC). The GC-MS system consisted of a Varian Saturn 3 GC/MS, with electron multiplier as detector and a SPI injector. A DB-5 fused silica capillary column (30 m x 0.25 mm) was used. The operating conditions were: the injector and detector temperatures 250 °C and 260 °C, respectively. The column temperature program was used: 90 °C, hold for 0 - 3 min, then a rise in temperature to 120 °C of 3 - 7.47 min at a rate of 6.7 °C/min, 250 °C of 7.47 - 23.92 min at a rate 7.9 °C/min, 280 °C of 23.93 - 26.92 min at a 10 °C/min and 280 °C of 26.92 - 41.92 min. Helium was used as a carrier gas at a flow rate of 1 mL/min; sample column 1 µl. Interpretation of the chromatographic data was carried out by an integration software (Varian, Saturn 3GC/MS version 4.1). The HPLC system consisted of Consta Metric 3200 pump (LCD Analytical), an UV Spectro Monitor 3200 detector (Thermo Separation Products) and a Chromoject integrator. The column was a C18 (Chromanetics, 150 x 4.6 mm, particule diameter 5µm). The solvents were acetonitrile - water (80:20 v/v) at a flow rate of 1.0 mL/min. The quantitation of PCB's was performed by standard procedure: calculation from the sum of characteristic peaks selected from chromatograms of samples and compared with the untreated contaminated soil.

Treatments. The treatments were grown in solid culture in 125 mL glass serum bottles sealed with teflon coated rubber stoppers. Contaminated soil was added at differents times (0, 7 and 15 days) on the inoculum. Each type of inoculum was put on PCB's contaminated soil at rate 15:85 respectively, and let to grow. Kirk's medium without glucose was added to adjust the moisture in the system (60%). The cultures were aerated every day until end of treatment to avoid the accumulation of CO_2. After 60 days, were done extractions and analysis of PCB's from each bottle. Duplicated were done. The control was a sample with contaminated soil and without inoculum.

RESULTS

pH and moisture. pH was in range from 4.4 to 5.5. These dates are good for fungus because its optimum value of growth (4.5). It is important to mention that addition of bagasse with *P. chrysosporium* maintain this pH, because the fungus produces acids when some lignocellulosic residues are present. The moisture was constant during all experiment (60 %).

CO$_2$ evolution. Effects of addition of contaminated soil at differents times (0, 7 and 15 days) on diverse ages of growth of *P. chrysosporium* were evaluated. In the treatments the rate of CO$_2$ evolution was inferior only on control treatment, because the microbial activity of soil was due only to native flora (18 mg CO$_2$/g initial dry matter [IDM]). The supply of inocula increase this metabolic activity. Both shown similar evolution on soil during 60 days; the maximum activity was 48 mg CO$_2$/g IDM approximately. None apparent variation was shown in using spores or mycelium with this determination. But this information is not enough to predict which treatment is better to transformation of PCB's.

Results of the biotransformation analysis. To establish the level of transformation of the treatments an extraction and analysis by HPLC and GC-MS were done for each sample. The results of HPLC analysis were used to estimate the PCB's biotransformation by comparison of final total area with respect to initial total area (Figure 1). The treatments showed a diminution in their PCB's content from 75.87 to 90.51 %. The samples with mycelia inoculum showed higher degradation than spore inoculum. In order to establish which treatment will be better for PCB's biotransformation, a comparison of media with the SAS System Program version 6 was done. The treatment which result better for our purpose and had a good level of degradation was cultivated with young inoculum of mycelia (90 % biotransformation). This treatment is very suitable to remediation process because the time necessary to produce inoculum is shorter (1.5 days) than others, with a good level of transformation. In addition, *P. chrysosporium* is a highly successfull competitor in nature, especially when the carbon source consists of lignocellulosic residues. If this kind of material is used in biotreatment systems, competition by non - lignin degrading organisms is likely to be minimal (Bumpus et al., 1985) as it shown in this experiment.

The change in composition of congeneres were determinated by GC-MS. A dehalogenation of PCB's most chlorinated congeners to least chlorinated congeneres was observed in general. The treatments with mycelium showed increase in the less chlorinated biphenyls (mono-, di- pentachlorobiphenyls) with respect to treatment without inoculum. The main transformation occurred in pentachlorobiphenyls. This difference probably due to transformation of most chlorinated biphenyls to least chlorinated (Eaton, 1985) by *P. chrysosporium*.

The mycelium show specialized and complex tissues that avoid effect of contaminant and its capacity to modify toxic wastes. The effectiveness of bioaugmentation depends on the survival and activity of inoculated microorganisms in soil. When introduced to a nonsterile contaminated soil, with active fungus resisted attack from indigenous soil bacteria and fungi. Likewise, easy manipulation of *P. chrysosporium* in this age produces a simple and cheap treatment. The optimization of using bioaugmentation with mycelium of *P. chrysosporium* on PCB's contaminated soil is the subject of our ongoing research.

ACKNOWLEDGMENTS

We are especially grateful to Elvira Ríos-Leal for technical assistance and for the financial support of Fondo de apoyo al desarrollo de proyectos de investigación básica y tecnológica en colaboración con las Instituciones de Educación Superior (FIES 95-106-VI).

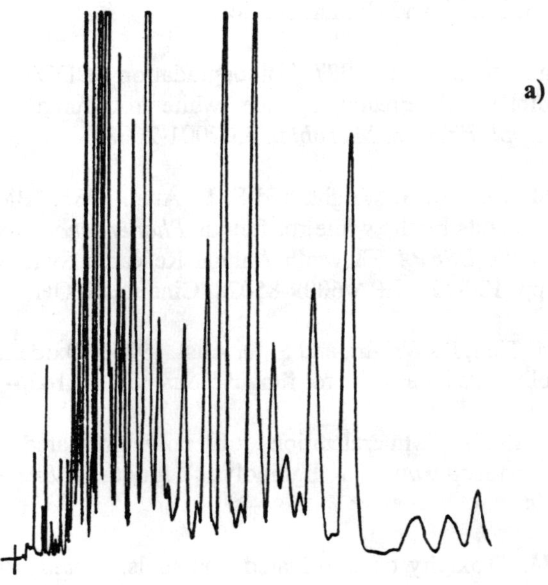

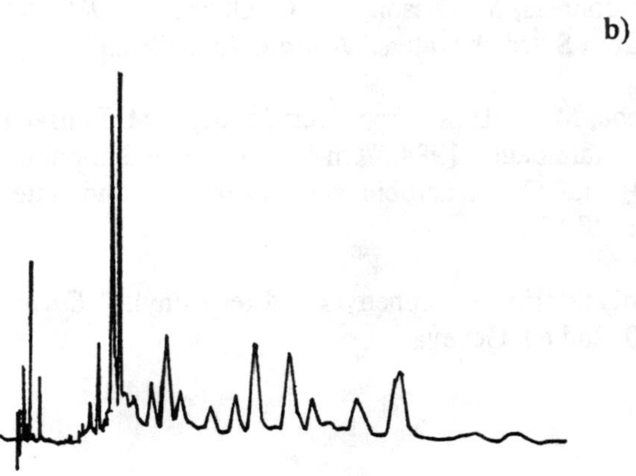

FIGURE 1. HPLC chromatograms of extract of treatment with mycelium inoculum. a) Initial sample. b) Final sample.

REFERENCES

ATSDR. 1996. *Toxicological profile for polychlorinated biphenyls.* Prepared for U.S. Department of Health and Human Services.

Bumpus, J. A., and S. D. Aust. 1987. "Biodegradation of DDT (1,1,1-trichloro-2,2-bis(4-chrorobiphenyls) ethane) by the white rot fungus *Phanerochaete chrysosporium*." *Appl. Environ. Microbiol. 53*:2001-2008.

Bumpus, J. A., M. Tien, D. S. Wright, and S. D. Aust. 1985. "Biodegradation of environmental pollutants by the white rot fungus *Phanerochaete chrysosporium*." *In Proceedings of the USEPA*. Eleventh Annual Research Symposium on Toxic Waste Disposal, pp. 120-126. EPA/6009-85028, Cincinnati, OH.

Bumpus, J. A., M. Tien, D. Wright, and S. D. Aust. 1985. "Oxidation of persistent environmental pollutants by a white rot fungus". *Science. 53*:1434-1436.

Eaton, D. C. 1985. "Mineralization of polychlorinated biphenyls by *Phanerochaete chrysosporium*: a ligninolytic fungus." *Enzyme Microbiol. Technol. 7*:194-196.

Fishbein, L. 1974. "Toxicity of chlorinated biphenyls." *Annu. Rev. Pharmacol. 14*:139-156.

Gutiérrez-Correa, M., and R. P. Tangerdy. 1997. "Production of cellulose on sugar cane bagasse by fungal mixed cultured solid substrate fermentation." *Biotechnol. Letters. 19*(7):665-667.

Jensen, S., A. G. Johnels, M. Olsson, and G. Otterlind. 1969. "DDT and PCB in marine animals from Swedish waters." *Nature. 224*:247-250.

Saucedo-Castañeda, M. R., B. K. Trejo-Hernández, J. M. Lonsane, S. Navarro, D. Roussos and M. Raimbault. 1994. "On-line automated monitoring and control systems for CO_2 and O_2 in aerobic and anaerobic solid-state fermentations." *Process Biochem. 29*:13-24.

WHO. 1993. "Polychlorinated biphenyls and terphenyls." *Environmental Health Criteria*. Vol.140. 2nd ed. Geneva.

TRANSFORMATION OF AROCLOR BY INDIGENOUS AND INOCULATED MICROBES IN SLURRY REACTORS

John P. Tharakan, **Ebenezar Sada, Raycharn Liou, and Ramesh Chawla**
(Howard University, Washington, DC)

ABSTRACT: In this work, we investigated the biotransformation of Aroclor 1242 in a bioslurry configuration. The soil was from Wurtsmith Airforce Base (MI), spiked with Aroclor 1242 (final concentration: 100ppm) and slurried (10% m/v) with culture media supplemented with biphenyl. Bioslurries were also operated with unautoclaved soils to investigate any indigenous microbial PCB biotransformative activity. Control abiotic slurries of autoclaved soil with no additional inoculum or cosubstrate demonstrated no reduction of spiked Aroclor, suggesting minimal abiotic PCB losses. Cosubstrate addition to unautoclaved soil only slightly enhanced PCB biotransformation extents, with 98% of the total initial spiked Aroclor remaining. When the slurry was amended with *R. erythropolis* (NY05) with no cosubstrate, only 51% of the total spiked Aroclor remained. Inoculation of the bioslurry with NY05 and amendment with biphenyl markedly increased total PCB transformation so that only 19% of spiked Aroclor remained after 10 days. These results point to the potential for *in situ* and on-site biodegradation of PCB contaminated media as long as suitable microbial and cosubstrate amendments are utilized.

INTRODUCTION

Polychlorinated biphenyl's (PCBs) are a family of nonpolar, aromatic chlorinated hydrocarbons in formerly widely used industrial fluids such as Aroclors [USA], Kanechlors [Japan] and Sovols [former USSR] (Erickson, 1997). Widespread PCB application as adhesives, transformer dielectrics, coolants, and hydraulics combined with lax disposal practices has resulted in almost ubiquitous contamination of various environmental media, ranging from soils and sediments to rivers and streams, with PCBs (Hutzinger et al. 1975). Health concerns arose from PCBs suspected toxicity, carcinogenicity and endocrine disruptive effects. PCBs also bioaccumulate in the food chain, with concentrations increasing by several orders of magnitude at succeeding trophic levels. These factors, combined with growing awareness of the fragility of the environment resulted in the banning of PCB production in the early seventies (Abramowicz et al. 1993)

Currently PCBs are regulated under the Toxic Substances Control Act (TSCA) and the Comprehensive Environmental Response and Liability Act (CERCLA). Widespread existing contamination at current and former industrial sites is the target of several Superfund remediation actions. Of the many potential PCB remediation technologies, incineration is the main EPA approved method. Alternate technologies have not yet demonstrated the destruction and removal efficiency (DRE) of 99.999 mandated by the EPA (Cookson, 1995). Biological treatment technologies, however, have shown potential for complete biodegradation of various halo-organics including PCBs. Bioremediation offers the possibility PCB mineralization while lowering energy requirements, problematic emissions and byproducts (Hicks and Caplan, 1993). Several barriers exist to the implementation of PCB bioremediation including high toxicity, low bioavailability, and lack of PCB metabolizing microbes (Barriault and Sylvestre, 1993). Successful laboratory PCB biodegradation has been achieved through use of cosubstrates to support cometabolic PCB biodegradation (Kohler et al. 1988).

In this study, we compare the aerobic biotransformation of spiked Aroclor 1242 under different biotic and abiotic conditions in a slurry reactor. Specifically, we first examine the effect biphenyl as cosubstrate amendment to stimulate indigenous microbes to biotransform PCB congeners. Second, we investigated the ability of inoculated NY05 to cometabolically biotransform PCB congeners. The impact of these varying culture conditions on the extents of PCB congener biotransformation was evaluated.

MATERIALS AND METHODS

Microbial Isolate. The microbial isolate, obtained from Dr. J. Tiedje at Michigan State University, is *Rhodococcus erythropolis* (strain NY05), a gram positive, aerobic cocci isolated from Hudson River (NY) sediments. It can cometabolically biotransform several different PCBs (Pellizari et al. 1996).

Soil and Indigenous Microbes. Uncontaminated soil, previously characterized as sandy with specific gravity of 2.65 and void ratio of 0.87, was obtained from Wurtsmith Air Force Base (MI). Soil was autoclaved thrice, spiked with Aroclor 1242 and then slurried with growth media. Unautoclaved soil was used to examine the activity of any microorganisms indigenous to the soil.

Growth Media. The growth media was comprised of minimal salt base, vitamin supplement, metal ion solution and mineral base, and has been described elsewhere(Sada et al. 1998). Biphenyl was added as the cosubstrate.

Batch Bioslurry Inoculation. Bioslurry experiments were conducted in single light-shielded 500-ml flasks maintained at 30°C. Bioslurry reactors were agitated with a magnetic stir-plate and immersed in a temperature controlled water-bath.

Flask Bioslurry Set-up. Fifteen gram aliquots of soil were autoclaved three times in succession, and then evenly spiked with Aroclor 1242 (dissolved in hexane) to a final concentration of 100 mg Aroclor/kg soil. The soil was then mixed with sterile culture media to form a 10%(m/v) slurry, while leaving 350-ml of headspace. Biphenyl was added to the appropriate reactors to a final level of 500 ppm. Biotic reactors were inoculated with NY05. Abiotic control bioslurry containing autoclaved, spiked soil with cosubstrate supplemented media and no inoculum. Cosubstrate-free biotic controls were established with Aroclor 1242 spiked soil in minimal media inoculated with NY05. Samples (2-ml) at various times were collected in triplicate for PCB congener concentration analysis.

PCB Measurement. Slurry samples were extracted with 1:1 hexane:acetone mixture, using a total of 2X sample volume. Extracts were concentrated and injected into a HP 5890 Series II Gas Chromatograph (GC) utilizing a 0.2mm i.d., 50m fused silica column with a 0.33μm film using flame ionization detection (FID). Octochloronaphthalene was the internal standard. Data are reported as a mole percent of each congener or as a percentage of total initial Aroclor spiked.

RESULTS AND DISCUSSION

Space limitations preclude the discussion of all the results on PCB congener biotransformation obtained in this study. Results pertinent to assessing the impact of indigenous microbes, cosubstrate amendments and NY05 on PCB congener transformation are shown in Fig.1 to 3. PCB congener biotransformation results using microbes indigenous to the Wurtsmith AFB soil will be discussed initially, followed by discussion of the effect of microbial and cosubstrate amendments on enhancing PCB biotransformation.

Figure 1a shows the relative mole percent of each congener in the abiotic control after 10 days in culture. The abiotic control consisted of thrice-autoclaved soil slurried with culture media with no cosubstrate or microbial supplementation. The data are depicted in bar chart form with each congener peak represented as a bar; the error bar indicates one standard deviation. The distribution of peaks and relative mole percent of each conforms to the profile obtained with direct GC-FID injection of Aroclor 1242. Further the difference between the peak areas for the individual congeners in the GC chromatogram of the extract from day 0 and day 10 of the abiotic sample was insignificant (less than 5%), indicating that abiotic losses of individual congeners are minimal. Figure 1b shows the relative mole percents of each detected congener in the biotic slurry culture using unautoclaved soil and amended with biphenyl after 10 days. The distribution of peaks indicates some biotransformation has occurred, primarily in the lower chlorinated congeners as indicated by the slight decreases in relative mole percent of the di-chlorbobiphenyls (peaks 2, 3, and 4) and the trichlorobiphenyls (peaks 5-10). Reductions in the higher chlorinated congeners, although detectable, are significantly lower.

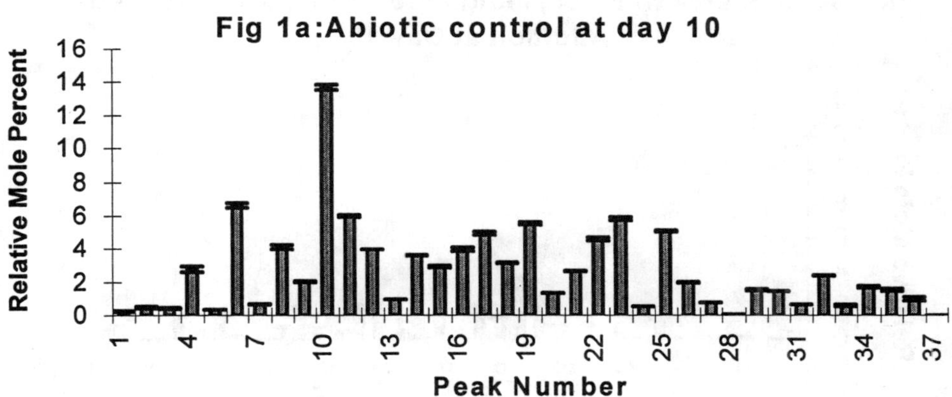

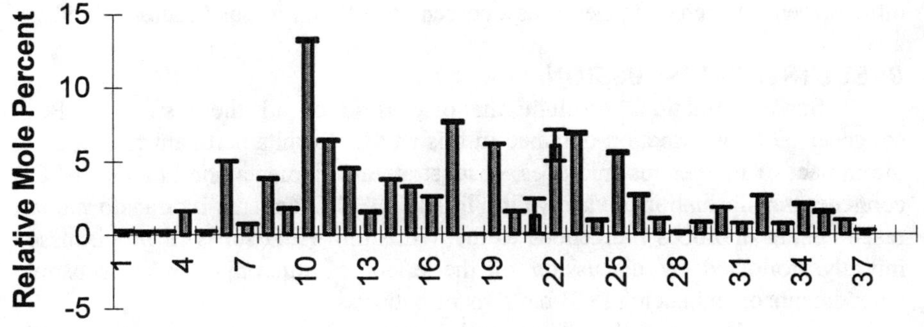

In Figure 2a, the relative mole percent of the various congeners are shown after 10 days in the bioslurry culture. In this bioslurry, the matrix was amended with an inoculum of NY05 but with no cosubstrate amendment. Figure 2b shows the relative mole percent of the various congeners in the spiked Aroclor 1242 amended with NY05 and biphenyl. In both cases, the reductions in peak height across the spectrum, ranging from dichlorobiphenyls to as high as hexachlorobiphenyls are consistent with results that have been obtained in biphenyl supported cultures of *R. erythropolis* (Pellizari et al. 1996; Sada et al. 1998). In those two studies, cosubstrate amendments, including biphenyl, terpenes and naphthalene, were added to both aqueous and slurry cultures. Although the data in Figure 2a correspond to a bioslurry with no amended cosubstrate, it is known that Aroclor 1242 contains small quantities of unchlorinated biphenyl, which may serve as cosubstrate. In addition, the soil used could potentially provide a cosubstrate from its organic content. When the

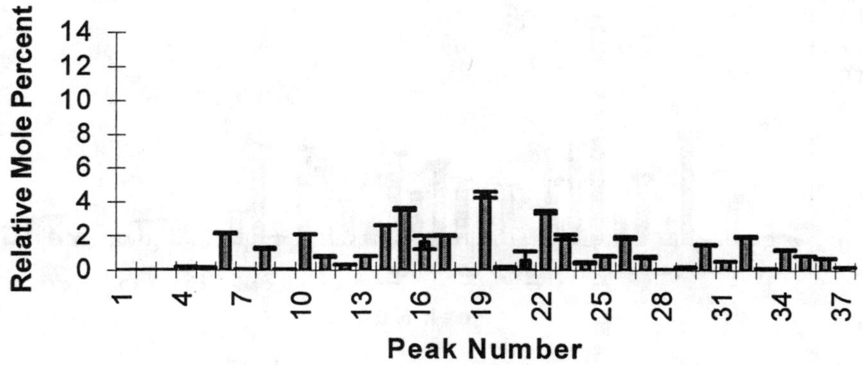

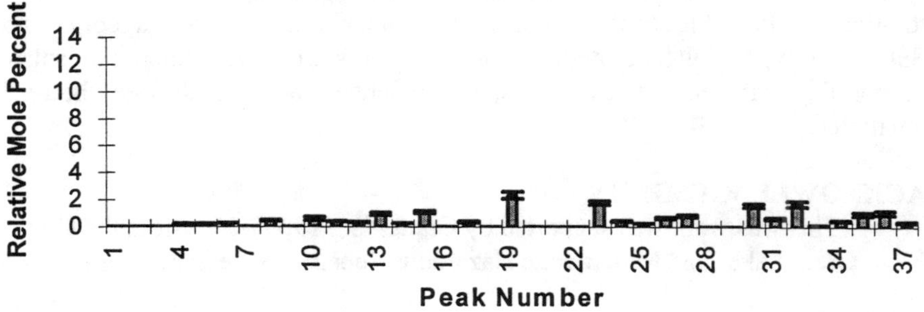

Fig 2b: Bioslurry w/ NY05, Indigenous soil microbes w/ Biphenyl at Day 10

bioslurry is amended with cosubstrate, the reduction of the various congeners is markedly increased, consistent with earlier studies.

Figure 3 shows the percentage reduction in total Aroclor; the data are shown as the percentage of initial total Aroclor that was spiked onto the soil prior to slurrying. This demonstrates the potential effect of amending a PCB contaminated soil bioslurry with a specific PCB-degrading microorganism and its attendant cosubstrate.

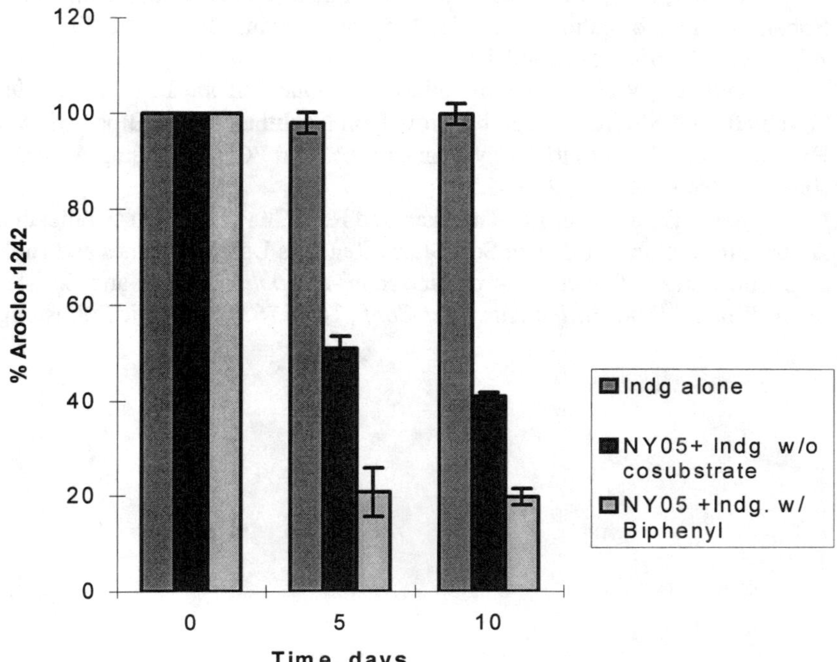

Fig 3: Percentage of Total Initial Aroclor 1242 Remaining

SUMMARY

For *in situ* or on-site PCB bioremediation process development, site-specific analysis must initially examine the potential for cosubstrate amendments to stimulate PCB biotransformation by any capable indigenous microorganisms. Subsequently, inoculation of contaminated media with specific microbial isolates, or possibly with engineered strains, and amended with cosubstrate should be attempted.

ACKNOWLEDGEMENTS

This was work was supported by Grant No. 919605-01-7 from the U.S. EPA, Great-Lakes and MidAtlantic Hazardous Substances Research Center.

REFERENCES

Abramowicz, D.A., M.J. Brennan, H.M. VanDort and E.L. Gallagher (1993). "Factors Influencing the Rate of PCB Dechlorination in Hudson River Sediments," *Env. Sci.Tech.*, 27, 1125-1131.

Cookson, J.T. (1995). *Bioremediation Engineering: Design and Application,* McGraw-Hill Inc., New York, NY.

Erickson, M.D. (1997). *Analytical Chemistry of PCBs,* Lewis Publishers, New York, NY.

Hicks, B.N. and J.A. Caplan (1993). "Bioremediation: A natural Solution," *Pollution Engineering*, 25, 30-33.

Hutzinger, O., S. Safe, and V. Zitko (1975). *The Chemistry of PCBs*, CRC Press, Cleveland, OH.

Kohler, H.P., D.K. Staub, and D.D. Focht (1988) "Cometabolism of PCB: Enhanced Transformation of Aroclor 1254 by Growing Bacterial Cells," *Appl.Env. Microbiol.* 54, 1940-1945.

Pellizari, V.H., S. Bezborodnikov, J.F. Quensen and J.M. Tiedje (1996) "Evaluation of Strains Isolated by Growth on Naphthalene and Biphenyl for Hybridization of Genes to Dioxygenase Probes and PCB Degrading Ability," *Appl.Env.Microbiol.* 62, 2503-2508.

Sada, E., R. Liou, J.P. Tharakan and R.C. Chawla (1998) "Cometabolic Biotransformation of PCBs in Soil-Slurry Reactors Using Terpenes and Biphenyl as Cosubstrates with *C.testosteroni* and *R. erythropolis*," in Suri and Christensen (eds), *Proc.30thMid-Atl.Ind.Haz.Wast.Conf.*, 167-176. Technomic Publishing, PA.

COMPUTER MODEL FOR PREDICTION OF PCB DECHLORINATION AND BIODEGRADATION ENDPOINTS

K. Thomas Klasson (Oak Ridge National Laboratory, Oak Ridge, Tennessee)
Eric M. Just (Northwestern University, Evanston, Illinois)

ABSTRACT: Mathematical modeling of polychlorinated biphenyl (PCB) transformation served as a means of predicting possible endpoints of bioremediation, thus allowing evaluation of several of the most common transformation patterns. Correlation between laboratory-observed and predicted endpoint data was, in some cases, as good as 0.98 (perfect correlation = 1.0).

INTRODUCTION

Although PCBs are relatively inert, biological degradation by anaerobic dechlorination and aerobic oxidation is possible (Abramowicz, 1990). The enzymes involved in the aerobic ring cleavage have been studied and reviewed extensively (Abramowicz, 1990; Furukawa, 1982); however, details are not known about the enzymes responsible for anaerobic dechlorination, primarily due to the difficulty in isolating the microorganisms responsible (Bedard and Quensen, 1995). The anaerobic dechlorination follows distinct dechlorination patterns, which refer to the type of chlorines removed, and is dependent on environmental conditions. As the knowledge of microbial dechlorination and degradation increases, it may soon be possible to control these bioremediation processes for optimal results.

METHODOLOGY

This work presents the development of a predictive modeling tool to aid in evaluation of PCB degradation outcomes. This tool is a computer model based on the susceptibility of individual PCB compounds (congeners) to undergo bacterial transformation (Bedard and Quensen, 1995; Bedard et al., 1986; Williams, 1994). The model was developed for use on a personal computer using Microsoft Excel (Microsoft Corporation, Redmond, WA); and in order to utilize the various built-in functions offered in Excel, the construction of a new nomenclature for PCBs was incorporated. The nomenclature name was 11 characters wide, and both the position and the numeric value of a character in the name mark the chlorine substitution type.

The susceptibility of individual congeners to undergo anaerobic transformation has been postulated and demonstrated by Williams (1994), who found that the chlorine position on the PCB carbon backbone determined sensitivity to attack. These dechlorination systems were adapted, modified, and used in the computer model; the susceptibility for attack is based on simple rules related to the structure of the PCB congeners.

As previously mentioned, dechlorination processes do not result in destruction of PCBs — only alteration. The most common pathway of biological destruction is through aerobic cometabolism. There are two known enzymatic oxidative processes, one that oxygenates the C-C bond in the 2,3- and 5,6-positions (Furukawa, 1982) and another that oxygenates the bond in the 3,4- and 4,5-positions (Bedard et al., 1986).

Rules for dechlorination systems and degradation systems used in predictions may be expressed as the Excel formulas shown in Figure 1. The structure of the congener under attack is found in column A, and the end-product structure in cells is found in column C. In this manner, any PCB congener can be evaluated for susceptibility, and the resulting end product can be determined.

	A	B	C
1	23450-20450	=SUBSTITUTE(A1,"234","204")	=SUBSTITUTE(B1,"456","406")
2	23450-20450		=SUBSTITUTE(A2,"345","305")
3	23450-20450	=SUBSTITUTE(A3,"340","300")	=SUBSTITUTE(B3,"045","005")
4	23450-20450	=SUBSTITUTE(SUBSTITUTE(A4,"230","200"),"034","004")	=SUBSTITUTE(SUBSTITUTE(B4,"450","400"),"056","006")
5	20406-20050	=SUBSTITUTE(SUBSTITUTE(A5,"03056","23000"),"03006","20000")	=SUBSTITUTE(SUBSTITUTE(B5,"3050","3000"),"20050","20000")
6	20406-20050		=SUBSTITUTE(A6,"2040","2000")
7	20400-00400	=IF(LEFT(A7,5)="00000",A7,IF(MID(A7,8,3)="040",REPLACE(A7,8,3,"000"),A7))	=IF(RIGHT(B7,5)="00000",B7,IF(MID(B7,2,3)="040",REPLACE(B7,2,3,"000"),B7))
8	20406-20050	=IF(OR(MID(A8,1,2)="00",MID(A8,4,2)="00"),"",A8)	=IF(OR(MID(B8,7,2)="00",MID(B8,10,2)="00"),"",B8)
9	20400-00400	=IF(OR(MID(A9,2,2)="00",MID(A9,3,2)="00"),"",A9)	=IF(OR(MID(B9,8,2)="00",MID(B9,9,2)="00"),"",B9)

FIGURE 1. Excel spreadsheet with underlying functions corresponding to dechlorination systems and oxidative degradation. Column A contains examples of congeners. The formulas describe doubly flanked meta chlorine (DFM), doubly flanked para chlorine (DFP), singly flanked para chlorine (SFP), singly flanked meta chlorine (SFM), unflanked meta chlorine on di- or tri-substituted ring (UFM), unflanked para chlorine on di- or tri-substituted ring (UFP), lone para chlorine on ring opposite substituted ring (LP), 2,3-dioxygenase attack, and 3,4-dioxygenase attack.

The dechlorination systems described above and by Williams (1995) occur both in nature and in laboratory experiments in the form of combinations — not as isolated individual systems. The more complex of these systems are often referred to as dechlorination activities or processes, which are denoted by letters such as C, H, H′, M, N, P, Q, etc. (Bedard and Quensen, 1995). For example, activity M is described as the removal of flanked and unflanked *meta* chlorines, and activity Q is described as the removal of flanked and unflanked *para* chlorines. Activity C can be described as a combination of activities M and Q (Bedard and Quensen, 1995; Quensen et al., 1990).

RESULTS

Figure 2 shows the predicted dechlorination of Aroclor 1242 through process C. The individual charts were constructed by entering published values for Aroclor 1242 congener composition (Schulz et al., 1989) and evaluating each congener's susceptibility to dechlorination in a sequence of dechlorination systems:

1. doubly flanked *meta* (DFM), followed by
2. singly flanked *meta* (SFM), followed by
3. unflanked *meta* on di- or tri-substituted ring (UFM), followed by
4. unflanked *para* on di- or tri-substituted ring (UFP), and followed by
5. lone *para* on ring opposite substituted ring (LP).

After each sequence step, the concentration of each of the 209 possible PCB congeners was recalculated to account for the products created, and the individual congeners were combined into "peak concentrations" for comparison with experimental data (Quensen et al., 1990).

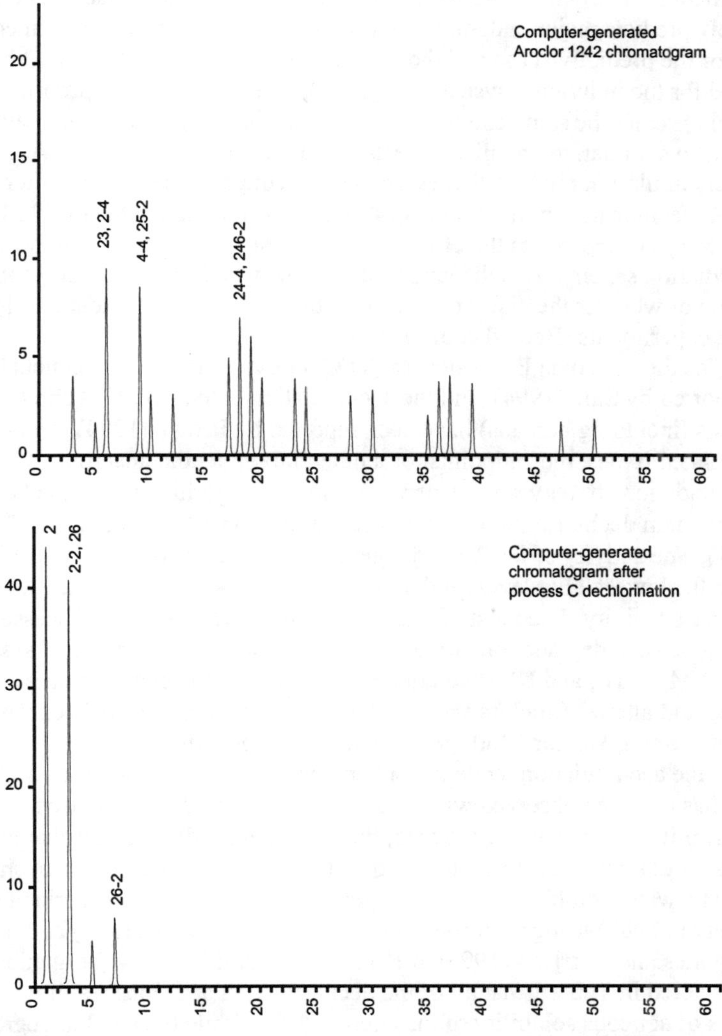

FIGURE 2. Dechlorination of Aroclor 1242 modeled via DFM+SFM+UFM+ UFP+LP. Values on the Y-axis refer to mole percent. The congeners in the three largest peaks have been identified.

The construction of the charts (displaying chromatograms) in Figure 2 was made by generating these congener-containing peaks. Quensen et al. (1990) obtained the laboratory data by combining Aroclor 1242 with Hudson River (NY) sediment organisms, inducing dechlorination activity. The correlation between starting values in the computer model and the sterile control in the experimental case was 0.95. The correlation of the final values in the predictive case and the experimental case was 0.98.

A computer model prediction via DFM+SFM+UFM dechlorination of Aroclor 1242 compares well with experimental data (Quensen et al., 1990) for dechlorination of Aroclor 1242 with Silver Lake sediment organisms. The model accurately predicts major endpoint products, and the correlation between end values for the predictive case and the experimental case was 0.92. Algorithms proposed for the individual systems (Figure 1), when combined appropriately, accurately predict the same end products as those listed by Bedard and Quensen (1995), who summarized published dechlorination activities (processes). The computer simulation predicts that essentially all congeners remaining after a process C dechlorination of Aroclor 1242 would be biodegradable (Table 1), which perhaps overpredicts the effectiveness of a sequential anaerobic-aerobic bioremediation scheme. It is difficult to determine whether the model is too optimistic or whether the experimental conditions have not been ideal during reported experiments (Bedard et al., 1987).

The dioxin Toxic Equivalences (TEQs) of various PCB congeners have been reported by Safe (1994), and the accumulation extents of PCB congeners in organisms (including humans) have been reported by Brown (1994). To assess health-related issues, the endpoints for a potential sequential anaerobic-aerobic PCB degradation strategy were generated using the algorithms described earlier. All of the main dechlorination activities result in a substantial reduction of TEQ (Table 1). The activity of the 2,3-dioxygenase is predicted to be sufficient in reaching the lowest TEQ levels and does not have to be complemented with 3,4-dioxygenase activity. The most efficient anaerobic dechlorination processes preceding aerobic degradation are the ones that include DFM and SFM systems (activities M, C, H′, and N). Prediction shows that aerobic degradation of unaltered and altered Aroclors should be effective in reducing TEQs except in the case of unaltered Aroclor 1260 (not shown). The most effective activity in reducing the accumulation tendency in humans is dechlorination process C. This activity has not been observed with Aroclor 1260 (Bedard and Quensen, 1995); however, if it was or could be induced, the computer model predicts that all end products would have half-lives in humans of less than 0.1 year (data not shown).

The water solubility of PCB congeners was estimated as described by (Voice et al., 1983) using partition coefficients from Mackay (1982) and data from Holmes and Harrison (1993) at 1% dissolved solids with 4% organic carbon. It can be noted in Table 1 that all of the dechlorination activities result in increases of aqueous solubility of the altered PCBs, while the aerobic degradation decreases PCB solubility.

TABLE 1. Summary results of predicted anaerobic dechlorination[a] followed by aerobic degradation (via ring cleavage) by 2,3-dioxygenase, 3,4-dioxygenase, and a combination of both enzyme activities.

		Process for Aroclor 1242 (100 ppm)					
		None	M	Q	C	H'	H
Anaerobic only	Half-life in humans						
	ppm<0.1y	31.0	66.9	71.4	76.2	52.3	39.5
	0.1y<ppm<1.0y	51.5	4.8	12.6	0.0	30.1	45.4
	1.0y<ppm<10y	15.2	15.2	0.0	0.0	9.2	9.8
	10y<ppm	0.9	0.0	0.0	0.0	0.0	0.0
	TEQ as ppb TCDD	10.1	0.0	0.0	0.0	0.0	0.0
	Aqueous solubility, ppb	3.3	6.3	10.2	15.9	4.6	3.9
2,3-Attack	Half-life in humans						
	ppm<0.1y	0.7	0.1	3.8	0.1	3.8	0.7
	0.1y<ppm<1.0y	9.2	0.0	6.0	0.0	6.0	9.6
	1.0y<ppm<10y	0.0	0.0	0.0	0.0	0.0	0.0
	10y<ppm	0.7	0.0	0.0	0.0	0.0	0.0
	TEQ as ppb TCDD	0.0	0.0	0.0	0.0	0.0	0.0
	Aqueous solubility, ppb	0.1	0.0	0.1	0.0	0.1	0.1
3,4-Attack	Half-life in humans						
	ppm<0.1y	0.0	0.0	0.0	0.0	0.0	0.0
	0.1y<ppm<1.0y	4.9	3.5	0.0	0.0	1.6	1.6
	1.0y<ppm<10y	15.2	15.2	0.0	0.0	9.2	9.8
	10y<ppm	0.9	0.0	0.0	0.0	0.0	0.0
	TEQ as ppb TCDD	10.1	0.0	0.0	0.0	0.0	0.0
	Aqueous solubility, ppb	0.4	0.6	0.0	0.0	0.4	0.4
2,3- & 3,4-Attack	Half-life in humans						
	ppm<0.1y	0.0	0.0	0.0	0.0	0.0	0.0
	0.1y<ppm<1.0y	0.0	0.0	0.0	0.0	0.0	0.0
	1.0y<ppm<10y	0.0	0.0	0.0	0.0	0.0	0.0
	10y<ppm	0.7	0.0	0.0	0.0	0.0	0.0
	TEQ as ppb TCDD	0.0	0.0	0.0	0.0	0.0	0.0
	Aqueous solubility, ppb	0.0	0.0	0.0	0.0	0.0	0.0

[a]Activity M was modeled as DFM+SFM+UFM; activity Q was modeled as DFP+SFP+ SFM+UFP+LP; activity C was modeled as DFM+SFM+UFM+UFP+LP; activity H′ was modeled as DFP+DFM+SFP+SFM; and activity H was modeled as DFP+DFM+SFP

REFERENCES

Abramowicz, D. A. 1990. "Aerobic and Anaerobic Biodegradation of PCBs: A Review." *Crit. Rev. Biotechnol. 10*: 241–251.

Bedard, D. L., and J. F. Quensen III. 1995. "Microbial Reductive Dechlorination of Polychlorinated Biphenyls." In L. Y. Young et al. (Eds.), *Microbial Transformation and Degradation of Toxic Organic Chemicals*, pp. 127–216. Wiley-Liss, New York.

Bedard, D. L., R. Unterman, L. H. Bopp, M. J. Brennan, M. L. Harber, and C. Johnson. 1986. "Rapid Assay for Screening and Characterizing Microorganisms for the Ability to Degrade Polychlorinated Biphenyls." *Appl. Environ. Microbiol.* *51*(4): 761–768.

Bedard, D. L., R. E. Wagner, M. J. Brennan, M. L. Haberl, and J. F. Brown, Jr. 1987. "Extensive Degradation of Aroclors and Environmentally Transformed Polychlorinated Biphenyls by *Alcaligenes eutrophus* H850." *Appl. Environ. Microbiol.* *53*(5): 1094–1102.

Brown, J. F., Jr. 1994. "Determination of PCB Metabolic Excretion, and Accumulation Rated for Use as Indicators of Biological Response and Relative Risk." *Environ. Sci. Technol.* *28*(13): 2295–2305.

Furukawa, K. 1982. "Microbial Degradation of Polychlorinated Biphenyls." In A. M. Chakrabarty (Ed.), *Biodegradation and Detoxification of Environmental Pollutants*, pp. 33–57. CRC Press, Boca Raton, FL.

Holmes, D. A., and B. K. Harrison. 1993. "Estimation of Gibbs Free Energy of Formation for Polychlorinated Biphenyls." *Environ. Sci. Technol.* *27*(4): 725–731.

Mackay, D. 1982. "Correlation of Bioconcentration Factors." *Environ. Sci. Technol.* *16*(5): 274–278.

Quensen, J. F., III, S. A. Boyd, and J. M. Tiedje. 1990. "Dechlorination of Four Commercial Polychlorinated Biphenyl Mixtures (Aroclors) by Anaerobic Microorganisms from Sediments." *Appl. Environ. Microbiol.* *56*(8): 2360–2360.

Safe, S. H. 1994. "Polychlorinated Biphenyls (PCBs): Environmental Impacts, Biochemical and Toxic Responses, and Implications for Risk Assessment." *Crit. Rev. Toxicol.* *24*(2): 87–149.

Schulz, D. E., G. Petrick, and J. C. Duinker. 1989. "Complete Characterization of Polychlorinated Biphenyl Congeners in Commercial Aroclor and Clophen Mixtures by Multidimensional Gas Chromatography-Electron Capture Detection." *Environ. Sci. Technol.* *23*(7): 852–859.

Voice, T. C., C. P. Rice, and W. J. Weber, Jr. 1983. "Effect of Solids Concentration on the Sorptive Partition of Hydrophobic Pollutants in Aqueous Systems." *Environ. Sci. Technol.* *17*(9): 513–518.

Williams, W. A. 1994. "Microbial Reductive Dechlorination of Trichlorobiphenyls in Anaerobic Sediment Slurries." *Environ. Sci. Technol.* *28*(4): 630–635.

UPTAKE OF CHLOROPHENOLS BY RYE GRASS: ROLE OF PHYSICO-CHEMICAL PROCESSES

Cynthia E. Crane and *John T. Novak* (Virginia Tech, Blacksburg, VA 24061)

ABSTRACT: Rye grass was grown in the presence of 2,4,6-trichlorophenol (TCP) and pentachlorophenol (PCP). The grass removed substantial amounts of TCP and PCP from solution. The majority of the removed contaminant could not be recovered by ethanol extraction. This unrecoverable fraction is considered to be "internalized" by the plant. The mass fractions removed and internalized were much greater than the fraction of water transpired, indicating that other mechanisms besides translocation were important in plant-contaminant interactions. The relative root surface area was correlated with both chlorophenol removal and internalization, suggesting that sorption of the contaminants onto the roots from solution was an important factor in plant-contaminant interactions.

INTRODUCTION

The use of plants to enhance in situ remediation appears to be a promising technology. Researchers have demonstrated that some plant species are able to directly take up organic contaminants from soil and transform them (Burken and Schnoor, 1997). Past research has identified pathways by which organic compounds enter the root endoderm from the surrounding solution and has explored the effect of contaminant characteristics on plant uptake (Briggs, *et al.* 1982). In terms of maximizing the rate and extent of remediation, there are many issues that require further research (Cunningham, *et al.*, 1996). One such issue is how to promote the transfer of the contaminant from the surrounding soil solution into the roots. The goal of this research was to study the mechanisms for removal of contaminants from solution by plants.

MATERIALS AND METHODS

In order to isolate the interactions between the plants and the test compounds, all experiments were conducted in a hydroponic system. The experimental compounds were 2,4,6-trichlorophenol (TCP) and pentachlorophenol (PCP), and the experimental plant was rye grass. Seeds were germinated in jars filled with semisolid bactoagar (6 grams bactoagar/liter nutrient solution). After germination, the seedlings were grown under cool, white lights with a 14 hour photoperiod for three weeks prior to experimental use.

Fifty mL of a chlorophenol/nutrient solution were pipetted into 70 mL, straight-sided glass jars. The initial concentrations were 0.0242 mM PCP and 0.0377 mM TCP. Seedlings were removed from agar, rinsed and then anchored into place in the jars using cotton, parafilm and tape. The number of plants in each jar was varied in order to determine which variables, such as root mass and transpired volume, were most closely related to compound removal. Each sampling episode consisted of jars containing from ten to forty seedlings and three jars containing no plants to monitor system loss.

Replicates were sacrificed at week one and week three. The shoots were clipped from the roots and weighed. The roots were weighed, placed in screwtop vials and extracted with ethanol for 24 hours. After extraction, the root surface area index was determined. The chlorophenol content of the root extract was analyzed on a Hewlett Packard 5890A Gas Chromatograph fitted with an electron capture detector. The aqueous samples were extracted with hexane and the extracts were analyzed by gas chromatography.

The relative surface area of the roots was measured by adsorption of rhodamine red (Sorenson and Wakeman, 1996). The roots were placed into 125 mL Erlenmeyer flasks containing 30 mL of a 2 µM rhodamine red solution at pH 4.9. The flasks were agitated for 18 hours on a shaker table and then the solution was centrifuged at 10,000 rpm for 10 minutes. The absorbance of the centrate was measured on a spectrophotometer at a wavelength of 550 nm. The adsorbed mass, expressed as umoles, is referred to throughout the paper as the root surface area index and reflects the relative adsorptive capacity of the different samples.

RESULTS AND DISCUSSION

Control losses were minimal during the first week: 6.3% of the PCP and 11.6% of the TCP. By three weeks, the control losses were more substantial: 38.5% for PCP and 50.5% for TCP. To normalize the sample values, mass fractions were calculated relative to the mass remaining in the corresponding controls.

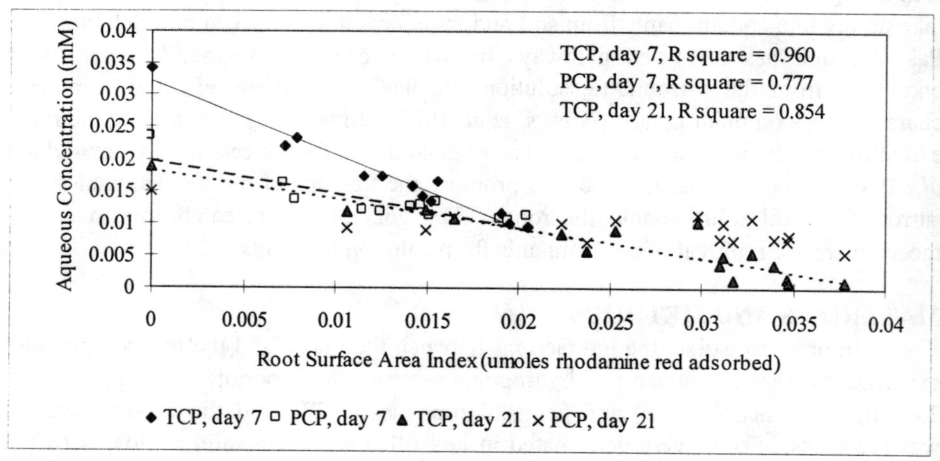

FIGURE 1. Decrease in aqueous concentration with increasing root surface area.

Within one week, the amount of TCP and PCP in solutions containing grass decreased substantially (Figure 1). Depending on the amount of roots present, up to 59% of the PCP and 75% of the TCP were removed. The mass remaining in solution after one week was correlated with root mass (correlation coefficients of -0.684 for PCP and -0.870 for TCP) and root surface area index (correlation coefficients of -0.834 for PCP and -0.969 for TCP). After three weeks, up to 95% of the TCP and 69% of the PCP had been removed from solution. The samples with the highest root surface area indices contained as little as 0.021 mg TCP/L (0.0011 mM, Figure 1). After 21 days, the mass of PCP remaining in solution yielded correlation coefficients of -0.727 with respect to root mass and -0.607 with respect to root surface area index. For TCP, the correlation coefficients were -0.905 for root mass and -0.897 for root surface area index. At both sample times, TCP was preferentially removed from solution.

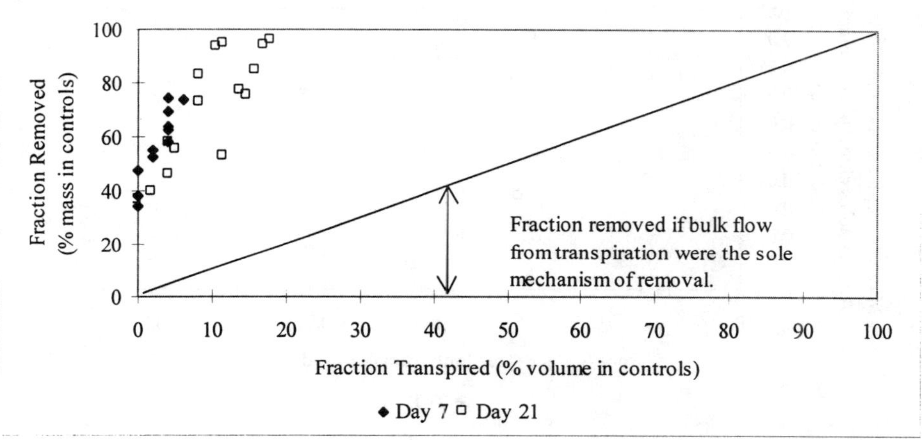

FIGURE 2. The mass fraction of TCP removed was consistently greater than the fraction of water transpired.

Transpiration is thought to be an important mechanism by which plants remove a compound from solution. While water removal was correlated with the removal of both TCP and PCP (correlation coefficients ranging from 0.615 to 0.854), the mass fraction removed was not commensurate with the volume fraction of water removed (Figure 2). For example, in one of the day 7 samples, no transpiration was measured while 38.1% of the TCP was removed from solution. In another day 7 sample, a TCP mass fraction removal of 74.5% coincided with a transpired fraction of 4.08%.

Most of the TCP and PCP that had been removed from solution could not be recovered by extraction with ethanol. The chlorophenol mass extracted from the roots by ethanol was divided by the chlorophenol mass removed from solution to yield the extractable fraction. This fraction was very small for TCP, less than 3.52% at day 7 and less than 1.01% at day 21, indicating that very little of the TCP was desorbed. On the other hand, at 7 days, between 39.9% and 83.7% of the PCP removed from solution was recovered by ethanol extraction of the roots. This result suggests that the majority of the PCP either was reversibly associated with the root surface or had reversibly diffused into the epiderm. Between day 7 and day 21, the extractable PCP fraction decreased to less than 5.7%, indicating that the majority of the PCP had been rendered unrecoverable to the external system by the plant between day 7 and day 21.

In this paper, the amount of each compound that could not be recovered is referred to as "internalized", a purely operational definition. This mass was calculated by subtracting the chlorophenol mass extracted from the roots and the mass of chlorophenol remaining in solution from the mass present in the controls. The difference in the TCP and PCP extractable fractions at day 7 and the substantial decrease in the PCP extractable fraction between day 7 and day 21 indicate that removal and internalization follow different kinetics.

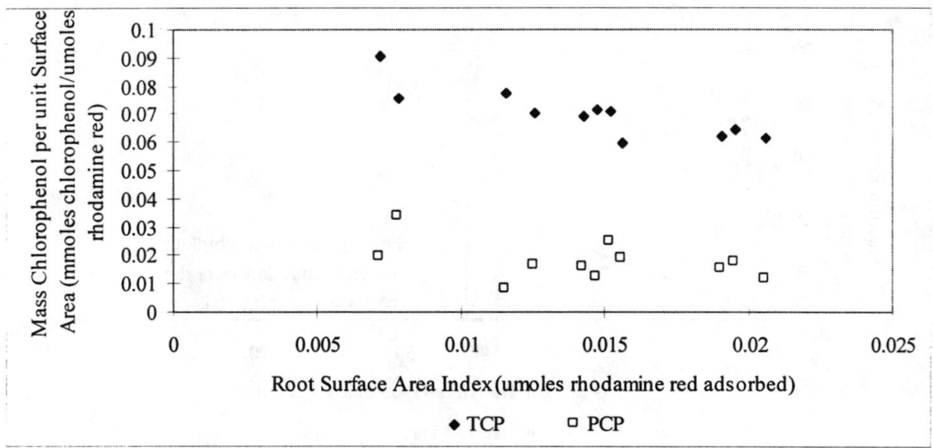

FIGURE 3. The relative propensity of TCP and PCP to be internalized on day 7.

Particularly at day 7, a greater mass of TCP was internalized on a per root surface area basis as compared to PCP (Figure 3). At 7 days, as much as 32.7% of the PCP and 72.1% of the TCP was internalized. The mass of internalized PCP was not correlated with either root mass or root surface area index, while the mass of internalized TCP had correlation coefficients of 0.858 with respect to root mass and 0.970 with respect to root surface area index. After three weeks, up to 68.2% of the PCP and 95.2% of the TCP was internalized. The mass fraction of internalized PCP was correlated with root mass (correlation coefficient of 0.733) and root surface area index (coefficient of 0.615). The internalized TCP fraction exhibited strong relationships with both root mass (correlation coefficient of 0.906) and root surface area index (correlation coefficient of 0.897). Thus, TCP was preferentially internalized over PCP, and the internalization of TCP was more strongly related to the amount of roots present than was the internalization of PCP.

One of the mechanisms by which plants may internalize contaminants is through translocation. As shown in Figure 4, the fraction internalized was not commensurate with the fraction transpired. For example, one of the day 21 samples did not exhibit measurable transpiration. In this sample, the grass had internalized 37.7% of the PCP and 36.0% of the TCP. Thus, although translocation appears to play a role in the internalization process, it is not the only mechanism.

From the data, it is clear that TCP was preferentially removed and preferentially internalized over PCP. In addition, the much greater removal and internalization of both chlorophenols indicates that the compounds concentrate either in the vicinity of the roots or on the roots. To assess the extent of this concentration effect, a concentration factor was calculated in the following manner:

Concentration Factor = (mg internalized/L transpired)/(initial mg/L bulk solution)

The resulting number is the amount by which the initial concentration would be multiplied in order to obtain the concentration in the transpiration stream if transpiration constituted the sole means by which the plant internalized the chlorophenol. This factor is similar in concept to the transpiration stream concentration factor (TSCF) described by Burken and Schnoor (1998).

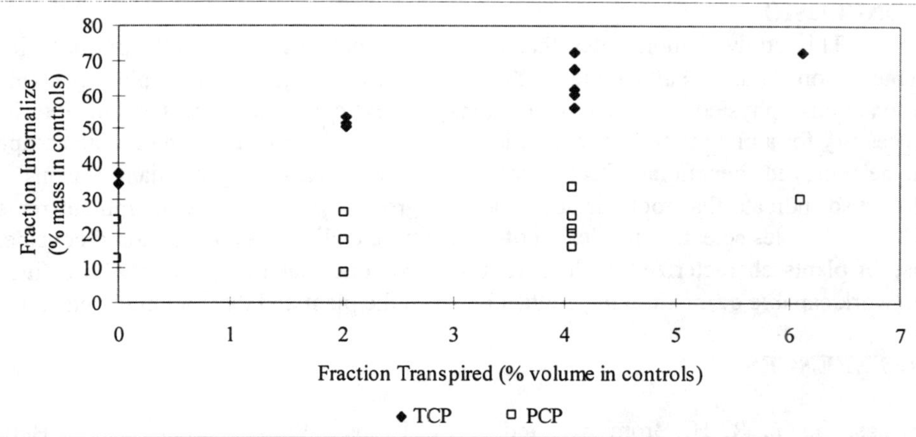

FIGURE 4. The internalized fractions of TCP and PCP were substantially greater than the fraction of water transpired on day 7.

Consistent with the preferential removal and internalization of TCP over PCP, the concentration factors were consistently higher for TCP than PCP. At day 7, the concentration factors ranged from 9.16 to 23.8 for PCP and from 11.4 to 25.4 for TCP. At day 21, these values were between 2.56 and 18.4 for PCP and between 4.47 and 18.3 for TCP.

These data suggest that, during removal from solution, the chlorophenol accumulates on or in the vicinity of the roots. Within the aqueous system, roots provide an organic surface with which a hydrophobic compound may associate. Thus, it is likely that the first step of a contaminant being directly remediated by a plant is the adsorption of the contaminant onto the root tissue and absorption into the epiderm. Based strictly on polarity, PCP should exhibit a greater affinity with the roots than TCP. However, because PCP's pKa is 4.74 while TCP's is 5.99 (solution pH was 5), a greater fraction of PCP would be in the phenolate form as compared to TCP. Ionization reduces the force driving the compound to move out of an aqueous matrix (Schwarzenbach, *et al.*, 1993).

Another explanation for the preferential removal of TCP over PCP lies in the preferential internalization of TCP. Once equilibrium between the chlorophenol in solution and the chlorophenol on the roots is reached, further sorption should not occur unless the aqueous concentration is augmented or the solid concentration is reduced. The proposed internalization mechanisms, by moving or transforming the compound, would effectively reduce the chlorophenol concentration in the parts of the root exposed to the aqueous system. Thus, internalization would allow sorption, or removal, to continue. The slower rate at which PCP was removed was due to its slower internalization. Briggs, *et al.* (1982) hypothesized that very nonpolar compounds, although readily able to diffuse across membranes, could not redissolve into the symplast. Thus, it is likely that the TCP was more readily moved into the endoderm than the PCP, increasing the rate of internalization and thus promoting removal.

CONCLUSION

This study demonstrates that plants have potential to directly promote in situ remediation. The results indicate that, as the first stage in direct plant-contaminant interactions, physico-chemical interactions are extremely important. Since it is not necessary for a plant to transpire in order for substantial amounts of an organic compound to be removed, beneficial effects may be achieved even during dormant seasons. The data also indicate that root surface area is of great importance in contaminant removal. Thus, in species selection for design of an in situ remediation system, should consider the use of plants characterized by high root surface areas, such as grasses with a fine root network capable of maximizing contact between the plant and the contaminated soil.

REFERENCES

Briggs, G. G., R. H. Bromilow, and A. A. Evans (1982). "Relationships Between Lipophilicity and Root Uptake and Translocation of Non-ionized Chemicals by Barley." *Pestic. Sci. 13*, 495 - 504.

Burken, J. G. and J. L. Schnoor (1997). "Uptake and Metabolism of Atrazine by Poplar Trees." *Environ. Sci. Technol. 31*(5), 1399 - 1405.

Burken, J. G. and J. L. Schnoor (1998). "Predictive Relationships for Uptake of Organic Contaminants by Hybrid Poplar Trees." *Environ. Sci. Technol. 32*(21), 3379 - 3385.

Cunningham, S. D., T. A. Anderson, A. P. Schwab, and F. C. Hsu (1996). "Phytoremediation of Soils Contaminated with Organic Pollutants." *Advances in Agronomy 56*, 55 - 114.

Schwarzenbach, R. P., P. M. Gschwend, and D. M. Imboden (1993). *Environmental Organic Chemistry.* John Wiley and Sons, Inc., New York.

Sorenson, B. L. and R. J. Wakeman (1996). "Filtration Characterization and Specific Surface Area Measurement of Activated Sludge by Rhodamine B Adsorption." *Wat. Res. 30*, 115 - 121.

DECONTAMINATION OF TNT-POLLUTED WATER BY MODIFIED FENTON REAGENT

M. Arienzo (Università degli Studi di Napoli, Federico II, Italy)

ABSTRACT: A modified Fenton reagent, using solid pyrite in the place of soluble ferrous iron salt, has been used to clean-up pure water spiked with 2,4,6-trinitrotoluene. Treatment was conducted at pH 3 with 1.8-28.6 mM Fe (II) (as pyrite) + 0.015-0.029 M of H_2O_2 and compared with Classic Fenton (CF) reaction: 1.8 mM Fe (II) (as ferrous sulfate) + 0.029-0.29 M H_2O_2. The pseudo-first order rate constant, k, did not significantly increased as a function of peroxide addition like in CF reagent, where all Fe (II) was in solution. When both reagents were used at the concentrations of 1.8 mM Fe (II) + 0.029 M H_2O_2 complete oxidation of TNT with modified Fenton reagent was slower respect to classic Fenton reagent (48 h vs. 24 h). However mineralization of TNT with 1.8 mM Fe (II) from pyrite + 0.029 M H_2O_2 under Dark/Light conditions was of the same order of magnitude (42/85 %) as with CF using a 10 times higher concentration of peroxide (0.29 M H_2O_2). Even though reuse of oxidized pyrite after a first oxidation treatment, prolonged the time of TNT disappearance from a second solution (48 h vs 24 h with 5.4 mM Fe^{2+} from pyrite + 0.029 M H_2O_2) the extent of mineralization remained unvaried ($\sim$ 45 %).

INTRODUCTION

The manufacture, processing, and packaging of 2,4,6-Trinitrotoluene (TNT) and other explosives at military ammunition plants across central U.S. have severly contaminated numerous acres of soil. At some location TNT contamination is widely dispersed resulting in contaminated ground and surface waters. Current technology for treating explosives-containing waste-waters are activated carbon adsorption and advanced oxidative processes such as UV-H_2O_2 which generate powerful oxidants (hydroxyl radicals) (Venkatadri and Peters, 1993). Other studies have successfully used the classic Fenton or Fenton-like reaction for removing explosives in water (Li et al., 1997a, 1997b, 1997c, 1998; Hundal et al., 1997). In the classic Fenton reaction, a ferrous salt, typically ferrous sulfate, is mixed with hydrogen peroxide to produce the hydroxyl radical. Optimization of the amounts of H_2O_2 in the Fenton reagent can keep low the cost of the treatment respect to UV/H_2O_2 process. As an alternative to using addition of soluble iron as in classic Fenton reagent, another means to generate $OH°$ radicals is to use a Fe (II)-bearing mineral that has the ability to act as surface catalyst. Pyrite, an iron (II) sulfide mineral, is a highly reactive mineral due to its high specific surface and high porosity, which makes the mineral highly susceptible to oxidation by O_2, Fe^{3+}, or H_2O_2 leading to the generation of Fe^{2+}. By introducing H_2O_2 into the system, peroxide can react with pyrite to produce Fe^{3+} (Eqn. 3) or react with Fe^{2+} and initiate the Fenton reaction

Objective. Our objective was to use pyrite as a catalyst for H_2O_2 decomposition and compare TNT destruction between pyrite/H_2O_2 and $FeSO_4$/H_2O_2 systems.

MATERIALS AND METHODS

The classic Fenton reaction was prepared by adding 1 mL of 89.2 mM Fe^{2+} (as $FeSO_4$:$7H_2O$ acidified with H_2SO_4) and 0.175 or 1.75 mL of 30% H_2O_2 to a 125-mL flask containing 49.825 or 47.25 mL of TNT solution yielding an initial concentration of 0.029 or 0.29 M H_2O_2 (v/v) and 1.8 mM Fe^{2+}. Solution pH (ca pH 6) was lowered to pH 3.0 with H_2SO_4.

The pyrite/H_2O_2 suspension were prepared by mixing varying concentrations of pyrite with H_2O_2. Specifically, treatments included varying concentrations of H_2O_2 at 0, 0.015, 0.029 and 0.29 M ad pyrite: 0.22 g pyrite L^{-1} (equivalent to 1.8 mM Fe(II)), 0.44 (3.6), 0.64 (5.4), 0.88 (7.2), 1.76 (14.3), 3.52 (28.6). For all treatments, duplicate flasks were agitated in the dark at room temperature (23 ± 2 ° C). on a shaker table at 130 circular rpm. At selected times, two 0.5 mL aliquots were removed from each reaction vessel. One was mixed with 6 mL Ultima Gold Cocktail (Packard, Meriden, CT) and ^{14}C-activity determined on a Packard 1900TR liquid scintillation counter (Packard Instruments Co., Downers Grove, IL). In the other aliquots Fenton reaction was terminated by adding one drop of concentrated H_2SO_4 (Watts et al., 1991).TNT concentration was determined by high performance liquid chromatography (HPLC) with a Keistone Betasil NA column (Keystone Scientific Inc., Bellefonte, PA). NO_3^- concentrations were determined by ion chromatography with conductivity detection (Dionex Corp., Sunnyvale, CA). The Fe (II) was quantified using the modified 1,10-phenantroline colorimetric method (Tamura et al., 1974; Zuo, 1992). Changes in peroxide concentrations in solutions were monitored with time using E.M. Quant test-strips (EM Science, Gibbstown, NJ) and a Reflectance Meter. Reactions were either conducted in the dark for 2 d (Dark Fenton) followed by exposure to light for 2 d (Light Fenton).The light source was two incandescent lamp (400-1000 nm) located approximately 1 m from the reaction vessels in a fume hood.

In order to evaluate if solid pyrite is sufficiently reactive and insoluble for a reasonable length of time, pyrite, following a treatment of an initial solution of TNT, was reused in another experiment.

RESULTS AND DISCUSSION

TNT Destruction. Monitoring changes in TNT concentrations over the time indicated that pyrite/H_2O_2 mixtures cam mediate a favorable degradation of TNT with complete destruction of the contaminant occurring within 24 and 48 h, Figure 1. However when modified and classic Fenton (CF) reagents were used at the same concentration of reactants, 1.8 mM Fe^{2+} + 0.029 M H_2O_2 the destruction of TNT was more rapid in CF, (24 vs 48h) than in pyrite sospension, Figure 1. Unlike results obtained when $FeSO_4$ was used as iron source, doubling the H_2O_2 concentration (0.015 to 0.029 M) in the pyrite suspensions had no effect on the rates of TNT transformation.

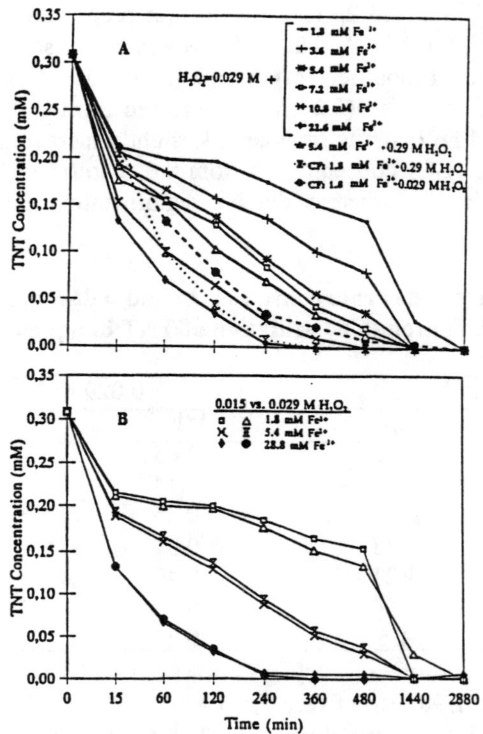

FIGURE 1. TNT oxidation by modified and Classic Fenton (CF) reagent with time (A). The initial TNT concentration was 0.31 mM. Comparison between 0.015 and 0.029 M H_2O_2 concentration on TNT destruction for varying pyrite concentrations (B).

Considering that initial soluble Fe (II) concentration before peroxide addition were low and similar for all pyrite concentrations (0.11 to 0.15 mM Fe(II)) Table 1, the observed slow rates of TNT loss were mainly due to catalytic decomposition of H_2O_2 on mineral surface and by soluble Fe(II) generated by pyrite oxidation.

TABLE 1. Fe(II) equilibrium concentrations in TNT solutions at pH 3.0 with varying pyrite concentration before H_2O_2 addition.

Pyrite Concentration (g L^{-1})	Equilibrium Conc. (mM Fe(II))
0.22	0.11
0.43	0.11
0.64	0.13
0.86	0.15
1.72	0.15
3.44	0.14

The first-order rate constant k for TNT degradation was calculated from the linear regression ln (Co/Ct) vs. time (t) with all regression coefficients more than 0.9 as summarized in Table 2. The pseudo-first order rate constant of CF reaction (Table 2) using 1.8 mM Fe^{2+} 0.29 M H_2O_2 was about seven times higher respect to the corresponding modified reagent (0.43 vs 0.06 h^{-1}). Increasing of ten times the amount of peroxide (0.29 M) k increased from 0.43 to 0.85 h^{-1} in CF, whereas for 5.4 mM Fe^{2+} pyrite treatment, k slightly increased from 0.23 to 0.35, Table 2. Reducing H_2O_2 concentration from 0.029 to 0.015 M, k. reduced from 0.43 to 0.23 h^{-1} in CF, whereas did not significantly changed in pyrite treatments.

TABLE 2. Pseudo first-order rate constants (k) and half-live times $\tau_{1/2}$ for modified and classic Fenton (CF) oxidation of TNT in aqueous solution.

mM Fe^{2+}	0.015 M H_2O_2		0.029 M H_2O_2	
	$k(h^{-1})$	$\tau_{1/2}$	$k(h^{-1})$	$\tau_{1/2}$
1.8	0.06	11.5	0.05	13.8
3.6	0.13	5.33	0.14	4.95
[a]5.4	0.23	3.01	0.25	2.77
7.2	0.29	2.38	0.31	2.23
14.4	0.57	1.21	0.49	1.41
28.8	0.82	0.84	0.78	0.88
[b]CF	0.43	1.6	0.23	2.77

k= pseudo-first order rate constant (k=ln[Co/Ct]/t); $\tau_{1/2}$ = 0.693/k
[a]k and $\tau_{1/2}$ were 0.35 and 1.98 respectively for 5.4 mM Fe^{2+} + 0.29 M H_2O_2
[b]k and $\tau_{1/2}$ were 0.85 and 0.81 respectively for 1.8 mM Fe^{2+} + 0.29 M H_2O_2

Moreover whereas the use of a higher dose of peroxide in CF reduced the time for complete disappearance of TNT in solution from 24 h (0.1 %) to 8 h (1 %), in modified Fenton reaction this time remained unvaried (24 h), Figure 1. These data indicate that for pyrite using reagent there is no practical convenience to increase the amount of peroxide, since this has a scarce effect on the kinetic of the reaction.

TNT mineralization. Complete mineralization of TNT was not achieved by any of the experimental treatments, but cumulative ^{14}C-loss was greater at lower pyrite concentration, 1.8 mM Fe^{2+}, for both peroxide concentrations (0.015 and 0.029 M), Table 3. Fortytwo percent mineralization was achieved with 1.8 mM Fe^{2+}as pyrite + 0.029 M H_2O_2 after 48 h of reaction in the dark respect to 30 % mineralization of the correspondent CF treatment, (Table 3). Adding more pyrite (above 5.4 mM Fe^{2+}) reduced the amount of TNT mineralized up to 32 % with 0.015 mM % H_2O_2 in the dark, Table 3. This could be due to OH° excavenging by Fe(II)/Fe(III) on the surface of the oxidized mineral.

Mineralization of TNT was greatly accelerated by light. In fact, application of light after 48 h of reaction in the dark, caused higher TNT mineralization rates (Table 3). As observed in the dark treatments, the highest rate of TNT mineralization was observed in the 1.8 mM Fe^{2+} pyrite treatment. This pyrite concentration + 0.029 M H_2O_2 mineralized approximately the same amount of TNT (~88 %) as with Classic Fenton treatment using 0.29 M H_2O_2.

TABLE 3. Cumulative percent ^{14}C-loss of TNT from pyrite/H_2O_2 suspensions and $FeSO_4$/H_2O_2 solution following 48 h of dark or 24 h of light treatment for two initial H_2O_2 concentrations.

Fe^{2+} Conc. mM	0.015 M H_2O_2		0.029 M H_2O_2		0.29 M H_2O_2	
	48h dark	24h light	48h dark	24h light	48h dark	24h light
1.8	41	72	42	85	nd*	nd
3.6	39	68	45	72	nd	nd
5.4	35	65	42	71	47	84
7.2	32	65	41	64	nd	nd
14.4	32	56	37	60	nd	nd
28.8	32	48	36	55	nd	nd
[a]CF	22	40	30	62	45	88

* not determined

[a]Classic Fenton reagent applies 1.8 mM Fe^{2+}.

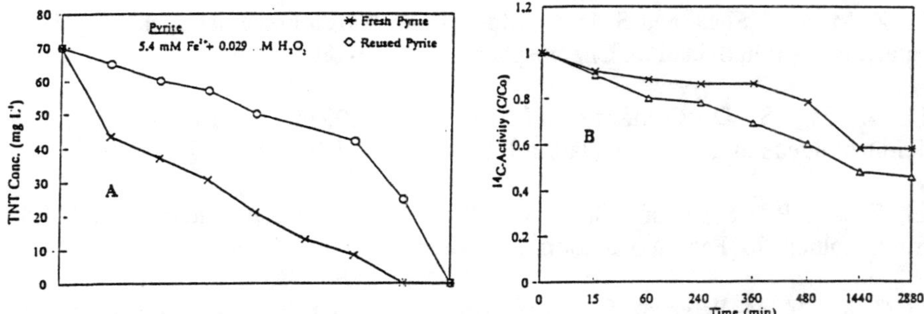

FIGURE 2. Changes in TNT destruction (A) and ^{14}C-loss (B) using fresh and reused pyrite. Pyrite, 5.4 mM Fe^{2+} was mixed with 0.29 M H_2O_2.

When pyrite was used a second time, TNT completely disappeared from solution within 2 d, instead of 1 d Figure 4. Although temporal changes in TNT were slower with the reused pyrite, loss of ^{14}C- from solution was similar for the fresh and reused pyrite (~50 %).

CONCLUSIONS

Results demonstrated that pyrite can be used in place of soluble ferrous iron to transform and mineralize TNT in water. The pyrite/H_2O_2 suspension produced consistently better results in terms of lower oxidant requirements and more extensive destruction of TNT. However, in the pyrite-peroxide system destruction rates are slower than the Fenton reaction and consequently require longer treatment times, at least 24 h. An additional advantage of using pyrite/H_2O_2 suspensions is that the pyrite can be easily separated from the treated solution by sedimentation/filtration and reused.

AKNOWLEDGEMENTS

Support for the first author was provided by the Organization for Economic Cooperation and Development (OECD), for research supporting "Surface and Groundwater Quality and Agricultural Practices" and by NATO-CNR (National Council of Research of Italy), Advanced Fellowship Program, area Environmental Technologies.

REFERENCES

Hundal, L. S., J. Singh, E. L. Bier, P. J. Shea, S. D. Comfort, and W. L. Powers. 1997. Removal of TNT and RDX from water and soil using iron metal. Environ. Pollu. 97:55-64.

Li, Z. M., P. J. Shea, and S. D. Comfort. 1998. Nitrotoluene destruction by UV-catalyzed Fenton oxidation. Chemosphere. 36:1849-1865.

Li, Z. M., S. D. Comfort, and P.J. Shea. 1997a. Destruction of 2,4,6-Trinitrotoluene by Fenton Oxidation J. Environ. Qual. 26: 480-487.

Li, Z. M., P.J. Shea, and S. D. Comfort. 1997b. Fenton oxidation of 2,4,6-trinitrotoluene by Fenton oxidation. J. Environ. Qual. 26:480-487.

Li, Z. M., M. M. Peterson, S. D. Comfort, G. L. Horst, P. J. Shea, and B. T. Oh. 1997. Remediating TNT-contaminated soil by soil washing and Fenton oxidation. Sci. Total Environ. 204:107-115.

Tamura, H., K. Groto, T. Yotsuyanagi, and N. Nagayama. 1974. Spectroscopic determination of iron (II) with 1,10-phenantroline in the presence og large amounts of iron (III). Talanta 21: 314-318.

Venkatadri R., and R. W. Peters. 1993. Chemical Oxidation Technologies: Ultraviolet Light/Hydrogen Peroxide, Fenton's Reagent, and Titanium Dioxide-Assisted Photocatalysis. Hazard. Waste Hazard. Mater. 2:107-149.

Watts, R. J. SW, Leung, and M. D. Udell. 1991. Treatment of contaminated soils using catalyzed hydrogen peroxide In J.A. Roth (ed) Chemical Oxidations: Technology for the '90s Technomic, Lancaster, PA.

Zuo, Y. 1992. Photochemistry of iron (II)/iron (III) complexes in atmospheric liquid phases and its environmental significance. Ph.D. diss. Swiss Fed. ed. Inst. Techn., Zurich (Diss. ETH No. 9727).

OXIDATION OF 2, 4, 6-TRINITROTOLUENE IN WATER AND SOIL SLURRY BY CALCIUM PEROXIDE

Michele Arienzo (Università degli Studi di Napoli Federico II, Portici, Italy)

ABSTRACT: The degradation of 2,4,6-trinitrotoluene was examined in pure water and contaminated soil slurry using calcium peroxide as a source of solid hydrogen peroxide and oxygen. The extent of TNT oxidation was compared with that obtained by using hydrated lime, which is normally generated by slurrying CaO_2 in water and contained in CaO_2 technical formulation (~50 %, w/w). Complete TNT degradation occurred between 280 min, 0.1 % $CaO_2/Ca(OH)_2$ and 20 min, 1 % $CaO_2/Ca(OH)_2$. A large part of the generated oxidation products, 80-90 %, were adsorbed on the solid calcium hydroxide, whereas the remaining 10-20 % was detected in solution until 48 h. Removal of nitro groups was extremely effective in CaO_2 slurry, where all the nitrogen (3 mol per mole of TNT) was removed from TNT within 240 min. Respect to calcium hydroxide, the peroxy compound liberated H_2O_2 in solution, 370 mg L^{-1} at 0.2 % CaO_2, w/v, which then decomposed within 480 min. Most of the ^{14}C-TNT was retained more strongly on the calcium hydroxide generated by slurrying CaO_2. The treatment of a contaminated soil slurry was effective at low concentration of $CaO_2/Ca(OH)_2$, ~0.2 %,w/w. Both oxidants do not lead to soil sterilization as the phosphorus added to neutralize the pH serves as a source of nutrient for the soil biomass.

INTRODUCTION

Bioremediation of munitions-contaminated soil is often slow and may be incomplete (Boopathy et al., 1994) because of the electron-withdrawing nitro groups that impede electrophilic attack by oxygenases of aerobic bacteria (Thiele et al., 1988). TNT resistance to biological oxidation indicates that effective destruction requires stronger oxidative treatment.

In recent years there has been a growing interest toward the use of oxygen releasing compounds (ORC), such as sodium percarbonate and metal peroxides. ORC technology is being used on over 2,000 sites in 46 states of U.S. and several foreign countries. ORC are intended to promote direct oxidation of contaminant and enhance in situ aerobic microbial degradation.

Some of the ORC, like peroxide metals, beside oxygen, release $Ca(OH)_2$ or $Mg(OH)_2$ causing a significant rise in pH. This creates the proper environment to accelerate oxidation and reactivity of recalcitrant chemicals like nitroaromatics. In such alkaline environment TNT, as a polynitroaromatic compounds, react easily with bases to form brightly colored solids or solutions (Strauss, 1970).

Calcium peroxide, CaO_2, is one of the most versatile and safest to handle of the family of solid inorganic peroxy compounds which can be considered a "solid form" of hydrogen peroxide (Tieckelmann, 1991). The commercially available material is a powder containing approximately equivalent amounts (50 % wt) of

CaO_2 and $Ca(OH)_2$. The byproducts of slurrying calcium peroxide in water are lime CaO, hydrated lime $Ca(OH)_2$, and hydrogen peroxide. The pH of a 1 wt % aqueous slurry is 12-12.5.

Objective. This study explores the potential to use CaO_2 as an alternative oxidant to promote chemical degradation (i.e. primary degradation or complete mineralization) of TNT in water and soil slurry. The effect of the treatment on soil biomass was also studied.

MATERIALS AND METHODS

TNT oxidation in pure water was initiated by adding various amount of CaO_2 fine dust, 0.1-1 % w/v, to a 125-ml flask containing 50 ml of TNT solution. Because technical CaO_2 contains about 50 % of $Ca(OH)_2$ and in order to evaluate the effect of CaO_2 respect to that of $Ca(OH)_2$ over TNT oxidation, $Ca(OH)_2$ was added to reaction vessels at the same dosage as for CaO_2 treatments. Duplicate flasks were agitated in the dark at room temperature (23 ± 2 °C) on a shaker table at 130 circular rpm. At selected times, two 0.5 ml aliquots were removed from each vessel. One aliquot was mixed with 6 ml Ultima Gold Cocktail and ^{14}C-activity determined on a Packard 1900 TR liquid scintillation counter, LSC. The other aliquot was used for chemical analysis. In these sample TNT oxidation was terminated by adding one drop of 10 % H_3PO_4.

Sequential extraction of adsorbed ^{14}C-TNT. Sorption of TNT to the solid peroxy or hydroxyl surface, was determined by equilibrating 40 mL of ^{14}C-spiked TNT solutions with 0.2 % CaO_2 or $Ca(OH)_2$ (w/v) in 50 mL Teflon tubes at 25 ± 1 °C for 24 h. At the end of equilibration time, suspensions were centrifuged at 3500 x g for 30 min and the supernatant removed. The residue was washed with deionized water and sequentially extracted with water, CH_3CN (sonication for 18 h at 30 °C) and 1 M NaOH defining respectively the readily-potentially-alkali-hydrolyzable ^{14}C-TNT. Carbon-14 in each pool was determined by LSC. Unextractable ^{14}C residue was determined by combustion of the solid phase to $^{14}CO_2$ with a biological oxidizer (Packard Tri-Carb B306) and $^{14}CO_2$ trapped in a mixture (3:2 v/v) of Carbosorb and Permafluor and quantified by LSC. The ^{14}C mass balance calculations indicated an average recovery of approximately 95 %.

Oxidation of TNT in soil-slurry. The soil was a Sharpsburg silty clay loam (20 % sand, 44 % silt, 36 % clay and 1.9 % OC, pH 8.0) contained 700 mg TNT kg^{-1}. A soil slurry was prepared by shaking 3 g of soil with 15 mL of H_2O (1:5 w/v) for 24 h in a 30-mL Teflon tube. CaO_2 and $Ca(OH)_2$ were separately added to two different set of tubes at concentrations of 0.1-0.2-0.3-0.4-0.5-1 and 2 % (w/v). Experiments were conducted in duplicate with experimental units agitated continuously on a reciprocal shaker at room temperature (23 ± 2 °C). Following treatment residual TNT in the soil was extracted by sonicating with 15 mL CH_3CN (1:5 w/v) for 18 h at 30 °C.

Viable biomass. The content of microbial biomass in 0.2 % CaO_2/$Ca(OH)_2$ munition treated soil was determined by lipid phosphate method (Findlay and Dobbs, 1993). The soil slurry was prepared by mixing 100 g of contaminated soil with 500 mL of tap water. The slurry was added of 0.2 % CaO_2 or $Ca(OH)_2$ w/w and shacked at room temperature for 24 h. At the end of the 24 h the pH of the soil suspension was adjusted to 7.0 with 0.7 mL of diluted H_3PO_4 and the slurry shacked for 2 more hours. Then the slurry was centrifuged at 5000 rpm for 20 min, supernatant was withdrawn and soil was air dried to 20 % humidity and sieved through 2 mm sieve. Soil was placed into Petri dishes, watered by deionized water till 34 % humidity and incubated at 24 °C for 6 days. A control experiment without CaO_2 was carried out simultaneously. For determination of soil microbial biomass the soil was sampled (1 g in 3 replicates) over 7 days every 2 days.

RESULTS AND DISCUSSION

TNT oxidation in CaO_2/$Ca(OH)_2$ water slurries. Complete TNT oxidation in CaO_2 and $Ca(OH)_2$ water slurry occurred rapidly for all concentrations experimented. In fact, TNT disappeared from solution between 270, 0.1 %-CaO_2/$Ca(OH)_2$, and 20 minutes, 1%-CaO_2/$Ca(OH)_2$, Figure 1a.

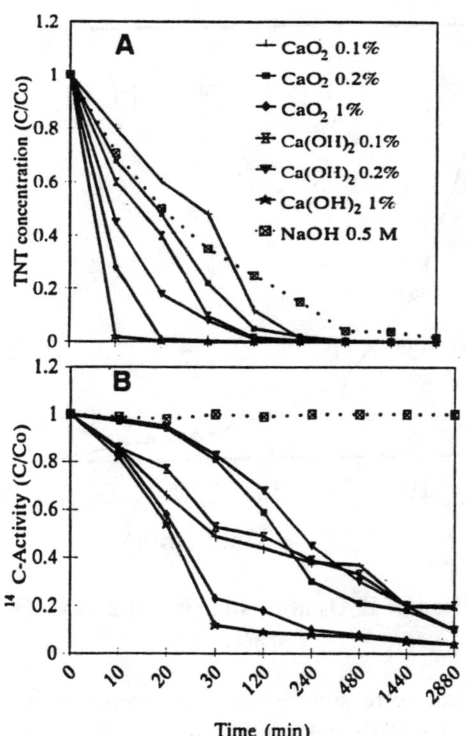

FIGURE 1. Disappearance of TNT (A) and of ^{14}C-TNT oxidation products (B) from solution by addition of CaO_2/$Ca(OH)_2$.

TNT disappeared from solution more rapidly in Ca(OH)$_2$ slurry: in the case, in fact, of the 0.1 % treatment TNT oxidation was 92 % in Ca(OH)$_2$ slurry respect to 52 % in the CaO$_2$ slurry at 30 min.

Because both CaO$_2$ and Ca(OH)$_2$ caused a consistent rise in pH, from 5.8 to 12.0 which however, even at the lowest concentration, remained below TNT pKa, 14.5, TNT oxidation was also studied in 0.5 M NaOH solution at pH 12.0. Results indicate slower TNT destruction rates in 0.5 M NaOH solution, with complete oxidation of TNT occurring only within 48 h, Figure 1a.

[14]C-data, Figure 1b, show that, even though all TNT, depending on oxidant concentration, disappeared from solution within 20-270 min, ^{14}C-TNT oxidation byproducts still persist in solution until 48 h. In fact at 48 h, ^{14}C-TNT percentages in solution were 20 % at 0.1 % CaO$_2$/Ca(OH)$_2$ and 10 % at 1 % oxidant's concentration. For all the systems, CaO$_2$, Ca(OH)$_2$ and NaOH, the ^{14}C mass balance, however, indicated no TNT mineralization to CO$_2$.

Generation of peroxide. Tracking peroxide concentration with time as in the case of 0.2 % CaO$_2$ water slurry, Figure 2, revealed that the peroxy-compound liberates approximately 370 mg L^{-1} of H$_2$O$_2$, which then completely disappeared within 480 min.

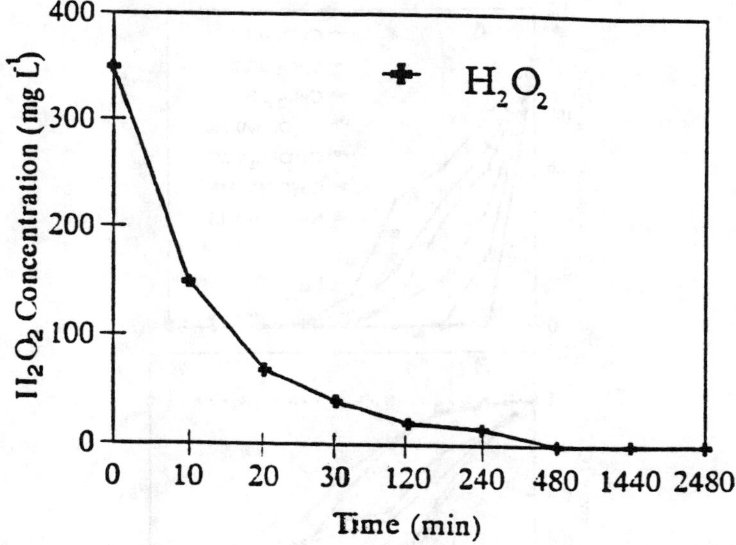

FIGURE 2. Production of H$_2$O$_2$ after slurring 0.2 % CaO$_2$ (w/v) powder in water.

TNT-NO$_2$ groups were stoichiometrically removed in the three oxidative systems studied: CaO$_2$, Ca(OH)$_2$ and NaOH, Figure 3. However removal of strongly electron withdrawwing-NO$_2$ groups was much more effective in CaO$_2$ than Ca(OH)$_2$ and NaOH slurry. In CaO$_2$ slurry all the nitrogen was removed from TNT within 4 h and detected in CaO$_2$ suspension as NO$_3^-$.

Totally, per mole of TNT, 3 mol of NO_3^- were found in CaO_2 slurry against 2.4 mol of NO_3^- detected in $Ca(OH)_2$ and NaOH solution within 24 h. This difference might be due to the more oxidative conditions existing in the CaO_2 slurry generated by the contemporary generation of oxygen, hydrogen peroxide and calcium hydroxide. Results from TNT-NaOH solution seems to be consistent with those showed by Cuta and Beranek (1974) who described the stoichiometric release of nitrite from TNT degradation in concentrated NaOH solution through the isolation by TLC of 2,4,6 trinitrophenol and 3,5 dinitrophenol.

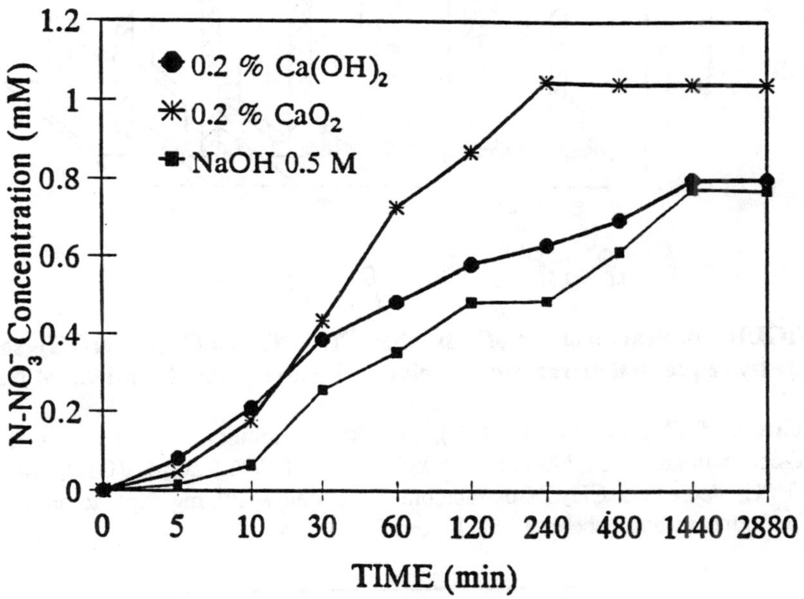

FIGURE 3. Production of NO_3^--N (mM) during oxidation of TNT in 0.2 % CaO_2/$Ca(OH)_2$ slurry and in 0.5 M NaOH solution at pH 12.0.

Fractionation of adsorbed ^{14}C-TNT after CaO_2/$Ca(OH)_2$ treatment. As displayed by Figure 1b, approximately, 80-90 % of the ^{14}C added as ^{14}C-TNT was adsorbed by the Ca-hydroxide surface within 24 h of equilibration with 0.1-1 % CaO_2/$Ca(OH)_2$ (w/v). Fractionation of sorbed ^{14}C-TNT after 24 h showed that less than 10 % of the adsorbed ^{14}C was extractable with deionized water ('readily available' pool) and about 2 % was removed with CH_3CN ('potentially available' pool) independently of the CaO_2-$Ca(OH)_2$ concentration, Figure 4. Among the two oxidants, the alkali-hydrolizable fraction predominates in $Ca(OH)_2$ system (> 60 % of adsorbed ^{14}C for all tested concentrations). In fact, subsequent extraction with 0.5 N NaOH removed about 55.6 and 84.8 % of the adsorbed ^{14}C, in 0.2 % CaO_2-$Ca(OH)_2$ slurry, respectively. In the same systems, about 32.5 and 4.5 % of the adsorbed ^{14}C was unextractable after these exhaustive extractions. Thus, even though, calcium hydroxide represents the predominant solid generated after slurrying CaO_2 and $Ca(OH)_2$ in water, ^{14}C-TNT was more strongly retained on the solid calcium hydroxide generated by slurrying CaO_2.

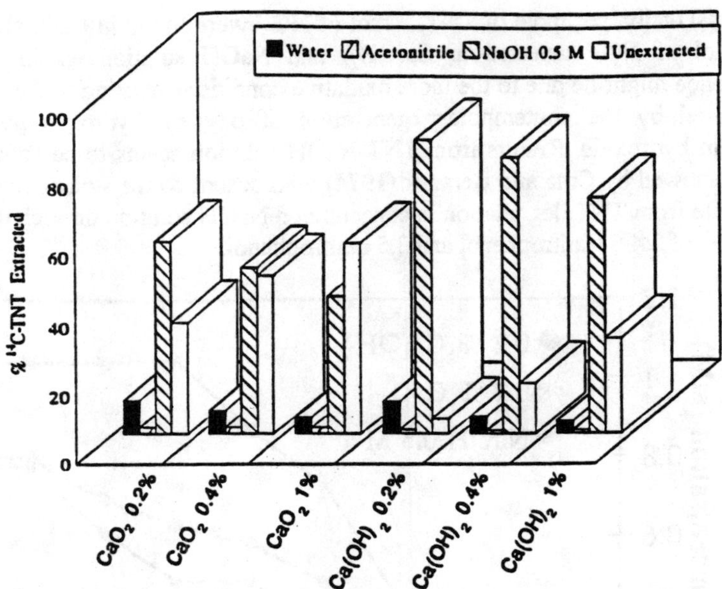

FIGURE 4. Fractionation of adsorbed ^{14}C-TNT on $CaO_2/Ca(OH)_2$ solid phase by sequential extraction by deionized water, CH_3CN and 0.5 M NaOH.

Oxidation of TNT in $CaO_2/Ca(OH)_2$ soil slurry. Equilibrating for 24 h slurries of TNT-contaminated soil, 700 mg TNT kg^{-1}, with 0.1-2 % CaO_2-$Ca(OH)_2$ (w/w soil) at 23 °C, decreased CH_3CN-extractable TNT below 20 mg kg^{-1} at an effective oxidant concentration above 0.1 %, Figure 5.

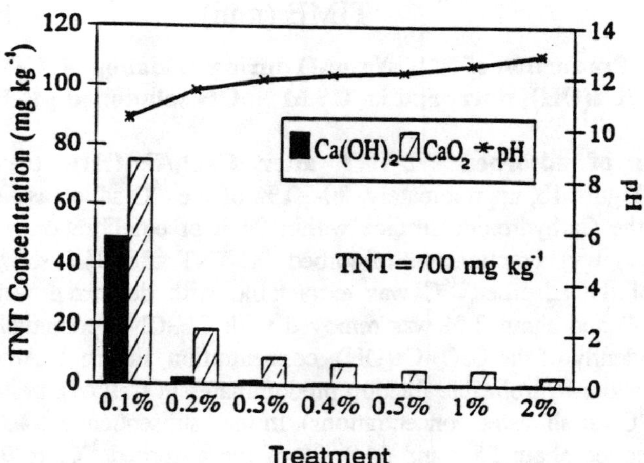

FIGURE 5. Destruction of TNT in contaminated soil slurry utilizing various concentration of $CaO_2/Ca(OH)_2$.

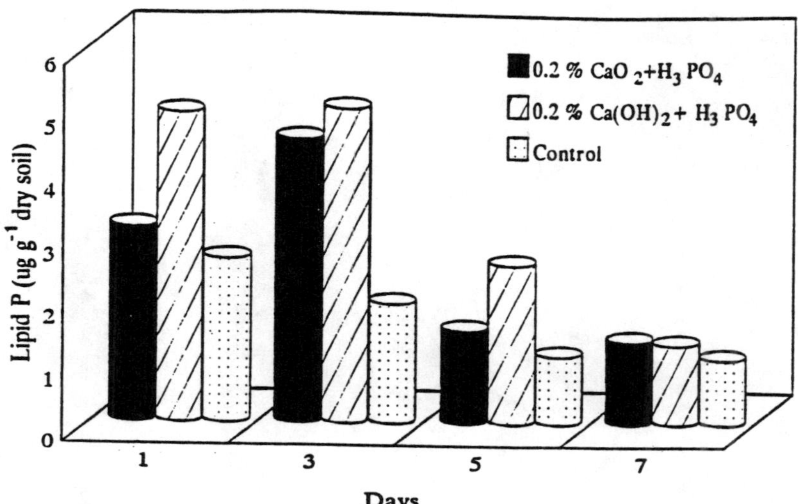

FIGURE 6. Influence of 0.2 % CaO$_2$/CaOH)$_2$ treatment on microbial biomass in TNT polluted soil.

Effect of the oxidants on microbial biomass. The results, Figure 6, show that treatment with low concentrations of CaO$_2$ and Ca(OH)$_2$ (less than 0.2 %) does not lead to soil sterilization. Content of total microbial biomass was higher than control in CaO$_2$/Ca(OH)$_2$-treated soil within 7 days of incubation. The addition of a readily available form of phosphorus from phosphoric acid to neutralize alkalinity might have generated optimum conditions and sustained microbial growth.

REFERENCES

Boopathy, R., M. Wilson, C. D. Montemagno, J. F. Manning, and C.F. Kulpa. 1994."Biological transformation of TNT by soil bacteria isolated from TNT-contaminated soil". *Bioresour. Technol.* 47: 19-24.

Cuta, F., and E. Beranek. 1974. *Collect. Czech. Chem. Commun.* 39:736.

Findlay, R. H., and F.C. Dobbs. 1993. *Quantitation Description of Microbial Communities Using Lipid Analysis.* In Handbook of Methods in Aquatic Microbial Ecology (Edited by P. F. Kemp, B. F. Sherr, E. B. Sherr and J. J. Cole), Chap. 32, Lewis Publisher.

Strauss, M. J. 1970. "Anionic Sigma Complexes". Chem. Rev. 70(6): 667-712.

Tieckelmann, R. E., and R. Steele. 1991. "Higher Assay Grade of calcium peroxide improves properties of dough". Food Technol. 45(1): 106-112.

Thiele, J., R. Muller, and F. Lingens. 1988. "Enzymatic Dehalogenation of Chlorinated Nitroaromatic Compounds". Appl. Environ. Microbiol. 54: 1199-1202.

COUPLED ABIOTIC/BIOTIC MINERALIZATION OF TNT

P. Schrader, T. F. Hess, and T. Renn (University of Idaho, Moscow, Idaho)
R. J. Watts (Washington State University, Pullman, Washington)

ABSTRACT: Coupled abiotic and biotic reactions were investigated for the mineralization of 2,4,6-trinitrotoluene, TNT. Modified Fenton reactions (with Fe^{+3} catalyst) were used as a chemical pretreatment of TNT prior to biological mineralization of the Fenton degradation products by unclassified activated sludge cultures. Using a hydrogen peroxide concentration of 0.9 %, a Fe^{+3} concentration of 13 mM, and biomass added 12 hours after initiation of the abiotic reaction, the observed extent of mineralization with the coupled abiotic/biotic system, 81%, was approximately 7 % greater than with the abiotic, Fenton system alone. Results of this study showed that, if properly optimized, the use of a coupled abiotic/biotic system may be a viable alternative for the treatment of waters and soils containing TNT.

INTRODUCTION

Bioremediation of water and soils contaminated with nitroaromatic chemicals has received substantial attention as a cost-effective clean up process alternate to incineration (Roberts *et al.*, 1993; Funk *et al.*, 1993). The general consensus of recent research is that anaerobic biological processes hold the most promise for stand-alone bioremediation, due to the problems of recalcitrance, build-up of lethal metabolic intermediates, and polymerization of intermediates inherent with the aerobic biodegradation of nitro-substituted compounds (Funk *et al.*, 1993). The research has indicated, however, that substantial mineralization of these compounds may not be achieved by the initial biological cultures even though the parent compound can be entirely transformed to intermediary metabolites including mixtures of simple organic acids (Crawford, 1995).

In situ chemical oxidations have the potential for rapidly treating soils or ground waters contaminated with toxic and persistent organic wastes. One mechanism for introducing strong oxidants into contaminated sites is the catalyzed decomposition of hydrogen peroxide to form hydroxyl radical ($\cdot$OH), commonly known as Fenton's reagent (Haber and Weiss, 1934). The standard Fenton procedure involves adding dilute hydrogen peroxide to a degassed solution of iron (II), which results in nearly stoichiometric generation of hydroxyl radicals:

$$H_2O_2 + Fe^{+2} \rightarrow OH\cdot + OH^- + Fe^{+3} \qquad 1)$$

However, most environmental applications of Fenton chemistry have some modifications, including use of higher concentrations of hydrogen peroxide, phosphate buffered medium, iron (III) or heterogeneous catalysts. These conditions, although not as stoichiometrically efficient as the standard Fenton's reaction, are often necessary to treat industrial waste streams and sorbed contaminants in soils and groundwater (Tyre et al., 1994).

Hydroxyl radicals generated by modified Fenton reactions react with most environmental contaminants at near diffusion controlled rates ($>10^9 M^{-1} sec^{-1}$). The degradation of xenobiotic chemicals by hydroxyl radicals then proceeds via either hydroxylation or hydrogen atom abstraction:

$$\cdot OH + R \rightarrow \cdot ROH \qquad\qquad 2)$$
$$\cdot OH + RH_2 \rightarrow \cdot RH + H_2O \qquad\qquad 3)$$

Many biorefractory compounds such as perhalogenated alkenes, dienes, and benzenes are effectively destroyed by hydroxyl radicals within minutes (Watts et al., 1994). Additionally, TNT has been shown to be susceptible to degradative attack by hydroxyl radical (Li et al., 1997). Based on the relatively high transformation rates, there has been an increased interest in oxidation processes for soil and groundwater treatment.

Combined abiotic and biological technologies for the destruction of hazardous wastes have been the subject of recent attention (Carberry and Benzing, 1991; Koyama et al., 1994; Scott and Ollis, 1995). In these studies, the authors investigated sequential processes, using abiotic reactions as a pretreatment step for a separate, following biological reaction. Such technologies were developed to overcome the biorecalcitrance of a particular compound inherent with stand-alone biological processes. In coupled Fenton/biological processes, chemical oxidation first destroys or desorbs the parent compounds via Fenton-like reactions making them more susceptible to subsequent electrophilic attack by the enzyme systems of microorganisms which then can mineralize the altered compound aerobically. Such remediation methods hold the promise of rapid degradation kinetics, hence fast clean up times, and low overall implementation costs if the process is properly optimized. The benefits of such a technology result from both improvement of environmental quality and safety to human life and lower implementation costs for site remediation technologies.

Objective. The objective of this study was to investigate coupled Fenton/biological processes to promote mineralization of TNT.

MATERIALS AND METHODS

Chemicals. Compounds used in the experimentation were the following: 2,4,6-trinitrotoluene (TNT), ^{14}C [2,4,6]-TNT (259.5 µCi/gram) (both synthesized by the IMAGE laboratory at the University of Idaho, Moscow, ID); hydrogen peroxide (30%), $Fe_2(SO_4)_3$ (Fisher Scientific, Fair Lawn, NJ); Ecolite scintillation cocktail (ICN Pharmaceuticals, Costa Mesa, CA)

Biomass. Biomass used in all coupled Fenton/biological experiments was obtained from a continuous culture, sequencing batch reactor (SBR) containing waste activated sludge grown on synthetic sewage. The synthetic sewage recipe was similar to that used by Kennedy et al. (1990).

Analytical techniques. Analysis of TNT was performed using a liquid chromatograph equipped with a diode array detector (Hewlett Packard 1090). Reversed phase liquid chromatography using a C-18 column (Phenomenex) was employed.

Response surface methodology. The experimental matrix used for determining optimal conditions for parent compound destruction was a central composite, rotatable design using the concept of response surfaces (Cochran and Cox, 1992). The method allows researchers to analyze the effects of multiple variables with a minimal number of experiments while keeping statistical significance and has gained popularity as an effective research tool.

Radiotracer experiments. An aqueous solution of ^{14}C-TNT containing approximately 100,000 dpm at a concentration of 50 mg/l was added to flasks equipped with biometer cups for the capture of $^{14}CO_2$ in 1ml NaOH. Appropriate concentrations of iron (III), hydrogen peroxide and biomass were added to the flasks and incubated at room temperature for the appropriate length of time with shaking at 200 rpm. A 1 ml NaOH sample was collected from each flask as well as two 1 ml H_2O rinsates and mixed with 15 ml of Ecolite scintillation cocktail. All samples were counted for radioactivity using a scintillation counter (Packard Tri-Carb 2100TR) via standard ^{14}C protocol.

RESULTS AND DISCUSSION

Optimization of TNT degradation. TNT degradation was investigated using modified Fenton reactions (pH 3.5) and response surface methodology for design of experiments (Cochran et al., 1992). Ranges of 0-20 mM iron (III) and 0-1.0% hydrogen peroxide were chosen to degrade the 50 mg/L solution of TNT. Analysis of the results and the response model indicated a maximum theoretical degradation of 100% which coincided with the maximum observed degradation (Figure 1, panel A). This was confirmed by the complete loss of detection of the parent compound by reversed phase HPLC analysis. The optimum reaction condition for the promotion of TNT degradation as calculated by the model and shown by the contour plot was an iron (III) concentration of 15 mM and a hydrogen peroxide concentration of 0.7 %.

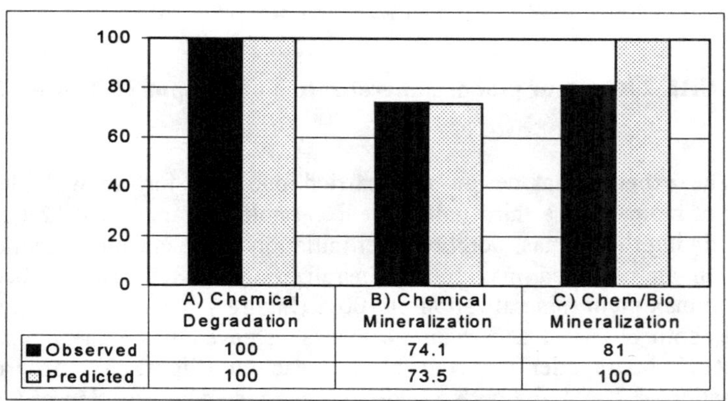

	A) Chemical Degradation	B) Chemical Mineralization	C) Chem/Bio Mineralization
■ Observed	100	74.1	81
□ Predicted	100	73.5	100

FIGURE 1. Comparison of extents of TNT destruction using various treatment processes.

Optimization of TNT mineralization. TNT mineralization experiments were designed using similar response surface methodology and reaction conditions as above. However, U-ring labeled ^{14}C-TNT was used so that mineralization could be quantified by capturing and measuring the production of $^{14}CO_2$. Results obtained from the experiments (Figure 1, panel B) showed an observed maximum mineralization of 74%, and a theoretical maximum mineralization of 73.5%. A contour plot of the response model (Figure 2) indicated the theoretical maximum response under the optimum reaction conditions of 13 mM iron (III) and 0.9% hydrogen peroxide, values similar to the optimum conditions found in the degradation experiments.

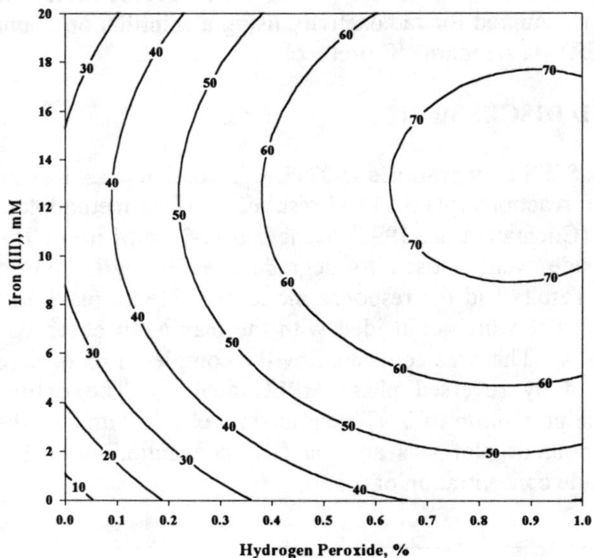

FIGURE 2. Contour plot of mineralization of 50 ppm TNT at pH 3.5.

The experimental design was carried one step further with the timed addition of biomass as a third independent variable. A range of 0-12 hours was used for timing the biomass addition after initiation of the chemical reaction. This resulted in an observed maximum mineralization of 81% and a theoretical, calculated maximum mineralization of 100% (Figure 1, panel C). The maximum response, both observed and theoretical, was achieved when the biomass was added at 12 hours after the start of the reaction (Figure 3). An iron (III) concentration of 7 mM and with a hydrogen peroxide concentration of 0.9% was required for the theoretical 100% mineralization.

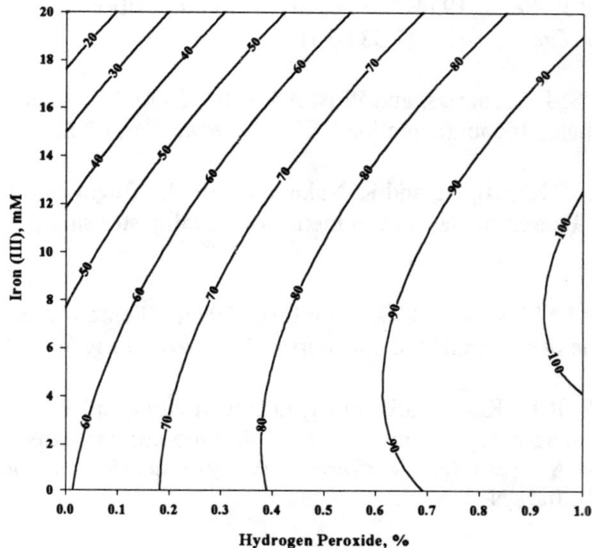

FIGURE 3. Contour plot of mineralization of 50 ppm TNT at pH 3.5 and biomass addition at hour 12 after reaction neutralization.

Future experiments to be performed include repetition of the above experiments at neutral pH with NTA chelated iron (III), both with and without biomass addition, as well as exploring nitroaromatic-acclimated biomasses. A suitable system for the measurement of reaction kinetics is also being explored. Results of this future data will be evaluated to produce an algorithm for the optimization of a chemical/biological treatment system for TNT.

REFERENCES

Carberry, J.B. and T.M. Benzing. 1991. "Peroxide pre-oxidation of recalcitrant toxic waste to enhance biodegradation." *Wat.Sci.Technol.* 23:367-376.

Cochran, W.G. and G.M. Cox. 1992. *Experimental Designs, Second Edition*, John Wiley & Sons, Inc., New York.

Crawford, R.L. 1995. "Biodegradation of nitrated munition compounds and herbicides by obligately anaerobic bacteria." In J. Spain (*Ed.*) *Biodegradation of Nitroaromatic Compounds*. Plenum Press, New York.

Funk, S.B., D.J. Roberts, D.L. Crawford and R.L. Crawford. 1993. "Initial-phase optimization for bioremediation of munitions compound-contaminated soils." *Appl. Environ. Microbiol.* 59:2171-2177.

Haber, F. and J. Weiss. 1934. "The catalytic decomposition of hydrogen peroxide by iron salts." *Proc.R.Soc.* 147:332-351.

Kennedy, M.S, J. Grammas, and W.B. Arbuckle. 1990. "Parachlorophenol degradation using bioaugmentation." *Wat. Environ. Res.* 62:227-233.

Koyama, O., Y. Kamagata, and K. Nakamura. 1994. "Degradation of chlorinated aromatics by Fenton oxidation and methanogenic digester sludge." *Wat.Res.* 28:885-899.

Li, Z.M., P.J. Shea and S.D. Comfort. 1997. "Fenton oxication of 2,4,6-trinitrotoluene in contaminated soil slurries." *Environ. Eng. Sci.* 14:55-66.

Roberts, D.J., R.H. Kaake, S.B. Funk, D.L. Crawford and R.L. Crawford. 1993. "Field-scale anaerobic bioremediation of dinoseb-contaminated soils." In M.A. Levin and M.A. Gealt (*Eds.*) *Biotreatment of industrial and hazardous waste.* McGraw-Hill, Inc., New York.

Scott, J.P. and D.F. Ollis. 1995. "Integration of chemical and biological oxidation processes for water treatment: review and recommendations." *Environ. Prog.* 14:88-103.

Tyre, B.W., R.J. Watts, and G. C. Miller. 1991. "Treatment of four biorefractory contaminants in soils using catalyzed hydrogen peroxide." *J. Environ. Qual.* 20:832-838.

Watts, R.J., S. Kong, M. Dipre, and W.T. Barnes. 1994. "Oxidation of sorbed hexachlorobenzene in soils using catalayzed hydrogen peroxide." *J. Haz. Mat.* 39:33-47.

BIOREGENERATION OF GRANULAR ACTIVATED CARBON CONTAMINATED WITH HIGH EXPLOSIVES COMPOUNDS

Matthew C. Morley, Samer Shammas, and Gerald E. Speitel Jr.
The University of Texas at Austin, Austin, Texas

ABSTRACT: Bioregeneration, an innovative groundwater treatment technique that combines adsorption to granular activated carbon (GAC) and biodegradation of contaminants, is being developed as a method for removing the high explosives RDX and HMX from contaminated groundwater. After removal from groundwater via adsorption, GAC is regenerated by desorbing RDX and HMX, which are subsequently biodegraded in a separate bioreactor. Initial batch biodegradation experiments with mixed cultures reduced initial RDX concentrations from 1 mg/L to non-detectable levels in 4 days. Over the same period, HMX was consistently reduced from 0.6 mg/L to 0.4 mg/L. Pseudo-first order rate constants for RDX and HMX were on the order of 7×10^{-4} L/mg TSS·day and 1×10^{-5} L/mg TSS·day, respectively. Continuous operation of a sequencing batch reactor (SBR) has effectively treated both contaminants. In an SBR that was operated with 4 day cycles for 72 days, RDX and HMX were biodegraded at similar rates and extents as in batch experiments. No persistent metabolites were produced. When coupled to removal from groundwater by adsorption, and subsequent desorption from GAC, the overall process is expected to effectively increase the GAC service life and provide a permanent treatment method for RDX and HMX.

INTRODUCTION

Due to past disposal practices, groundwater at the U.S. Department of Energy (DOE) Pantex Plant near Amarillo, Texas is contaminated with high explosive (HE) compounds. The HE compounds of concern are RDX (hexahydro-1,3,5-trinitro-1,3,5-triazocine) and HMX (octahydro-1,3,5,7-tetranitro-1,3,5,7-tetrazocine). To control contaminant migration and to treat groundwater, DOE has installed a pump-and-treat system. Contaminants are removed from pumped groundwater by adsorption to GAC, which is currently the most widely used treatment method for RDX- and HMX-contaminated waters because it is a simple and proven technology. Average influent concentrations to the groundwater treatment system were about 1.2 mg/L RDX and 0.26 mg/L HMX between July 1996 and January 1998 (personal communication, Jimmy Rogers, Battelle Pantex). Treatment goals are 0.026 mg/L for RDX and 5.11 mg/L for HMX (Battelle Pantex, 1997).

Although adsorption is effective, it has some disadvantages. Adsorbed contaminants are simply transferred from water and concentrated on the surface of the GAC, providing no permanent treatment or destruction of contaminants. Further, because GAC has a finite adsorption capacity, the carbon eventually becomes exhausted and must be replaced. Spent GAC may be classified as a hazardous waste, adding costs for treatment and/or disposal. Adsorption to GAC can thus have relatively high operating and maintenance costs.

Bioregeneration is a possible alternate treatment method which could reduce or eliminate disposal and replacement of spent GAC. Bioregeneration combines the reliability of contaminant removal by adsorption to GAC with permanent contaminant destruction by biodegradation. To effectively regenerate GAC, the process must include desorption for adsorbed contaminants to become bioavailable. RDX and HMX can both be effectively adsorbed to GAC; however, RDX is more soluble, more toxic, and less adsorbable than HMX, and will thus control the design and operation of a treatment system. Little is known about desorption and biodegradation of these compounds. One of these mechanisms will be the rate-limiting step in the overall bioregeneration process., and this research seeks to develop methods for increasing the rate and extent of desorption and biodegradation.

Bioregeneration systems have been developed in two general process configurations: on-line systems, in which biodegradation and adsorption occur simultaneously in a packed or fluidized bed; and off-line systems, in which contaminants are desorbed from GAC, and the regenerant solution is in a separate bioreactor. Initial results indicate that an on-line bioregeneration system will not be feasible due to the complex nutritional requirements of the bacterial cultures. An off-line system would minimize fouling of GAC and allow the application of enhanced desorption techniques such as the use of cosolvents or surfactant solutions.

This paper includes results of batch RDX and HMX biodegradation experiments and continuous biodegradation of these compounds in bench-scale sequencing batch reactors. Additionally, results of batch enhanced desorption experiments are presented.

MATERIALS AND METHODS

Analytical Methods. High explosives and their metabolites were analyzed via high performance liquid chromatography (HPLC) using EPA SW-846 Method 8330. Analytical standards for RDX and HMX were obtained from Supelco. Total organic carbon (TOC) was measured with a Beckman 915-B Total Organic Carbon Analyzer; TOC standards were prepared with glucose. Total suspended solids (TSS), used as a measure of biomass, was determined by filtering a known volume of culture suspension through a pre-weighed 0.5-µm glass fiber filter (Gelman Sciences). The dry weight of biomass was determined after drying the filter at 102°C for a minimum of one hour.

Biodegradation of High Explosives. All cultures used in these experiments were originally isolated from Pantex environmental samples (Autenrieth, *et al.*, 1999). Prior to use in biodegradation tests, each culture was grown in 300 mL of 25 g/L Oxoid Nutrient broth for four days, which enabled rapid bacterial growth. The biomass was separated from the nutrient solution by centrifugation prior to use in biodegradation tests.

Initial batch biodegradation tests were performed in 500 mL flasks with a total culture volume of 300 mL. The cultures were supplied with excess biodegradable organic carbon (a mixture of glucose, glycerol, and succinate), as well as nitrogen as ammonium nitrate. An additional carbon sources (ethanol) was tested for its ability to support cometabolism of RDX and HMX. Initial concentrations of RDX and HMX in batch biodegradation experiments were about 1.1 mg/L and 0.6 mg/L, respectively, comparable to contaminant levels in site groundwater.

Continuous treatment of HE-contaminated water was tested in sequencing batch reactors (SBRs). SBRs were 1-L flasks with a culture volume of 500 mL. After initial seeding with the cultures, SBRs were operated on a 4 day cycle, with periodic sampling for RDX, HMX, metabolites, DO, pH, and TSS. At the end of each cycle, 150 mL of the culture were taken from the reactor; the remainder of the culture was discarded. After centrifuging the biomass from the 150 mL sample, the biomass was returned to the reactor, and makeup water, nutrients, carbon sources, and HEs were added to return the reactor to its initial configuration. Treatment with SBRs was better able to simulate the potential requirement for cyclic operation of the bioreactor.

Enhanced Desorption Experiments. Northwestern LB-830, which is the GAC being used in the Pantex treatment system, was used in all adsorption and desorption tests. For enhanced desorption experiments, 0.15 g of washed 200×325 mesh size GAC was loaded with RDX and HMX in 500 mL bottles. After loading for one week, the GAC was separated from the solution by filtering through a 0.5-µm glass fiber filter. The filter, with GAC, was placed in the bottle, and the bottle was filled with a solution for desorbing

HEs from the GAC. Different solutions that were tested for enhanced desorption included ethanol and methanol as mixtures with water, surfactants, and cyclodextrins.

RESULTS AND DISCUSSION

Batch Biodegradation Experiments. Initial batch biodegradation experiments were conducted separately with the different cultures to determine differences in their ability to biodegrade RDX and HMX and their production of persistent metabolites. These experiments demonstrated that each of the cultures could biodegrade mixtures of RDX and HMX. Figure 1 shows RDX biodegradation for one of the initial experiments conducted with three separate cultures. Because biodegradation of RDX and HMX proceeds via an uncertain metabolic pathway, a pseudo-first order kinetics model was used to calculate the relative rates of biodegradation under different batch conditions. The model for batch biodegradation is as follows:

$$\frac{dC}{dt} = -k_1 X C$$

where C is the substrate (HE) concentration; k_1 is the pseudo-first order rate constant; and X is the biomass concentration. Both C and X are functions of time. For this experiment, RDX degradation rate constants (k_1 values) ranged from 6×10^{-4} L/mg TSS-day for the APBS1 culture to 9×10^{-4} L/mg TSS-day for the H2O culture. Although the H2O culture had the highest RDX degradation rate, it also produced the highest levels of persistent intermediates; all intermediates for the GAC and APBS cultures were transitory. (HMX biodegradation over four days was virtually the same for all three cultures; 30-35% of the initial HMX was degraded.) Because the GAC and APBS cultures were not markedly different in their abilities to degrade HE compounds, these two cultures were mixed for additional biodegradation experiments.

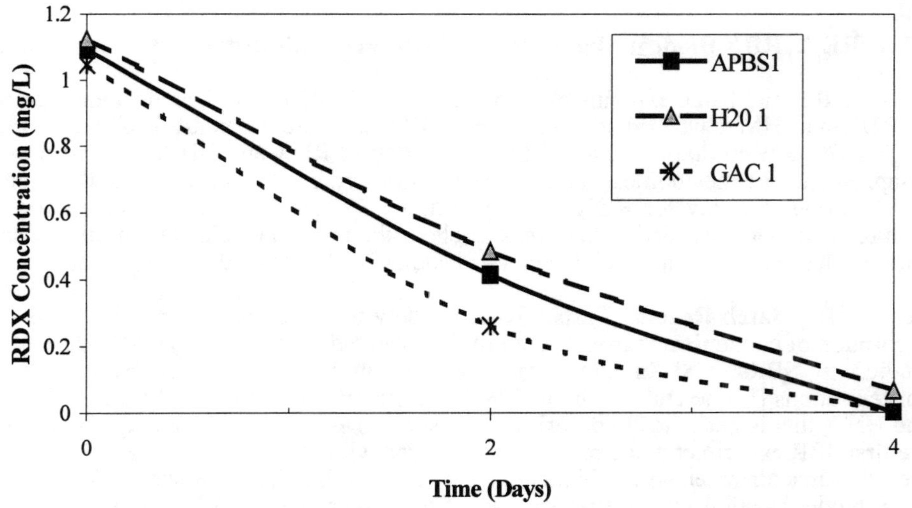

FIGURE 1. RDX Biodegradation in Batch Cultures.

Ethanol was tested as an alternate carbon source for supporting cometabolism of HEs (ethanol may also serve to promote desorption of HEs from GAC; Wilkie, 1994). One culture (GAC1) which had performed well in previous batch experiments was supplied with ethanol as a carbon source in one reactor, and a mixture of carbon sources in a different reactor. Figure 2 shows RDX degradation over eight days for this experiment. Over the initial 4 days of the experiment, the culture that was supplied with the mixed carbon sources reduced RDX to undetectable levels. With ethanol as a carbon source, RDX biodegradation was much slower, although about 85% of the RDX was degraded over 8 days.

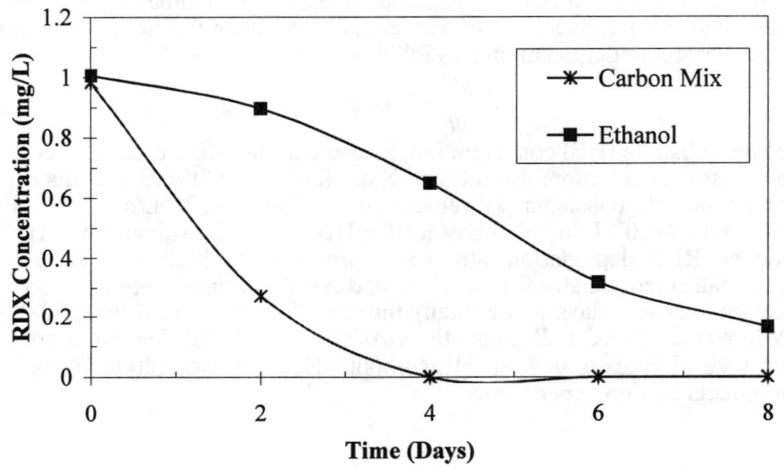

FIGURE 2. RDX Biodegradation by GAC Culture, with Different Carbon Sources.

All initial batch experiments showed that the cultures can best degrade a mixture of RDX and HMX when dissolved oxygen (DO) concentrations are less than 1 mg/L. Higher DO concentrations inhibited biodegradation of RDX and HMX, but did promote disappearance of metabolites. Higher pseudo-first-order rate constants (on the order of 7×10^{-4} L/mg TSS·day for RDX and 10^{-5} L/mg TSS·day for HMX) were observed in cultures that were supplied with excess levels of the biodegradable carbon mix, whereas cultures that were supplied with ethanol had somewhat lower biodegradation rates.

Sequencing Batch Reactor Tests. To adequately treat the periodic mass inflows from desorption of contaminants from spent GAC, biodegradation of RDX and HMX was next studied in SBRs. SBRs allow variations in substrate concentrations and oxygen concentrations (Irvine and Ketchum, 1988); thus, SBRs are well-suited for treating RDX and HMX that is periodically desorbed from spent GAC. Two SBR reactors were run in the first SBR experiment; one reactor was supplied with ethanol as a carbon source, while the other reactor received the biodegradable carbon mix. Figure 3 shows the results of RDX biodegradation for the first 16 days of this experiment, and Figure 4 shows HMX results over the same period. Although RDX and HMX removals during the initial cycles in both reactors were not as good as those achieved in batch tests, HE removal improved over time. By the end of the fourth cycle (day 16), RDX removal approached 100% for the SBR that received the carbon mix and 50% for the reactor that was fed ethanol. The experiment was run for a total of 72 days with similar results, although the reactors were subject to periodic upsets (unintended changes in operating conditions such as unstable pH or poor DO control) with subsequently poor HE removal for these cycles. Over each 4 day cycle, at least 85% of the initial RDX was consistently degraded in the culture that received the mixture of carbon sources, while about 50% of the initial RDX was degraded by the culture that received ethanol.

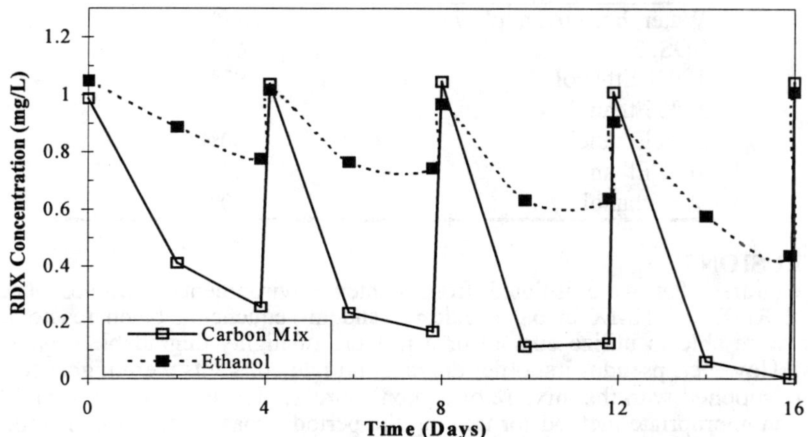

FIGURE 3. Initial RDX Degradation, SBR 1.

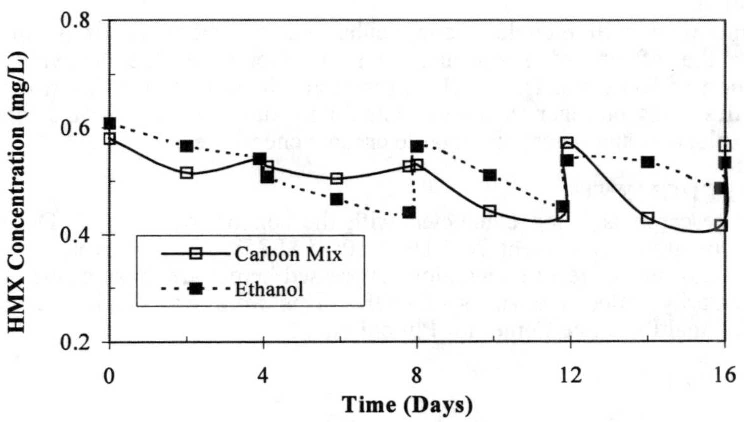

FIGURE 4. Initial HMX Degradation, SBR 1.

Enhanced Desorption Experiments.   Several chemical methods were tested for increasing the rate and extent of desorption of RDX and HMX from GAC and were compared to desorption using water alone.   Initial batch tests confirmed that cosolvents (e.g., ethanol and methanol) and surfactants (e.g., sodium dodecyl sulfate, or SDS) promoted the desorption of RDX from GAC. Table 1 summarizes some of the enhanced desorption results; the reported percent desorption is an average of two identical batch reactors. Data are given as percent of adsorbed RDX that was desorbed over a total of 10 days. Higher solvent concentrations desorbed more RDX, but the 5% ethanol solution still desorbed 20.3% of RDX. Methanol solutions were also effective, but desorbed slightly less RDX at similar solution concentrations. Surfactant solutions above the critical micelle concentration were also effective, desorbing nearly 70% of RDX.

TABLE 1. Percent Desorption of RDX by Enhanced Desorption Solutions.

Solution	RDX Desorbed
Water, buffered at pH 7	3.0%
SDS, 2.5 g/L	68%
100% Ethanol	93%
50% Ethanol	84%
25% Ethanol	60%
10% Ethanol	31%
5% Ethanol	20%

CONCLUSIONS

Cultures that were isolated from Pantex environmental samples successfully degraded RDX and HMX in batch cultures and in sequencing batch reactors. These cultures were able to utilize ethanol or a mixture of highly degradable organic carbon sources. However, pseudo-first order degradation rate constants were higher for cultures that were supplied with the mixture of carbon sources. Treatment of RDX and HMX in SBRs is an appropriate method for treating the periodic mass input of contaminants that will be generated by regenerating spent GAC via enhanced desorption. SBRs will also allow close control of dissolved oxygen conditions to promote RDX and HMX biodegradation (under low DO levels) and degradation of metabolites (at higher DO concentrations).

Future work will include treating enhanced desorption fluids in an SBR and determining the effects of surfactants and cosolvents on the bioavailability and biodegradation of RDX and HMX. This research will provide key information that is needed for designing bioregeneration systems for treating aqueous streams contaminated with high explosives and other semivolatile organic chemicals.

ACKNOWLEDGMENTS

This research is being conducted with the support of the U.S. Department of Energy, Cooperative Agreement No. DE-FC04-5AL85832. However, any opinions, findings, conclusions, or recommendations expressed herein are those of the authors and do not necessarily reflect the views of DOE. This work was conducted through the Amarillo National Resource Center for Plutonium.

REFERENCES

Autenrieth, R., M. Jankowksi, J. Bonner, and M. Kodikanti. 1999. "Optimizing the Biotransformation of RDX and HMX." *Proceedings, Fifth International Symposium on In Situ and On-Site Bioremediation.* Battelle, Columbus, OH.

Battelle Pantex. 1997. *1996 Environmental Report for Pantex Plant.* Technical Report DOE/AL/9704, Environmental Protection Dept., Battelle Pantex, Amarillo TX.

Irvine, R.L., and L.H. Ketchum, Jr. 1988. "Sequencing Batch Reactors for biological Wastewater Treatment." *Crit. Rev. Env. Control.* 18(4): 255-294.

Wilkie, J.A. 1994. "Biological Degradation of RDX." M.S. Thesis, University of California at Los Angeles, Los Angeles, CA.

BIOREMEDIATION OF NITROCHLOROBENZENE, NITROANILINE, CHLOROANILINE AND OTHER ORGANICS IN GROUNDWATER

Patrick Hicks (Regenesis, Kingwood, Texas)
Jerry Kubal (Kubal-Furr & Associates, Tampa, Florida)
Stephen Koenigsberg (Regenesis, San Juan Capistrano, California)

ABSTRACT: The uppermost aquifer beneath a chemical plant contains residual concentrations of o-nitrochlorobenzene from a tank car spill, as well as other plant-related organic compounds of concern (COC), including 2-nitroaniline and o-chloroaniline. The aquifer consists of low permeability clays and silts, and produces only a few gallons of groundwater per day from a shallow recovery well previously installed for remediation. Soil removal was the principal mechanism to treat the tank car spill. Subsequent recovery of residual groundwater contamination has been ineffective for more than 10 years due to low aquifer yield, and the recalcitrant nature of the COC. Aerobic treatment was facilitated using Oxygen Release Compound (ORC®). The treatment area was approximately 80 square meters and the depth of treatment was approximately 6 meters of aquifer profile. The DO and ORP measurements were highly variable, but trended up for the first three months following ORC injection. The DO and ORP trends began to decrease approximately six months following injection. The DO ranged between 3.5 mg/l and 10.7 mg/l, and the ORP ranged between 291 mV and 367 mV during this six month period. Maximum contaminant concentration reductions observed during the pilot test varied from 28% to 83%, and rebound was detected following the effective life of the ORC. The relatively low sorption of the compounds to soil, and the influx of contaminated groundwater into the pilot test area, indicate that a permeable bioremediation-barrier represents the most effective remedial option for this site.

INTRODUCTION

A field-scale pilot study to evaluate *in-situ* treatment of the compounds of concern (COC) was performed. Aerobic treatment of various COC in soil systems has been reported to be highly variable (Abou-Rizk et al., 1995; Myers et al., 1995; Zappi et al., 1995), and the goal of the pilot test was to establish that aerobic bioremediation of the COC was a viable option at this site. Oxygen Release Compound (ORC®), manufactured by Regenesis Bioremediation Products in San Juan Capistrano, California, is a patented formulation of magnesium peroxide that slowly releases molecular oxygen when hydrated, thereby facilitating aerobic bioremediation of the COC. Numerous remedial options were evaluated, and ORC was identified as the preferred method to oxygenate the aquifer. The ORC releases oxygen slowly over a period of 6 to 9 months, which corresponded to the scheduled testing period.

MATERIALS AND METHODS

The ORC design/loading rates were based on dissolved organic compound mass in the treatment area, as well as dissolved oxygen (DO) concentrations and oxidation-reduction potential (ORP) measurements. The ORC was injected in a grid pattern around the recovery well which continued to extract groundwater, thereby enhancing the rate of oxygen movement through the treatment area. Baseline COC sampling was performed prior to ORC injection, and the post-treatment monitoring protocol included DO and ORP in addition to the COC.

RESULTS AND DISCUSSION

Table 1 depicts the efficacy of the six month bioremediation process at this site for the COCs. Figures 1 to 3 illustrate the reductions over time in the COCs as measured in monitoring well F-2, which is located in the central portion of the pilot test area. Figure 4 illustrates the changes in dissolved oxygen (DO) and Figure 5 describes the oxidation-reduction potential (ORP) at the site during the pilot test. Interpretation of DO and ORP measurements must be made with reference to the fact that there is on-going demand in the system. Increases in DO and ORP are clearly evident at certain specific times in the fluctuating background pattern and cease altogether in conjunction with periods of rebound.

CONCLUSIONS

The bioremediation process effectively reduced the dissolved masses of o-nitrochlorobenzene, 2-nitroaniline and o-chloroaniline, and also reduced dissolved mass of other COCs. Rebound of the COC concentrations were observed after the oxygen from the initial ORC treatment was spent. The project will be expanded to include a long-term barrier design using ORC to continuously deliver oxygen to the aquifer, and eliminate the potential for off-site migration of dissolved COC.

TABLE 1. Maximum decrease in dissolved compounds of concern concentrations following ORC treatment.

Compound	% Concentration Reduction
1,2-dichloroethane	28.3
1,2-dichloropropane	61.0
chlorobenzene	63.5
naphthalene	54.5
2-chlorophenol	38.7
2-nitroaniline	82.1
nitrobenzene	54.5
o-chloroaniline	82.8

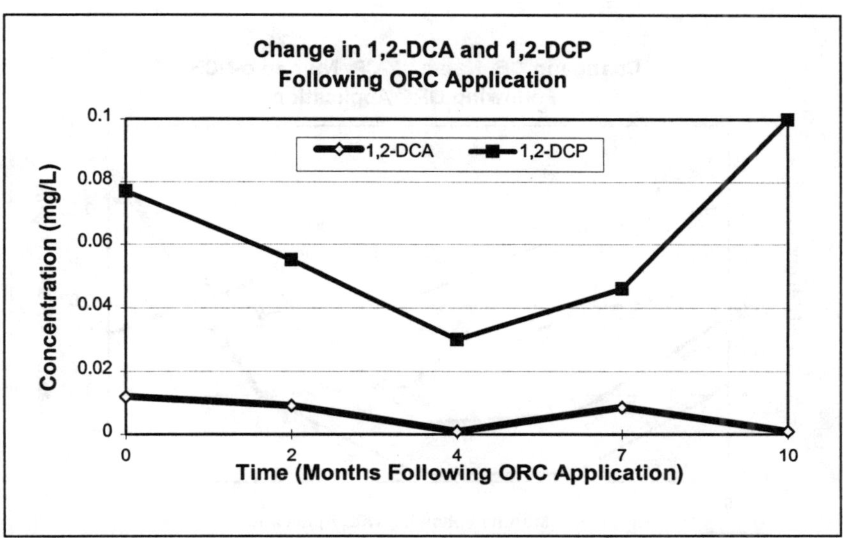

FIGURE 1. Change in 1,2-DCA and 1,2-DCP concentrations following ORC application.

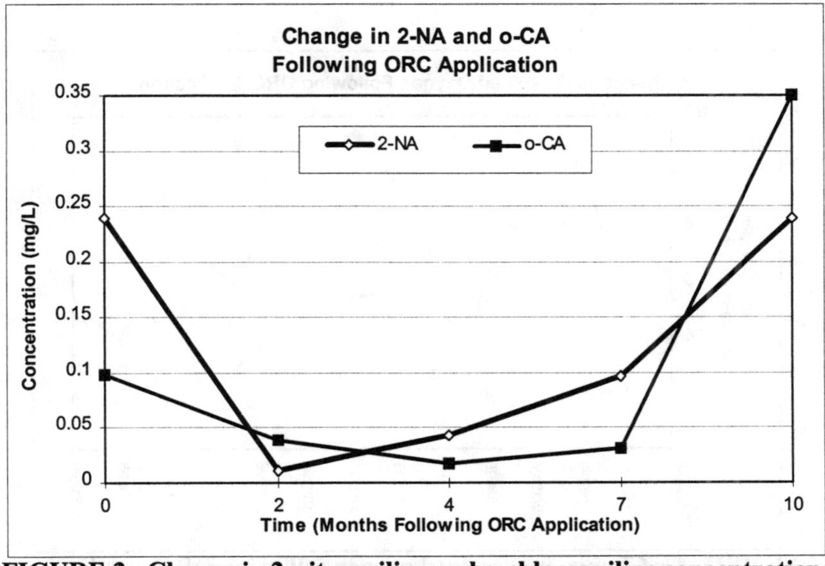

FIGURE 2. Change in 2-nitroaniline and o-chloroaniline concentrations following ORC application.

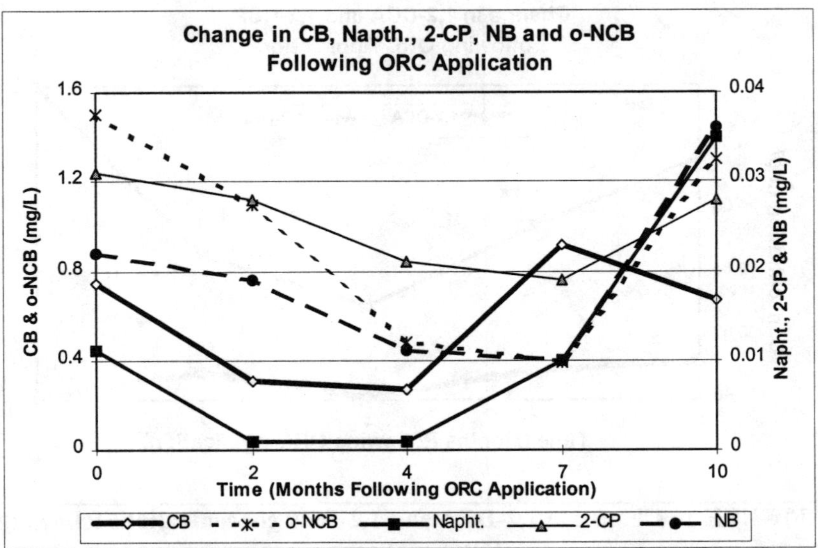

FIGURE 3. Change in chlorobenzne, nitrobenzene, o-nitrochlorobenzene, naphthalene and 2-chlorophenol concentrations following ORC application.

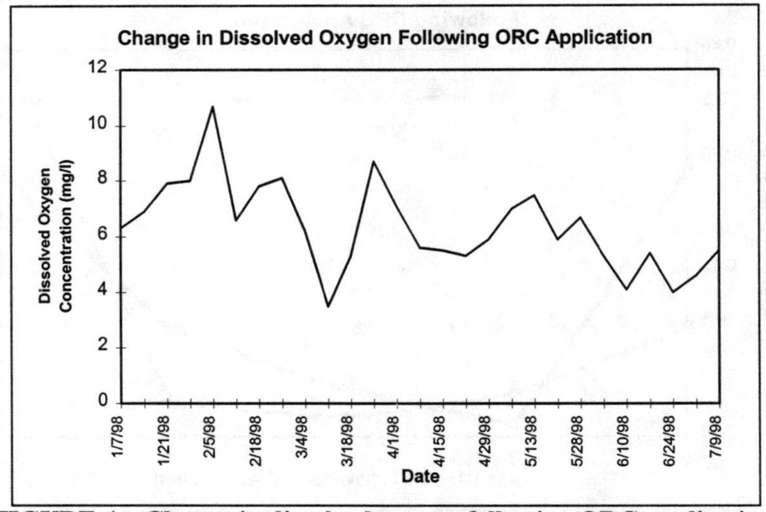

FIGURE 4. Change in dissolved oxygen following ORC application.

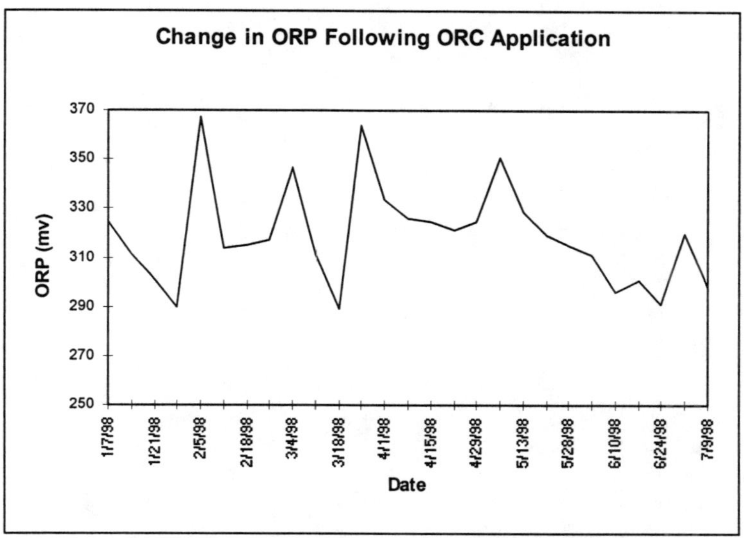

FIGURE 5. Change in ORP following ORC application.

REFERENCES

Abou-Rizk, J.A.M., M.E. Leavitt and D.A. Graves. 1995. In Situ Aquifer Bioremediation of Organics Including Cyanide and Carbon Disulfide. Applied Bioremediation of Petroleum Hydrocarbons. Vol. 3(6): 175-183. Battelle 1995.

Myers, J.M., B.S. Banipal and C.W. Fisher. 1995. Biodedgradation of Oil Refinery Wastes. Applied Bioremediation of Petroleum Hydrocarbons. Vol. 3(6): 445-451. Battelle 1995.

Zappi, M.E., D. Gunnison and H. Fredrickson. 1995. Aerobic Treatment of Explosives-Contaminated Soils Using Two Engineering Approaches. Bioremdiation of Recalcitrant Organics. Vol. 3(7): 281-288. Battelle 1995.

QUANTIFYING MASS AND TOXIC BIOAVAILABILITY RATES OF ORGANICS IN SATURATED POROUS MEDIA

Yves Dudal, Louise Deschênes and Rejean Samson
NSERC Industrial Chair in Site Bioremediation, École Polytechnique de Montréal
Montréal (Québec) Canada

ABSTRACT: Bioavailability is often defined as the fraction of contaminant available for microbial degradation. The definition of bioavailability is, in this study, extended to the rate of biological receptor exposure to the contaminant present in the environment. Regarding the type of biological receptor (contaminant-degrading or contaminant-sensitive), mass or toxic bioavailability rates are considered. A method was developed to assess these bioavailability rates from experimental results. An example of this is shown using a glass beads saturated column. Three elutions are performed on this column: non-reactive tracer, contaminant (pentachlorophenol) under abiotic conditions and in the presence of contaminant-degrading microorganisms. Contaminant concentrations and the associated toxicity are monitored at the exit of the column. Results obtained are expressed in non-dimensional numbers of mass and toxic bioavailability to characterize the dynamic of bioavailability within a flowthrough system as well as the studied medium in terms of intrinsic bioremediation potential and the contaminant-associated hazard. More saturated media are currently under investigations using this method, but results are already promising in terms of characterizing the fate of organics through their bioavailability rates.

INTRODUCTION

Bioavailability of organic contaminants in subsurface systems is usually understood to be the factor integrating the different limitations to bioremediation techniques (Ramaswami and Luthy 1997). It is widely accepted that sorbed contaminant is not fully accessible to microbes which results in a decrease of biodegradation rate and extent (Harms and Zehnder 1995; Pignatello and Xing 1996). The physicochemical mechanisms responsible for these limitations are of two kinds: partition of the contaminant between the aqueous phase and the organic phase and diffusion into the organic polymer-like matter or into the nanopores of the particulate phase (Nam and Alexander 1998).

The bioavailable fraction can also be regarded as the carrier of toxicity throughout the subsurface system. In this point of view, the mechanisms responsible for bioremediation limitations are now favoured as they tend to decrease the toxicity linked to the free contaminant (Kelsey and Alexander 1997). Target organisms susceptible to express toxicity when in contact with the free contaminant can then be used to monitor bioavailability, by, for example, the measurement of bioaccumulated contaminant in their organism (Umbreit *et al.* 1986).

In agreement to the up-to-date environmental policies, bioremediation is more and more monitored by measuring the decrease of contaminant mass as well as the resulting toxicity (Dasappa and Loehr 1991). When looking at bioavailability of organic contaminants in aquifers, it is also necessary to take into account the dynamic of the system. Here, the bioavailable contaminant can not be addressed as a fraction of the total contaminant. The groundwater flow brings some contaminant in and out of a section of the aquifer by constantly modifying the partition equilibrium as well as the diffusion of the contaminant. In this case, bioavailability should be expressed as the result of different kinetics (Bosma *et al.* 1997).

In this study, a definition of bioavailability is proposed as being: the rate of biological receptor exposure to the contaminant. In the case where the receptor is a contaminant-degrading microorganism, bioavailability leads to biodegradation. When the receptor is an organism susceptible to express toxicity when in contact with the contaminant, bioavailability leads to toxicity. Rates of mass and toxic bioavailability for organic compounds in aquifers are then introduced.

Bioavailability has been extensively studied in the field of pharmacokinetics. In these studies, the area under the elution curves (AUC) of a drug through an animal or human body are used. They are considered as a robust parameter, considering that its dependence towards experimental variations is rather low (Ritschel 1988). The AUC can also be used for studies of the fate of the contaminant through a section of an aquifer. Indeed, the derivative of the mass breakthrough curve of a step elution of the contaminant corresponds to the rate of contaminant recovery at the end of the section. Thus the area under this curve (AUCmass) represents the fraction of contaminant that was able to eluate through the section. This value reported to the elution reference time of the contaminant within the section leads to the mass bioavailability rate.

$$\tau_{massBioA} = \frac{1 - AUCmass}{\bar{t}_{mass}\sigma^2_{mass}} \qquad (1)$$

where t_{mass} is the average residence time and σ^2_{mass} is its standard deviation for the mass elution. The same approach can be used with the toxicity breakthrough curve. Here, the area under the curve (AUCtox) reported to the elution reference time leads to the toxic bioavailability rate.

$$\tau_{toxBioA} = \frac{AUCtox}{\bar{t}_{tox}\sigma^2_{tox}} \qquad (2)$$

This study shows how to use this approach and how the results can be obtained from a simple experimental set-up. For this experiment, a saturated porous medium consisting of contaminant-inert glass beads was used. Pentachlorophenol was the contaminant eluted through the saturated column under either abiotic or biotic conditions.

MATERIALS AND METHODS

Production of microorganisms. An amount of soil was sampled from a utility pole treatment contaminated site. Pentachlorophenol (PCP)-degrading microorganisms were grown from this sample by sequential addition of PCP under slurry conditions, using a basal salt medium (BSM) (Otte *et al.* 1994). This slurry was then filtered through 8μm and the remaining suspension was bioactivated the same way. PCP was monitored in the suspension by UV spectrophotometry and HPLC at 319 nm after filtration at 0.45μm.

Breakthrough experiments. The column used for the three different elutions (non-reactive tracer, contaminant under abiotic conditions and in the presence of degrading microorganisms) was a pre-chromatography Chromaflex[TM] column, made only of Kimax[TM] borosilicate glass and polytetrafluoroethylene (PTFE). The column was 15cm long with a 4.8cm inside diameter. Both endings included a 20μm PTFE filter as well as a PTFE grid for bed support and accurate flow distribution. The column was filled with glass beads and then saturated with BSM. All experiments were done with a saturation of 98±1%. Hydraulic conductivity was 9.2×10^{-4} cm/s. The hydrodynamic elution was a classic step tracer test using ^{3}H-BSM. The two other breakthrough curves were step elutions of PCP (25 mg/L) under abiotic conditions and in the presence of PCP-degrading microorganisms. After filtration at 0.45μm, the contaminant concentration was monitored by UV spectrophotometry at 319 nm and HPLC analysis. Prior to the biotic elution, a bacterial suspension (see above section) was inoculated in the column. This was done by recirculating the suspension during 20 pore volumes (Vp) followed by a flush during 2 pore volumes. Protein analysis using the micro-BCA (bicinchoninic acid) method (Pierce 1998)carried out on the samples taken during the flush showed no bacteria leakage from the column after inoculation.

Toxicity measurement. Toxicity of elution samples was assessed by using the Microtox® test. This test uses a bioluminescent marine bacterium and measures the inhibition to luminescence caused by the presence of contaminant in the sample. Results are collected in terms of number of dilutions necessary to reach the minimum concentration with an observable effect (LOEC).

Parameters Estimation. Mass elution curves were mathematically modelled using the classic transport equation for a reactive tracer in a saturated porous media, which includes a retardation factor for sorption equilibrium, dispersion and diffusion and first-order biodegradation, according to the non-dimensional form:

$$R \frac{\partial C}{\partial T} = \frac{1}{Pe} \frac{\partial^2 C}{\partial Z^2} - \frac{\partial C}{\partial Z} - \zeta C \qquad (3)$$

with R, the retardation factor, C, the concentration of contaminant, T and Z, the temporal and spatial variables, Pe, the Peclet number and ζ, the first-order biodegradation constant. The numerical solution was obtained using the finite difference method on this distributed parameter equation. Curve fitting for parameter evaluation from experimental data was done using a non-linear least square method based on the Marcquardt-Levenberg algorithm. The fitted mass elution curves were then derived to allow AUCmass calculation with the trapeze method. The experimental toxicity elution curves were derived and the AUCtox was calculated using the same trapeze method.

RESULTS AND DISCUSSION

Figure 1 shows the results obtained from the three different breakthrough experiments performed on the glass beads column: non-reactive tracer test, PCP elution under abiotic conditions and PCP elution under biotic conditions. The two first curves appear to be identical, proving that PCP eluted through the glass beads column behave as a non-reactive tracer. The biotic curve is clearly underneath the abiotic one, showing that PCP was partially degraded throughout the column. This also shows that the microorganisms inoculated in the column, not only stayed inside the column but also kept their metabolic capacities towards PCP. The estimated first-order biodegradation constant obtained from curve fitting is 0.74 mg/(L.h).

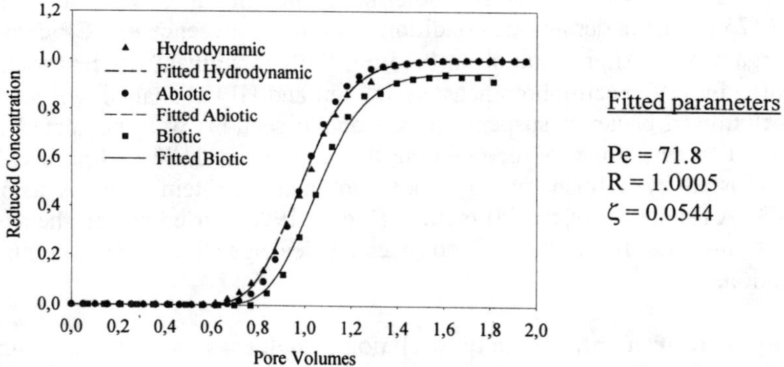

FIGURE 1: PCP breakthrough curves from the glass beads saturated column

Figure 2 shows the mass and toxic breakthrough curves as well as their derivatives obtained under biotic conditions. The toxicity is directly associated with the presence of the contaminant. The sensitivity of the Microtox® method towards PCP is observable as toxicity breaks through much faster than the actual mass of contaminant. The AUC and the reference times for the two curves are then calculated to obtain the two bioavailability rates: 0.4 Vp^{-1} for the mass bioavailability rate and 10.64 Vp^{-1} for the toxic bioavailability rate.

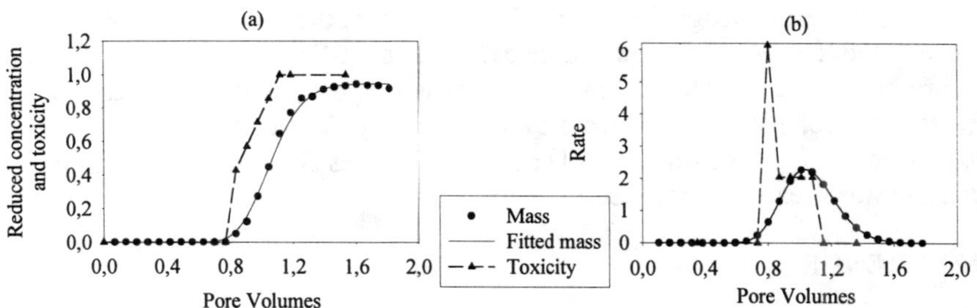

FIGURE 2: PCP mass and toxicity elution curves under biotic conditions. (a) Reduced concentration and toxicity and (b) their derivatives

These results can be reduced to a non-dimensional form. For the mass bioavailability, the reduction is made to the optimal case where the whole contaminant would have been degraded. This calculation leads to a non-dimensional number, massBioA, between 0 and 1, here 0.053. Concerning the toxic bioavailability, the reduction is carried out by comparing the rate obtained under biotic conditions to the rate obtained under abiotic conditions. This time, the non-dimensional number, toxBioA, is generally between 0 and 1 but may be superior to 1 in the case where the presence of the microorganisms responsible for biodegradation leads to an increase in toxicity. For the example of the glass beads column, the non-dimensional toxic bioavailability is 1, indeed, no reduction in toxicity has been associated with the partial degradation of PCP observed. This result is linked to the sensitivity of the method used to measure toxicity: figure 2 shows that the toxicity reaches a peak before PCP concentrations reach their maximum. The results in terms of mass and toxic bioavailability are in agreement to what could be expected from a glass bead medium. First, glass beads do not represent the best medium for microorganisms attachment. Second, the residence time of the contaminant within the column, knowing that no retardation is observed, is the shortest possible. Third, the contaminant elution was quite fast due to the fact that it is not sorbed on the medium, which leads to a small residence time. Thus, a very low mass bioavailability and a very high toxic bioavailability are understandable.

CONCLUSION

The method developed here to quantify mass and toxic bioavailability rates leads to interesting results concerning the characterisation of the medium towards the contamination in terms of bioremediation potential and hazard index. If hazard analysis is carried out along the bioremediation process, the approach to bioavailability presented here allows the calculation of these two index: first, the intrinsic bioremediation potential, IBP, expressed as the ratio of mass over toxic bioavailability and second, the hazard Index, HI, expressed as its inverse. More matrices are currently under investigation with promising results for integrating the dynamics of bioavailability

ACKNOWLEDGEMENTS
The authors acknowledge the financial support from the industrial Chair partners: Alcan, Bodycote/Analex, Bell Canada, Browning-Ferris Industries, Cambior, Centre québécois de valorisation de la biomasse et des biotechnologies (CQVB), Hydro-Québec, Natural Science and Engineering Research Council (NSERC), Petro-Canada and SNC-Lavalin. The help of Jacques Bureau in regard to toxicity analysis was greatly appreciated.

REFERENCES

Bosma, T. N. P., P. J. M. Middeldorp, G. Schraa and A. J. B. Zehnder (1997). "Mass transfer limitation of biotransformation: quantifying bioavailability." *Environmental Science and Technology 31*(1): 248-252.

Dasappa, S. M. and R. C. Loehr (1991). "Toxicity reduction in contaminated soil bioremediation processes." *Water Research 25*(9): 1121-1130.

Harms, H. and A. J. B. Zehnder (1995). "Bioavailability of sorbed 3-chlorodibenzofuran." *Applied and Environmental Microbiology 61*(1): 27-33.

Kelsey, J. W. and M. Alexander (1997). "Declining bioavailability and inappropriate estimation of risk of persistent compounds." *Environmental Toxicology and Chemistry 16*(3): 582-585.

Nam, K. and M. Alexander (1998). "Role of nanoporosity and hydrophobicity in sequestration and bioavailability: tests with model solids." *Environmental Science and Technology 32*(1): 71-74.

Otte, M.-P., J. Gagnon, Y. Comeau, N. Matte, C. W. Greer and R. Samson (1994). "Activation of an indigenous microbial consortium for bioaugmentation of pentachlorophenol/creosote contaminated soils." *Applied Microbiology and Biotechnology 40*: 926-932.

Pierce (1998). Micro BCA protein assay reagent kit. *Instructions catalogue, Pierce.*

Pignatello, J. J. and B. Xing (1996). "Mechanisms of slow sorption of organic chemicals to natural particles." *Environmental Science and Technology 30*(1): 1-11.

Ramaswami, A. and R. G. Luthy (1997). 'Measuring and modeling physicochemical limitations to bioavailability and biodegradation'. *Manual of Environmental Microbiology.* C. J. Hurst. Washington, D.C., ASM Press.

Ritschel, W. A. (1988). 'Pharmacokinetic and biopharmaceutical aspects in drug delivery'. *Drug Delivery Devices. Fundamentals and Applications.* P. Tyle. New York, Marcel Dekker Inc. 32: 17-79.

Umbreit, T. H., E. J. Hesse and M. A. Gallo (1986). "Bioavailability of dioxin in soil from a 2,4,5-T manufacturing site." *Science 232*: 497-499.

BIOREMEDIATION OF URANIUM MINE DRAINAGE USING METHYLOTROPHIC SULFATE-REDUCING BACTERIA

Barbara C. Hard (UFZ, Leipzig, Germany)
Wolfgang Babel (UFZ, Leipzig, Germany)

ABSTRACT: The presence of high concentrations of sulfate and metals in mining water is one of the major problems connected to the closure of uranium mines. Large volumes of acidic water from the mines and dumps collect in lakes. The water cannot be disposed off until it has been treated in some way as it poses a direct threat to drinking water and agriculture. A bioremediation process is proposed in which sulfate-reducing bacteria are used to decontaminate acid mine drainage (AMD).

A number of acidotolerant, methylotrophic sulfate-reducing strains have been isolated which are suitable. Lab scale experiments with different reactor types were carried out in order to find the optimum design for this bioremediation process. Comparisons were made between methanol and other electron donors with regards to their suitability as substrate for this process. Methanol was found to be most suited. Laboratory data suggest that immobilizing the bacteria on pumice particles increases the sulfate-reduction rate (SRR) up to three fold to 18 mg/l.h, compared to the rates of free flowing cells of between 3.7 and 6.8 mg/l.h. Preliminary experiments on a larger scale (15 l) using acid mine drainage pH 2.5 show SRR of 0.71 mg/l.h. In biosorption processes up to 140 mg of aluminium per g biomass was removed from the water. The application of the proposed process with regards to bioremediation in situ are discussed.

INTRODUCTION

In eastern Germany uranium mining was a very important industry and a large number of mining sites, uranium processing plants and waste dumps had been in use since the second world war. Acid mine drainange from these abandoned uranium mines and waste dumps poses a serious threat to the country site surrounding these mines. Treatment of this water using conventional methods is very expensive, therefore it is desirable to find an alternative method which is very effective and low in costs. Decontamination of this water requires a number of steps: sulfates have to be removed, (heavy) metals and radionuclides have to be collected and eliminated and the water has to be neutralized. A microbiological approach in which sulfate-reducing bacteria are used is a feasible solution (Hard et al., 1997): reduction of sulfated results in the precipitation of metal sulfides and neutralization of the water. Metals such as aluminium which don´t precipitate as sulfides but stay in solution can be removed by biosorption. Uranium can be reduced from the soluble form to the insoluble form thus eliminating it from the water. Costs for this process result largely from the need for a carbon and energy source for the bacterial metabolism. It is therefore important to use a cheap substrate: methanol. Methanol is an inexpensive and widely available substrate. A number of methylotrophic strains of sulfate-reducing bacteria were isolated. In order to find optimum conditions for sulfate-reduction and neutralization, a number of different reactor types were used. This paper reports on the suitability of the different cultivation methods tested, the optimization of the process, biosorption of metals and discusses its application for in situ remediation.

MATERIALS AND METHODS

Five strains of acidotolerant, methylotrophic sulfate-reducing bacteria were isolated and cultivated as described by Hard et al. (1997) using methanol (10 mM) as sole source of carbon and energy.

Strain descriptions. The strains of sulfate-reducing bacteria isolated were Gram-negative and belong to the genus *Desulfovibrio*. They are methylotrophic, but can also utilize a number of other substrates as sole source of energy and carbon, e.g. lactate, pyruvate, ethanol and others. All strains are acidotolerant as well as metal tolerant to a number of metals such as iron, uranium, aluminium and copper and were found to be suitable for biosorption processes (Hard et al., 1997).

Determination of growth rates. All strains were grown at 30 °C in batch, fed batch or continuous culture at pH 7. Growth was followed by protein measurements. Protein was determined according to the modified method of Bradford (1976). Cells were harvested by centrifugation at 10,000 g for 10 minutes. The pellet was resuspended in 1 ml 1 N NaOH and the samples were boiled for 10 minutes in the waterbath. 0.1 ml sample was taken and 1 ml Bradford solution was added. The absorbance at 595 nm was measured with a Hitachi 1100 spectrophotometer. Bovine serum albumin (BSA) was used as protein standard.

Chemical determinations. Sulfate was determined according to Schauder et al. (1986) whereby sulfate is precipitated with barium ions or with a sulfate assay kit "Spectroquant" (Merck, Darmstadt, Germany) based on the reaction of sulfate with barium ions. Methanol was determined using a Shimadzu GC 14 A gas chromatograph (Shimadzu, Duisburg, Germany) or a Dani GC 1000 head space gas chromatograph (Dani, Gera, Germany).

Reactors and Cultivation. Three different types of reactors were used for continuous cultivation under sterile conditions: 1. Fermentor (B. Braun Biotech International, Melsungen, Germany) with a working volume of 1 liter. 2. Airlift reactor, working volume 2 liter and 3. Fixed bed reactor (Jena Glass, Jena, Germany), working volume 2 liter. Pumice particles with a diameter of less than 0.5 mm served carriers for the bacteria. For fed batch cultivation with acid mine drainage (AMD) pH 2.5 an airlift reactor (non-sterile), working volume 15 l was used. Anaerobic conditions were maintained in all reactors using nitrogen gas. The concentration of methanol in the medium was 10 mM, yeast extract or vitamin solutions were not added.

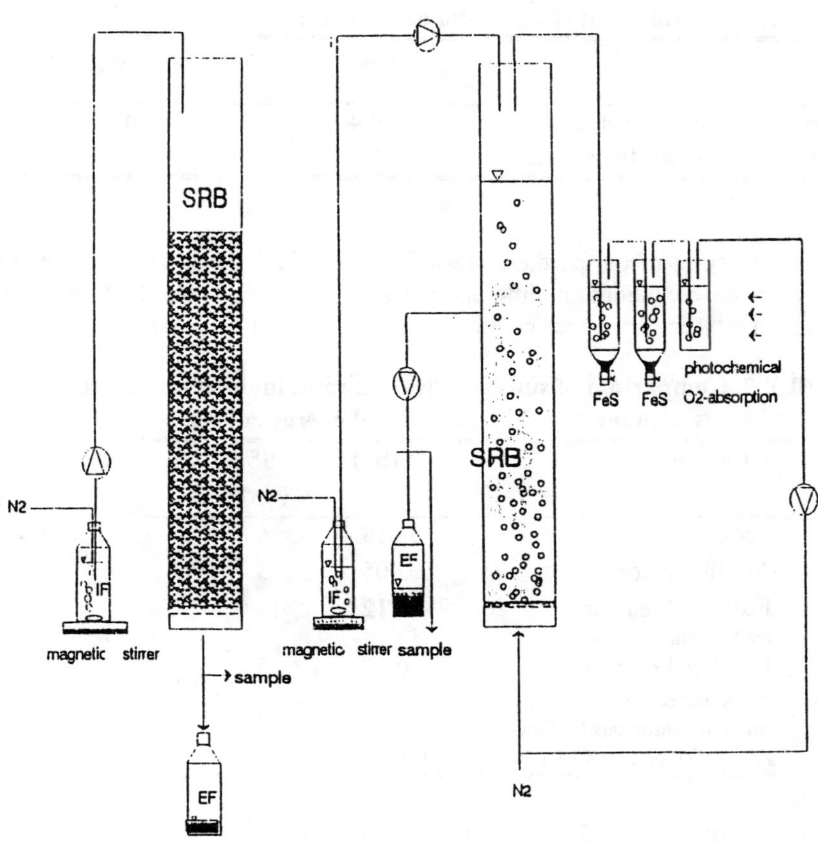

FIGURE 1. Experimental set-up of fixed bed reactor and air lift reactor

RESULTS AND DISCUSSION

The use of bacteria in bioremediation processes depends largely on the costs. Methanol is one of the cheapest substrate, however, sulfate-reducing bacteria generally show higher growth rates with other, much more expensive carbon and energy sources, e.g. lactate. In order to compare the growth yields of methanol and lactate, a number of preliminary growth experiments were carried out. In Table 1 the growth yields of one of the isolated strains are shown. In order to reduce sulfate and grow, a lot less methanol is needed than lactate. Therefore the costs per gram of sulfate is much lower with methanol used compared to lactate.

TABLE 1. Growth yield of strain UFZ B 406 in fed batch culture with different electron donors (10 mM)

		Methanol	Lactate	Pyruvate	Acetate
Ymx/s	BM/ mol substrate	9.4	1.1	14.5	8.45
YCs/SO_4	g substrate/ g sulfate	0.7	1.95	1.2	1.8

BM: biomass

For comparison, pure cultures of sulfate-reducing bacteria were cultivated under sterile conditions in different types of reactors. Table 2 shows sulfate-reduction rates in continuous culture.

TABLE 2. Comparison of sulfate-reduction rates in different reactor types with methanol as carbon and energy source

Reactor	$D\ (h^{-1})$	SSR (mg/l.h)
Fermentor	0.018	6.8
Air lift reactor	0.005	3.7
Fixed bed reactor (cells immobilized)	0.12	18.0
Fixed bed reactor (cells immobilized, without continuous N^2 flow)	0.12	5.4
Airlift fermentor (15 l)*	fed batch	0.71

* with AMD pH 2.5 and methanol (10 mM)

The data shows highest SSR by far in the culture growing in the fixed bed reactor with pumice particles with constant N_2 flow.

On a larger scale in a 15 l volume using acid mine drainage, the iron sulfide precipitates was used as sludge bed. The pH inside the reactor was at pH 7 and AMD pH 2.5 was added in first fed batch experiments. The culture conditions were non sterile and a seemingly stable mixed culture consisting of 2 or 3 different strains established itself in the reactor. The pH increased over a period of 40 days to pH 8 (Figure 2). In the first 25 days the formation of acetate was observed, which was completely utilized by day 40 (data not shown). The formation of acetate was probably due to the incomplete oxidation of methanol by one or more of the strains in the mixed culture, possibly not sulfate-reducers. The rate of sulfate reduction was fairly low compared to the rates of the other cultures. Possible reasons for this could be: FeS associated cells show less sulfate reducing activity than cells immobilized on particles or free living cells (Fukui and Takii, 1994) and competition for substrate between sulfate-reducing bacteria and other members of the mixed culture.

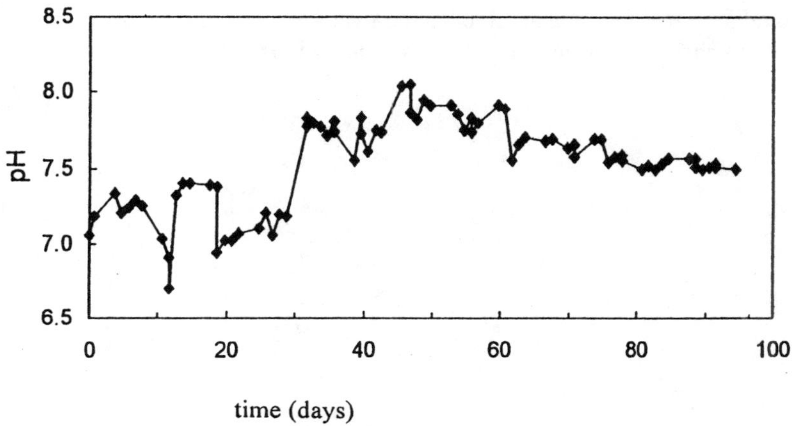

FIGURE 2. pH changes over time in the 15 l sludge bed reactor

Biosorption of metals. In addition to the removal of metals from the water as sulfides, biosorption processes took place. This is particularly important for metals which don´t precipitate as sulfides, e.g. aluminium. It is found in the AMD in high concentrations (85 mM). The mechanism of aluminium biosorption was found to be a passive one and deposition of aluminium ions took place exclusively on the surface of the cell. Up to 140 mg of aluminium per g biomass were removed from the water in this way.

Conclusions. The use of acidotolerant methylotrophic sulfate-reducing bacteria for an on site bioremediation process is feasible. Best sulfate-reduction rates are achieved when cells are immobilized on pumice particles. Neutralization of the AMD can take place in medium with slightly acidic to neutral pH, as protons are bound, acidic water can be pumped in. Biosorption of metals is an additional mechanism to eliminate metals from acid mine drainage.

REFERENCES

Bradford, M. M. 1976. "A rapid and sensitive method for the quantitation of microgram quantities of protein utilizing the principle of protein-dye binding." *Anal. Biochem. 72*: 248-254.

Fukui, M., and S. Takii. 1994. "Kinetics of sulfate respiration by free living and particle-associated sulfate-reducing bacteria." *FEMS Microbiol. Ecol. 13*: 241-248.

Hard, B. C., S. Friedrich, and W. Babel. 1997. "Bioremediation of acid mine water using facultatively methylotrophic metal-tolerant sulfate-reducing bacteria." *Microbiol. Res. 152*: 65-73.

Schauder, R., B. Eikmanns, R. K. Thauer, F. Widdel, and G. Fuchs. 1986. "Acetate oxidation to CO2 in anaerobic bacteria via a novel pathway not involving reactions of the citric acid cycle." *Arch. Microbiol. 145:* 162-172.

IN SITU MEASUREMENT OF PCB AVAILABILITY
IN UNSATURATED SOILS

Sean McNamara (Carnegie Mellon University, Pittsburgh, PA)
Richard G. Luthy (Carnegie Mellon University, Pittsburgh, PA)

ABSTRACT: A novel sampling device is being developed to measure the aqueous-phase availability of polychlorinated biphenyls (PCBs) in unsaturated field soils. The new sampling device integrates the mass of PCBs sampled from soil pore-water over extended sampling periods (weeks to months), and quantifies the volume of water from which that mass was collected. In this way, the new technique provides a *field* estimate of the time-averaged mass and volumetric flux rates of PCBs in the soil column. The new pore-water sampler is constructed from a porous stainless steel interface, granular activated carbon, and a fiberglass wick. The capillarity of the wick and elevation potential are used to draw water from the soil surrounding the sampler, though the cylindrical porous interface and activated carbon, and into the wick. PCBs are sorbed onto the activated carbon, while the sampled water is conveyed along the wick and collected at some distance from the sampling interface. The new device improves on previous wick sampler designs by easing field installation requirements and by providing a way to integrate low-level contaminants to analytical limits while minimizing target chemical losses.

INTRODUCTION

Numerous studies have been performed to investigate the reduced availability, mobility, and toxicity of extremely hydrophobic compounds, such as polychlorinated biphenyls (PCBs), in aged or biotreated soils (Linz and Nakles, 1997). Most investigations have used laboratory tests, such as SPLP or column leachate analysis, to help quantify the extent to which cost-effective remediation strategies, such as biotreatment or natural attenuation, may reduce soil risk. Currently, few field methods exist to quantify similar measures of aqueous-phase availability or mobility – especially in unsaturated soils. Existing unsaturated zone sampling techniques generally require extensive excavation to install devices that may only sample dilute saturated flows (e.g., pan lysimeters or tile drains), may only provide a point estimate of chemical availability (e.g., suction lysimeters), may not quantify the volume of water sampled (e.g., resin capsule samplers), and may have associated quantification problems due to sorptive losses or bioutilization (Skogley and Dobermann, 1996; Wilson et al., 1994).

A new sampling device is being developed which hopes to address these problems. The new sampling device builds on earlier investigations that have demonstrated the utility of using fiberglass wicks to overcome soil pore-water tension and to passively sample solutes and pore-water from soils at reduced saturation (Brown et al., 1988; Boll et al., 1992). In addition, the new sampling design adds a sorbent capacity to preserve and integrate low-level contaminants over extended sampling intervals (weeks, months).

Objective. The principal goal of the project is to develop a sampling method that allows for the in situ assessment of PCB mobility in soils and soil/sludge materials and to use this technique to help quantify the extent that field biotreatment may reduce PCB mobility. The key questions to address are:

1. What is the rate of release of residual PCBs in biotreated materials?
2. How do in situ measurements of PCB release compare with apparent biodegradation rates?
3. What are the implications of these factors for acceptable endpoints and risks?

The research approach is to design and evaluate a novel sampling device, and to deploy such devices in field soils and land biotreatment units at the Alcoa facility in Massena NY. The data from field investigations will be used to assess PCB mass flux and mobility in the land biotreatment units and to build on previous investigations relating to PCB mobility in aged or biotreated soils (Adeel et al., 1998; Smith et al., 1996).

MATERIALS AND METHODS

The general design concept for the new sampling device was to combine the sorbing properties of resin capsule samplers (Skogley and Dobermann, 1996) and the capillary action of wick sampler (Brown et al., 1988) into a single sampler design. The foreseen benefits of the new design would be to retain trace quantities of PCBs on a sorbent material over extended sampling intervals (weeks, months) in a manner that would limit their loss through bioutilization, sorption on sampling device materials, or other processes; to be able to sample soil pore-water across a range of reduced saturation states; and to be able to maintain a water balance such that time-averaged concentration of PCBs in the pore-water could be estimated from the volume of water sampled by the wick and the mass of target chemicals extracted from the sorbent material. Design considerations included size, geometry, materials, hydraulic performance, cost, ease of deployment and retrieval, and compatibility with existing extraction and analysis techniques for PCBs.

Sampler Design. The basic design of the new sampler is shown in Figure 1. The sampling end of the fiberglass wick is centered in a porous stainless steel cylinder. Between the wick and the permeable interface, a layer of sorbent material is packed. The sorbent material provides connectivity between the wick and the interface, and captures the aqueous phase chemicals of interest as water passes from the soil, through the interface and sorbent material, and into the wick. The cylindrical design of the sampler was chosen to minimize soil disturbance when deploying the sampler in field soils. The cylinder's outside diameter measures 3.2cm, which is comparable to a commercially available bore-size of a soil auger. The end of the cylinder that the wick extends through is threaded to accept standard pipefittings (1/2" NPT). The opposite end of the cylinder is fitted with a cone point to ease insertion into horizontal bore holes which will be drilled for installation into shallow land treatment units, as shown in Figure 2. A vertical installation design has been considered for deeper soils, in which a collection volume is added below the hanging wick length and water is pumped from that volume periodically. The development and optimization of the vertical design is currently beyond the scope of the study.

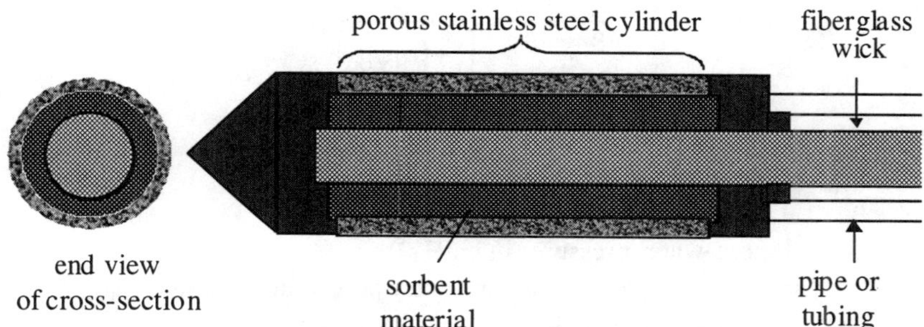

FIGURE 1. New sampler design for the capillary extraction of pore-water from unsaturated soils with retention of PCBs on sorbent media.

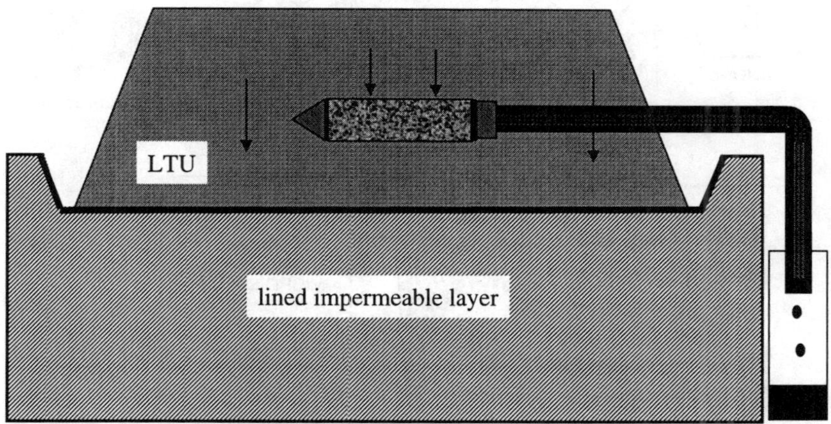

Figure 2. Horizontal sampler installation in shallow land treatment unit.

Hydraulic Tests. Laboratory experiments were performed to evaluate the hydraulic performance of the new sampling device under simulated wetting and drying cycles that might be expected in the field. The sampling device was installed in a 5-gallon bucket (18.9 L) of soil, and water was irrigated through the bucket to simulate rainfall. Soils (clean Massena fill and Monongahela River sand) of different hydraulic conductivities, different horizontal and vertical wick lengths, and different sorbent media (Amberlite XAD-4 resin and Calgon OL20x50 granular activated carbon), were included in the experimental matrix.

A data acquisition system was designed to measure the degree of unsaturation in both the soil and the wick throughout the wetting and drying experiments. Pressure transducers were fitted with porous stainless steel tips that were filled with water to act as tensiometers. A schematic of the data acquisition system and experiment setup is shown in Figure 3. Additional laboratory investigations were performed to translate the measured pore-water pressures to water content values using the retention curve relationship formulated by van Genuchten (1980).

$$\frac{\theta - \theta_r}{\theta_{sat} - \theta_r} = \frac{1}{\left[1 + \left(\alpha |\psi|\right)^n\right]^m}$$

Where θ = volumetric water content (cm^3 water/cm^3 soil)
     θ_{sat} = saturated water content, or porosity
     θ_r = irreducible water content
     ψ = pore-water pressure (in cmH$_2$0)
     α = positive inverse of the air-entry pore water pressure ($|\psi_a|^{-1}$)
    n, m = fitting parameters.

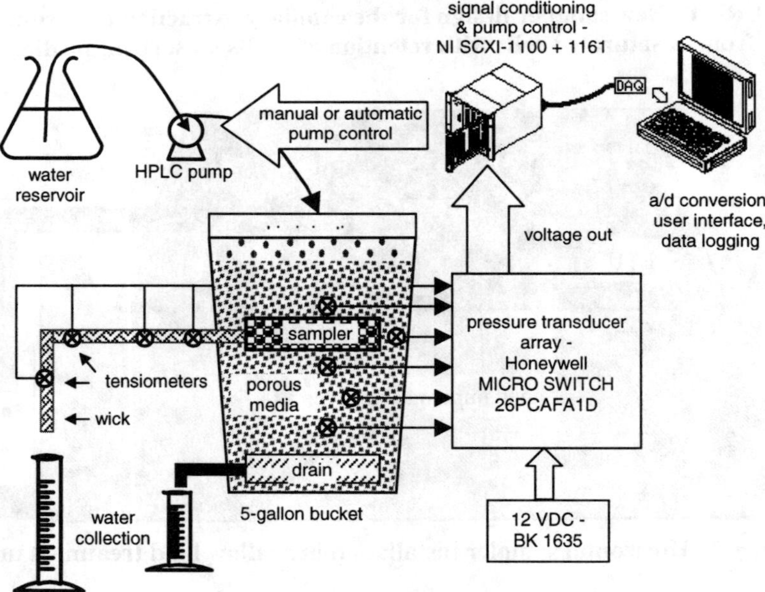

Figure 3. Schematic of hydraulic evaluation experiment design

RESULTS AND DISCUSSION

The hydraulic evaluation tests demonstrated that the new sampler can extract pore-water from soils at pore-water pressures ranging from saturation (≥ 0 cm H$_2$0) to less than -50 cm H$_2$0, using horizontal and vertical wick lengths between 50 and 100cm (up to 200cm combined length). For the Massena soil, this was equivalent to water contents ranging from saturation ($\theta_{sat} = 0.38$) to below 5 percent by volume. Typical results for hydraulic loading vs. volume of water sampled are shown in Figure 4. The device was able to sample pore-water under both wetting and drying conditions, and was able to recover even after several weeks of drying. This second observation represented a critical advance in the optimization of the sampler design. The sampler was initially evaluated using XAD-4 resin for the sorbent media. After several weeks of drying, the sampler failed to sample additional pore-water when the soil was returned to a wetting cycle. Later tests revealed that the physical properties of XAD-4 were altered during the drying cycle - enough that the polymeric resin became so hydrophobic that it would no longer transmit water. Granular activated carbon was hydraulically superior in this respect, and showed little hysteresis.

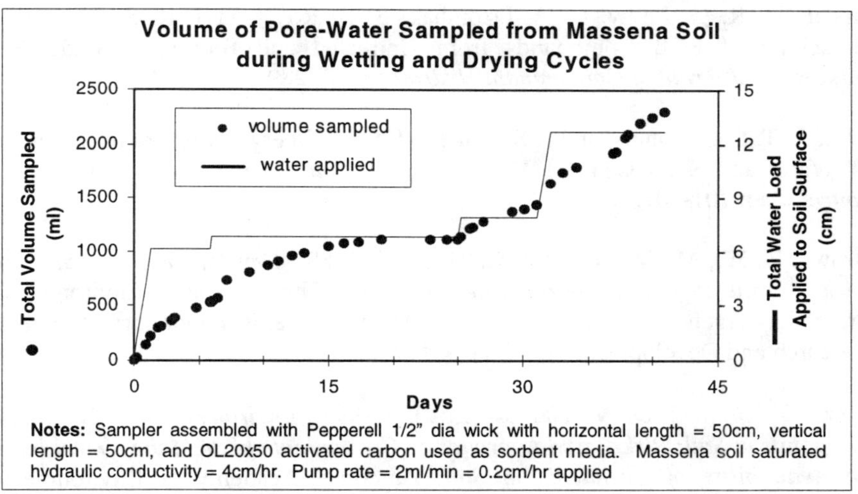

Notes: Sampler assembled with Pepperell 1/2" dia wick with horizontal length = 50cm, vertical length = 50cm, and OL20x50 activated carbon used as sorbent media. Massena soil saturated hydraulic conductivity = 4cm/hr. Pump rate = 2ml/min = 0.2cm/hr applied

Figure 4. Example hydraulic performance results for new sampler design.

Work Status. Currently, the design of the new pore-water sampler has been sufficiently refined to accomplish its intended objectives in the Massena land treatment units. Field trials are expected to begin in May 1999 at the Alcoa facility in Massena, NY. Several samplers will be installed in the land treatment units, as well as other impacted site soils.

Future work will concentrate on extending this design to a variety of field soils and sampler configurations through the completion of a finite difference model that will simulate the lengthy hydraulic tests previously described. The model will allow the user to input a desired sampler configuration in a characterized soil (i.e., van Genuchten parameters) and then run the hydraulic simulation for a specified precipitation profile. The model will be used to optimize the sampler for the given application by choosing wick type, diameter, or length combinations that will minimize sampler induced gradients, yet still allow a reasonable pore-water collection rate.

Work has also begun on a new method to extract PCBs from the activated carbon sorbent, as the EPA methods for soil (i.e., sonication or Soxhlet extraction with hexane and acetone) yielded poor recoveries. The new method uses benzene to elute PCBs from carbon in a pressurized column. Initial results using 100ml of benzene pumped through a 5g sample of spiked carbon at 3ml/min and 300psi have yielded repeatable recoveries of 85% or higher for Aroclor 1242 and several individual congeners.

Support for this work is funded by the DOE/EPA/NSF/ONR Joint Program on Bioremediation (EPA Grant Number R825365-01-0). Additional support is provided by a NSF Graduate Research Fellowship.

REFERENCES

Adeel, Z., R. G. Luthy, D. A. Dzombak, S. B. Roy, and J. R. Smith. 1998. "Leaching of PCB Compounds from Untreated and Biotreated Sludge-Soil Mixtures." *Journal of Contaminant Hydrology*. 28: 289.

Boll, J., T. S. Steenhuis, and J. S. Selker. 1992. "Fiberglass Wicks for Sampling of Water and Solutes in the Vadose Zone." *Soil Science Society of America Journal*. 56: 701-707.

Brown, K. W., M. W. Holder, J. C. Thomas. 1988. *Development of a capillary wick unsaturated zone pore water sampler*. United States Environmental Protection Agency, Environmental Monitoring Systems Laboratory, Office of Research and Development , Las Vegas, NV.

Linz, D. G., and D. V. Nakles (Eds.). 1997. *Environmentally Acceptable Endpoints in Soil: Risk-Based Approach to Contaminated Site Management based on Availability of Chemicals in Soil*. American Academy of Environmental Engineers, Annapolis, MD.

Skogley, E. O., and A. Dobermann. 1996. "Synthetic Ion-Exchange Resins: Soil and Environmental Studies." *Journal of Environmental Quality*. 25:13-24.

Smith, J. R., M. E. Egbe, T. C. Lightfoot, and W. J. Lyman. 1996. "Bioremediation of PCB- and PAH-Containing Sludge/Sediments in Land Treatment Units to Achieve Risk-Based Endpoints." *Hazardous and Industrial Waste : Proceedings of the Mid-Atlantic Industrial Waste Conference*. 28: 309.

van Genuchten, M. T. 1980. "A Closed Form Equation for Predicting the Hydraulic Conductivity of Unsaturated Soils." *Soil Science Society of America Journal*. 44: 892-898.

Wilson, L. G., L. G. Everett, and S. J. Cullen (Eds.). 1994. *Handbook of Vadose Zone Characterization & Monitoring*. Lewis Publishers, Boca Raton, FL.

LUX-BASED BIOSENSING OF 2,4-DICHLOROPHENOL BIODEGRADATION AND BIOAVAILABLILITY IN SOIL

Kenneth Killham (Aberdeen University, UK)
Liz J. Shaw & Andrew A. Meharg (ITE, Monks Wood, Camb., UK)
Tatiana Boucard, L. Anne Glover & Yvonne Beaton (Aberdeen University, UK)

ABSTRACT: Laboratory-based experiments were carried out to establish feasibility of *lux*-marked biosensors for assessing *in-situ* bioremediation potential of 2,4-DCP contaminated soil. Results showed that the decreases in 2,4-DCP concentration in soil water extracts from soil inoculated with a 2,4-DCP degrader, corresponded directly to decreases in toxicity to *lux*-marked *E. coli*. Response of the biosensor was, therefore, intimately linked to mineralisation activity in inoculated soil. In addition, light output of *P. putida* 12708 *luxCDABE* correlated with 2,4-DCP concentration in spiked sterile soil during ageing and in a field contaminated soil undergoing remediation. *Lux*-modified bacteria may be used as rapid, sensitive and reliable reporters of 2,4-DCP bioremediation and ageing in soil systems, complementing chemical analyses.

INTRODUCTION

Chlorophenols are used throughout the world in bleaching, in synthesis of herbicides and as biocides (Jensen, 1996). Soil is a major sink for chlorophenols and numerous soil micro-organisms are capable of mineralising various chlorophenols. Bioremediation is therefore an attractive option for cleaning contaminated environments (Jensen, 1996). Efficacy of bioremediation must, however, be established and Sousa et al. (1998) described a *lux*-modified biosensor to assess bioremediation potential and constraints of organic contaminants. *Lux* genes can be inserted randomly into non-luminescent bacteria and used to report on cell metabolic activity. Alternatively, *lux* genes can be fused to catabolic genes to report on xenobiotic concentration (Sousa et al. 1998). *Lux*-based biosensors can, therefore, complement chemical techniques for monitoring bioremediation.

Many factors affect the efficacy of bioremediation including bioavailability of the contaminant. Following release into soil, a xenobiotic will become aged i.e. degraded (biotically and/or abiotically), sorbed to soil organic matter or sequestered into soil nanopores. These processes are dynamic and will affect xenobiotic bioavailability and biodegradation potential.

The first aim of this study was to validate use of *lux*-marked *E. coli* HB101 pUCD607 (Rattray et al., 1990) as a sensor for mineralisation of 2,4-dichlorophenol (2,4-DCP) by *Burkholderia* sp. RASC c2 (Suwa et al., 1996) in spiked, autoclaved soil microcosms. The second aim was to determine the effects of 2,4-DCP ageing on toxicity to a chromosomally *lux*-marked bacterium, *Pseudomonas putida* 12708 *luxCDABE*. The final aim was to apply *P. putida* to investigate toxicity in soil historically contaminated with 2,4-DCP.

MATERIALS AND METHODS

2,4-DCP Mineralisation Microcosms. Autoclaved, replicate 4 g aliquots of brick earth soil Shaw *et al.* (1998) were prepared in glass EPA vials (40 ml). Replicate soil microcosms were spiked with 2, 4-DCP to give final concentrations of 49.3, 98.5, 197.0, 295.5 and 469.1 mg kg^{-1} and incubated at 25°C for 1 h. Batch cultures of *Burkholderia* were set-up in tenth strength PTYG-broth (Suwa et al., 1996) in a 125 ml Erlenmeyer flask and incubated with shaking (no. 6, Luckman R100 shaker) at 25 °C. Mid logarithmic phase cells were harvested in the presence of 0.1 M $CaCl_2$ by centrifugation (5 min, 11,600 g). Supernatant was discarded, cells were resuspended in an equal volume of sterile minimal salts media and the suspensions pooled. Aliquots (0.5 ml) of washed cell suspension were added to four replicate microcosms for each concentration treatment such that the final moisture content was 60 % of water holding capacity (WHC). Controls consisted of additional microcosms without ^{14}C-DCP or without cells, for determination of background ^{14}C and abiotic degradation/volatilisation, respectively. A test-tube containing an alkaline trap (1 ml of 1M NaOH) time intervals, NaOH solution was removed from the traps mixed 1:4 with Ultima Gold™ scintillation cocktail. Fresh NaOH solution was added to each trap. Radioactivity of NaOH samples was determined by scintillation counting for 180 s. Data was expressed as cumulative percent of the initial spike for each individual replicate.

2,4-DCP Toxicity Microcosms. Autoclaved soil, prepared as described above was spiked with 2,4-DCP to achieve a final concentration of 196.2 mg kg^{-1} and inoculated with mid-logarithmic phase (3.4 ± 0.25 x 10^7 cfu g^{-1} soil) *Burkholderia* to give a final moisture content of 60 % WHC. Control microcosms consisted of non-2,4-DCP spiked soil microcosms. Four replicate soil microcosms were randomly selected and destructively sampled from control and treatment samples at T = 3 h after spiking soil with 2,4-DCP and immediately after the addition of cells. Soil was slurried by adding 4 ml sterile distilled water, mixed and soil water extracts were obtained by horizontal shaking at 25 °C for 30 min (No. 6, Luckman R100 shaker). Soil was removed by centrifugation (268 g, 5 min, 25 °C), supernatant was decanted and the volume determined gravimetrically. HPLC analysis was carried out using extracts obtained from all sample points. At T = 1 d and subsequent sample points, triplicate 900 µl aliquots of soil water extracts were removed and added to scintillation vials for bioassays using *E. coli*.

2,4-DCP Ageing Microcosms. Soil microcosms were prepared as described above and spiked with 750 mg kg^{-1} 2,4-DCP, but samples were not inoculated with *Burkholderia*. Controls consisted of non-2,4-DCP spiked soil. Microcosms were incubated at 25 °C in darkness, destructively sampled at T = 1 h after soil was spiked with 2,4-DCP (day 0) as described earlier and toxicity of soil water extracts was determined using batch grown *P. putida* (final cell conc. in assay = 7.9 ± 0.64 10^6 cfu ml^{-1}).

Toxicity of Contaminated Field Soil Undergoing Remediation. Top soil was sampled from two sites (A and B) at a pesticide plant, São Paulo, Brazil. Soil water extracts were prepared as described earlier following addition of distilled water to an equivalent weight of oven dried soil, but samples were centrifuged at 760 g, 20 min. Triplicate samples of each soil were muffle furnaced, resuspended in an equal volume of distilled water and pH adjusted to 5.5 as described by Sousa et al. (1998). Toxicity of triplicate samples before and after muffle furnacing was determined using batch grown $P.$ $putida$ (final cell conc. in assay = $8.1 \pm 0.3 \ 10^6$ cfu ml^{-1}). 2,4-DCP in soil water extracts of samples A and B was measured by HPLC prior to muffle furnace. Soil water extracts were treated with methanol (50 % v/v, shaken at 25 °C for 10 min and 5 min centrifugation at 11,600 g) for HPLC analysis.

Preparation of Biosensors and Bioassays. Freeze dried cultures of $E.$ $coli$ were prepared and resuscitated by incubating with shaking at 25°C for 2 h in 10 ml LBG (Sousa et al., 1998). Cells were harvested by centrifugation (11,600 g), washed in an equal volume of 0.1 M KCl, diluted and 100 µl aliquots (final cell conc. in assay = $1.8 \pm 0.21 \times 10^5$ cfu ml^{-1}) were added to each soil water extract. Samples were mixed and luminescence was determined for randomly arranged samples by scintillation counting each sample for 15 s at room temperature. $P.$ $putida$ was chromosomally marked with $luxCDABE$ (originally isolated from $Vibrio$ $fischeri$) by transposon mutagenesis using a suicide delivery system. Six replicate logarithmic phase $P.$ $putida$ cultures were prepared in LB, harvested by centrifugation (11,600 g), washed once in phosphate buffered saline (PBS) and resuspended in 0.5 x Vol. sterile soil water extract (ageing experiment) or PBS (contaminated field soil study). Cell aliquots (50 µl,) were added to replicate 450 µl soil water extracts at 15 s intervals and mixed by aspirating with a pipette for 3 s. Light output of each sample was measured by luminometery for 1 s at 15 s intervals after 30 min exposure at room temperature. Mean percentage decrease in luminescence was calculated for all treatment samples against the non-treatment sample.

Data Analysis. Analysis of data was performed by ANOVA following log transformation.

RESULTS AND DISCUSSION

Figure 1 shows that the lag period before detection of ^{14}C-2,4-DCP mineralisation by $Burkholderia$ increased with increasing ^{14}C-2,4-DCP concentration. At 197.0 mg kg^{-1} soil, mineralisation occurred after a lag of 24 h, but 65.5 ± 1.0 % ^{14}C-2,4-DCP was released by day 6. The lag period increased further for samples spiked with higher ^{14}C-2,4-DCP concentrations and variability was greatest. Soil was therefore spiked with 2,4-DCP at ~ 200 mg kg^{-1} soil to determine toxicity of soil water extracts to $E.$ $coli$ during mineralisation by $Burkholderia$.

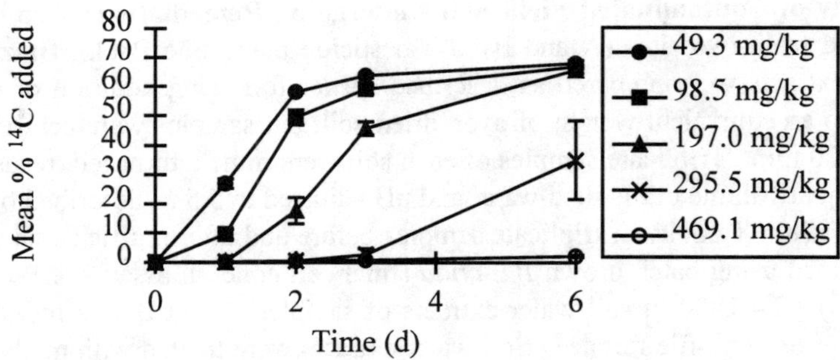

FIGURE 1. Mineralisation of ^{14}C-2,4-DCP by *Burkholderia* sp. RASC c2 in sterile soil microcosms. Bars represent standard errors.

Figure 2 shows that soil water extracts were initially very toxic to *E. coli*. HPLC analysis showed that 2,4-DCP concentration decreased on day 1 and by day 2 was near the limit of detection (0.1 mg L^{-1}). At low 2,4-DCP concentrations, soil water extracts stimulated *E. coli* light output on day 2, probably due to uncoupling of proton cycle. Luminescence of *E. coli* decreased to ~ 100 % values on day 3 and was relatively constant for the duration of the experiment. 2,4-DCP concentration in soil water extracts were below background values from days 3-7. Changes in 2,4-DCP concentration corresponded to the rapid phase of biodegradation as demonstrated in the first mineralisation experiment. Sterility of non-inoculated microcosms was confirmed by culture analysis. There was a slight, but significant (p < 0.001) decrease in 2,4-DCP concentration and toxicity in sterile spiked soil microcosms over time (data not shown). These effects were probably due to volatilisation or ageing of 2,4-DCP.

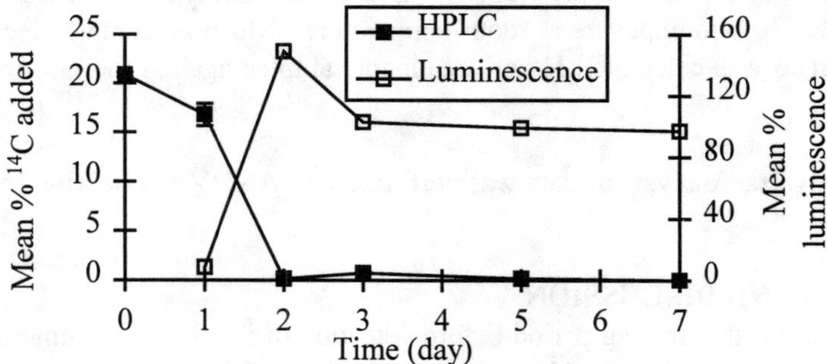

FIGURE 2. Mean concentration of 2,4-DCP (□) and mean toxicity of 2,4-DCP (■) in soil water extracts from 2,4-DCP spiked soil. Bars represent standard errors.

Soil water extracts from 2,4-DCP contaminated soil were initially very toxic to *P. putida* (Figure 3). Toxicity of soil water extracts significantly decreased (p < 0.001), however, with time, and on day 17, was 82.6 ± 4.3 % of the control. Rapid sorption of 2,4-DCP to soil organic matter would have occurred within the first hours of spiking soil (Shaw et al.). Soil was therefore spiked with a high 2,4-DCP concentration to allow measurable decreases in toxicity over time. Some 2,4-DCP may have been lost due to volatilisation, but the large decrease in toxicity with time (> hours) was likely due to intra-particle and micropore diffusion.

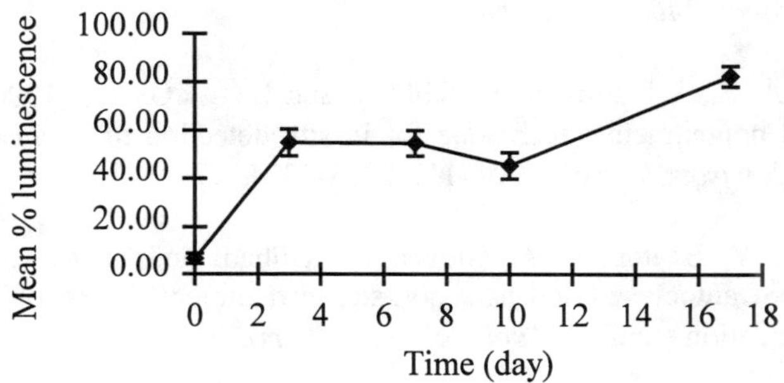

FIGURE 3. Mean toxicity of soil water extracts during ageing of 2,4-DCP in soil. Bars represent standard errors.

Muffle furnace treatment of samples A and B from a soil, historically contaminated with 2,4-DCP, was very effective in reducing toxicity (Table 1). Muffle furnacing would have destroyed any volatile and non-volatile organic compounds present in the samples. Treatment of sample B was most effective in stimulating light output, indicating that this sample was initially most toxic to *P. putida*. HPLC analysis of soil water extracts confirmed that sample B had highest concentration of 2,4-DCP.

TABLE 1. Concentration and mean toxicity of 2,4-DCP in soil water extracts from contaminated field soil. Toxicity after muffle furnacing was expressed relative to luminescence before treatment.

	Sample A	Sample B
Mean relative luminescence ± se	5.5 ± 1.3	22.3 ± 8.9
2,4-DCP conc. (mg kg^{-1} soil)	None detected	103.5

CONCLUSION

Results showed that decreases in 2,4-DCP concentration in soil water extracts of spiked soil inoculated with the 2,4-DCP degrader, *Burkholderia* sp. RASC c2, reflected a decrease in toxicity to *lux*-marked *E. coli*. Changes in toxicity of soil water extracts to the *lux*-based biosensor were therefore sensitive indicators of biodegradation of 2,4-DCP by *Burkholderia*. In addition, experiments showed that *lux*-marked *P. putida* was a rapid indicator of 2,4-DCP ageing in spiked soil and DCP remediation of a field contaminated soil.

ACKNOWLEDGEMENTS

The authors are grateful to the NERC Environmental Diagnostics Programme for funding this research and to Dr. G. I. Paton for HPLC analysis of the field contaminated soil.

REFERENCES

Jensen, J. 1996. "Chlorophenols in the terrestrial environment." *Rev. Environ. Contam. Toxicol. 146*(10): 25-51.

Rattray, E. A. S., J. I. Prosser, K. Killham and L. A. Glover. 1990. "Luminescence-based nonextractive technique for in situ detection of *Escherichia coli* in soil." *Appl. Environ. Microbiol. 56*(11): 3368-3374.

Shaw, L. J.; Y. Beaton, L. A. Glover, K. Killham and A. A. Meharg. "Reinoculation of autoclaved soil as a non-sterile treatment for xenobiotic sorption and biodegradation studies." *Appl. Soil Ecol. In press.*

Sousa, S., C. Duffy, H. Weitz, L. A. Glover, E. Bär, R. Henkler and K. Killham. 1998. "Use of a *lux*-modified bacterial biosensor to identify constraints to bioremediation of BTEX-contaminated sites." *Environ. Toxicol. Chem. 17*(6): 1039-1045.

Suwa, Y., A. D. Wright, F. Fukimori, K. A. Nummy, R. P. Hausinger, W. E. Holben and L. J. Forney. 1996. "Characterisation of a chromosomally encoded 2,4-dichlorophenoxyacteic acid/α-ketoglutarate dioxygenase from *Burkholderia* sp. strain RASC." *Appl. Environ. Microbiol. 62*(7): 2464-2469.

INTEGRATED ANAEROBIC/AEROBIC EGSB BIOREACTOR FOR AZO DYE DEGRADATION

Nico C.G. Tan, Joris L. Opsteeg, Gatze Lettinga and Jim A. Field
(Wageningen Agricultural University, Wageningen, The Netherlands)

ABSTRACT: A lab-scale integrated anaerobic/aerobic expanded granular sludge blanket (EGSB) bioreactor was tested for the biodegradation of the azo dye 4-phenylazophenol (4-PAP). The goal of this study was to investigate if the azo dye 4-PAP is degraded under the integrated anaerobic/aerobic conditions in continuous systems. The results from this bioreactor study indicated that the azo dye was completely decolorized. Furthermore, none of the expected biodegradation products (aromatic amines) were recovered in the effluent of the bioreactor. Batch tests indicated that aniline was aerobically degraded by facultative bacteria present in the granular sludge. Biodegradation for 4-aminophenol (4-AP) was difficult to prove since this compound was readily autoxidized. An anaerobic control bioreactor showed recovery of both aromatic amines, aniline and 4-AP, formed from the reductive cleavage of the azo dye. Therefore, integration of both anaerobic and aerobic conditions is beneficial for the complete removal of this azo dye.

INTRODUCTION

Aromatic compounds with electron-withdrawing substituents (azo, polynitro and polychloro) are not very susceptible for degradation under aerobic conditions (Field et al., 1995). On the other hand, most of these xenobiotic compounds are readily reduced under anaerobic conditions. However, the reduced products are generally not degraded further under anaerobic conditions. On the other hand, the reduced products are readily metabolized by aerobic microorganisms. Therefore, both anaerobic and aerobic conditions are required for the complete degradation of aromatic compounds with electron-withdrawing substituents such as azo dye.

Anaerobic and aerobic conditions can be integrated by creating an aerobic biofilm on anaerobic biomass (O'Reilly and Scott, 1995). For this purpose anaerobic granular sludge can be used, especially granular sludge which possesses high tolerance for oxygen (Kato et al., 1993). Previously, it was demonstrated in batch experiments that azo dye reduction by integrated anaerobic/aerobic bacterial cocultures and by granular sludge occurs in the presence of oxygen (Kudlich et al., 1996; Tan et al., 1999a). The resulting aromatic amines were mineralized aerobically (Kudlich et al., 1996; Tan et al., 1999b). The next challenge of this research is to demonstrate that these xenobiotic compounds can be degraded in continuous systems.

Objective. The objective of this study is to determine if the azo dye 4-PAP can be degraded continuously under integrated anaerobic/aerobic conditions in a lab-scale EGSB bioreactor. 4-PAP is expected to be reductively cleaved to aniline and 4-AP in anaerobic conditions. These reduction products are expected to be mineralized further by aerobic microflora.

The continuous experiments were conducted in an integrated anaerobic/aerobic EGSB bioreactor. An anaerobic control bioreactor was operated in order to investigate the influence of the addition of oxygen on the degradation of the azo dye and the aromatic amines. Batch experiments were preformed in order to check the bioreactor results.

MATERIALS AND METHODS

Reactor and Batch Media. The basal medium used in all batch experiments contained were previously described (Tan et al., 1999a). For the reactor experiments the amount of trace elements and carbonate were reduced 10-times and 5-times, respectively.

Analyses. The aromatic amines used in the aerobic and anaerobic biodegradation assays were analyzed spectrophotometrically with a spectrophotometer at their absorbance maxima; aniline at 230 nm; 4-AP at 230 nm. Liquid batch samples were centrifuged (7833 g, 10 minutes) and diluted in a 0.10 M sodium phosphate buffer solution (pH 7.0) and measured in a quartz cuvette.

The azo dye and their corresponding amines in the continuous integrated anaerobic/aerobic degradation experiments were analyzed with high performance liquid chromatography (HPLC). Samples from the bioreactor were centrifuged (7833 g, 10 minutes) and diluted in demineralized water and 10 µl samples were injected with a autosampler (Marathon, Separations). The azo dye and their corresponding aromatic amines were detected spectrophotometrically with a UV detector (Spectroflow 783, Kratos Analytical) at their absorbance maximum (347 nm for 4-PAP and the previously stated wavelengths for the aromatic amines).

Demineralized water (20 ml/L) in methanol (A) and acetic acid (5.0 ml/L) in demineralized water adjusted to pH 5.9 (B) were used as liquid phase and were pumped (High Precision Pump Model 104, Seperations) at a flow rate of 300 µl min^{-1} first through a degaser (GT-103, Separations) and afterwards through two reverse phase columns (Chromosphere C18, Chrompack). The following gradient program was used for the detection of 4-PAP and its corresponding aromatic amines aniline and 4-AP: 0 minutes 100% B; 5 minutes 100% B; 15 minutes 80% A : 20% B; 20 minutes 80% A : 20% ; 25 minutes 100% B.

Batch Experiments. The aerobic biodegradation of the aromatic amines was studied in 117 ml glass bottles filled with 22.5 ml of basal medium and inoculated with 2.0 g volatile suspended solids (VSS) per liter of Shell granular sludge (Refinery, Moerdijk, The Netherlands). Thereafter, the batches were closed and flushed with N_2 / CO_2 (70% / 30%) gas for 5 minutes. Different initial headspace oxygen percentages (IHOP) were arranged by first removing a given amount of gas from the bottles headspace and replenishing it with the same amount of oxygen. Anaerobic degradation was evaluated by omitting the addition of oxygen. Nedalco granular sludge (Alcohol distillery, Bergen op Zoom, The Netherlands,

1.0 g VSS/L) was used for the anaerobic degradation experiments. The concentrations of the aromatic amines applied were 200 mg/L and 100 mg/L in the aerobic and the anaerobic biodegradation experiments, respectively.

Bioreactor Experiment. Integrated anaerobic/aerobic conditions were created by aerating the recycle stream of the EGSB bioreactor (Figure 1). The bioreactor concept is a modified bioreactor described by Shen and Guiot (1995). In order to saturate the recycle stream with the added oxygen, gas from headspace of the aeration unit was recycled into the liquid phase. The lab-scale bioreactor was

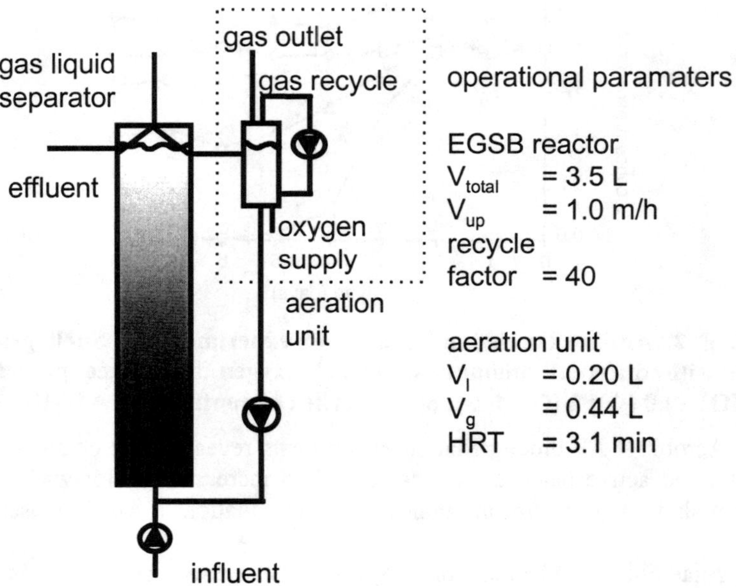

FIGURE 1. Lab-scale integrated anaerobic/aerobic expanded granular sludge blanket bioreactor.

seeded with granular sludge (20.0 g VSS/L$_{reactor}$ (L$_r$)) from Nedalco with a high tolerance for oxygen and Shell granular sludge (5.0 g VSS/L$_r$) with the ability to degrade aniline aerobically. An anaerobic control EGSB bioreactor (V = 2.5 L) without aeration unit was also operated for comparison.

The bioreactor experiments were divided into three periods. In the start-up period (0-14 days), only ethanol (960 mg/L) was fed and the hydraulic retention time (HRT) was 1.0 day. In the second period (15-44 days), aniline (200 mg/L), ethanol (720 mg/L) and oxygen (loaded at 17% to 25% of the influent chemical oxygen demand (COD)) were fed to the bioreactor. This was done in order to enrich a culture degrading aniline. The HRT of 1.1 day was applied in the second period. In the third period (45-63 days), azo dye 4-PAP (50 mg/L), ethanol (960 mg/L) and oxygen (loaded at 22% of the influent COD) were fed and the HRT was 1.4 day.

RESULTS AND DISCUSSION

Batch Experiments. Aerobic batch experiments were preformed to investigate the aerobic biodegradability of the compounds formed by the azo dye reduction. Figure 2 shows the results of the aerobic aniline degradation by granular sludge. The figure clearly indicates the necessity of oxygen for the biodegradation of aniline. The fact that aniline was degraded also indicates the occurrence of facultative bacteria in the anaerobic granular sludge.

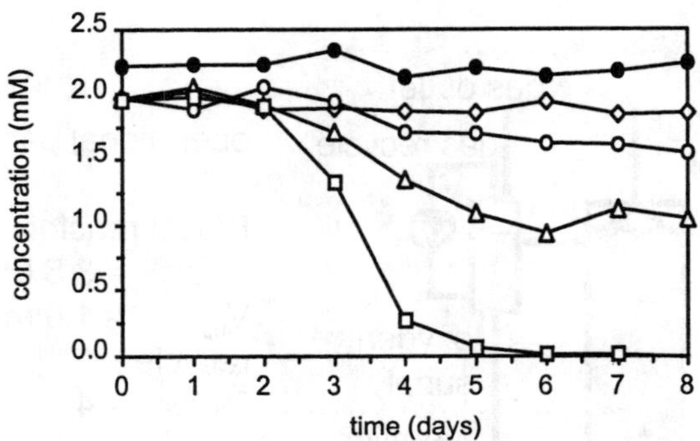

FIGURE 2. Aniline batch biodegradation experiment by Shell granular sludge with different amounts of initial oxygen headspace percentages (IOHP) 0 (◇); 2 (○); 4(△); 8(□); sterile (●) control with 4 IOHP.

Aerobic 4-AP biodegradation experiments revealed that degradation rate in sterile and active batches were same and an increase in color was observed (data not shown). These results indicate the autoxidation of 4-AP (Jensen et al., 1992).

Anaerobic biodegradation experiments were preformed in order investigate if the formed azo dye reduction products were degraded anaerobically. The results of the experiments showed no biodegradation of either aromatic amine during a test period of 100 days. In the literature there are also no reports of aniline degradation by methanogenic consortia. However, there are some reports of 4-AP degradation occurring under methanogenic conditions after long adaptation periods (Razo-Flores et al., 1996).

Bioreactor Experiment. The most important results form the integrated anaerobic/aerobic EGSB bioreactor and the control bioreactor are shown in Figure 3. Aniline was fed in period II and not recovered in the integrated anaerobic/aerobic bioreactor after a short lag phase. At the end of period II, the recovery of aniline increased due to limited oxygen addition which dropped to 17% of the influent COD. The completely anaerobic control bioreactor showed complete recovery of aniline in this period indicating no degradation. Therefore, these results indicate that the addition of oxygen in the integrated EGSB bioreactor was essential for the biomass to degraded aniline.

In period III, both bioreactors were fed with the azo dye 4-PAP. The azo dye was reduced in both systems. Generally, neither of the aromatic amines,

257

aniline nor 4-AP, could be detected in the effluent of the integrated anaerobic/aerobic EGSB bioreactor; whereas aniline and 4-AP were recovered in the effluent of the anaerobic control bioreactor. These results again indicate the necessity of oxygen addition for the removal of the formed reduction products.

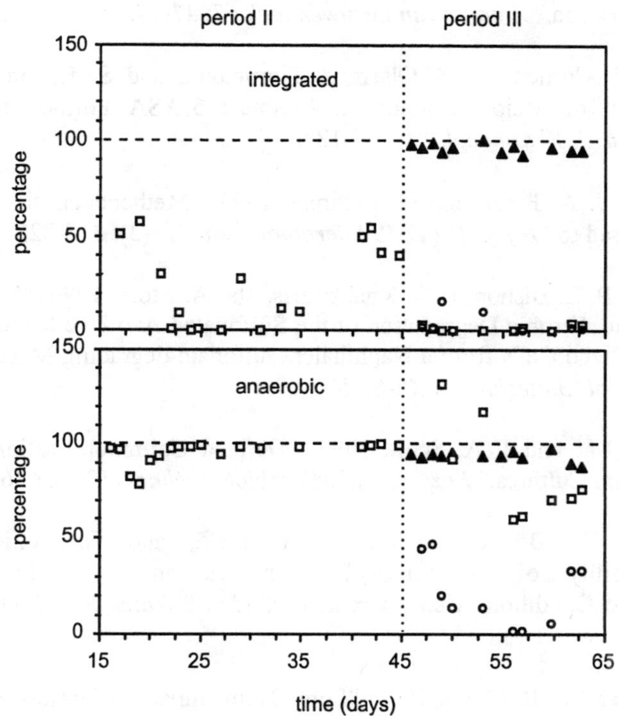

FIGURE 3. Bioreactor results period II and III bioreactor experiments; removal of 4-PAP (▲) and recovery of aniline (□) and 4-AP (○).

Since autoxidation of 4-AP was already observed in the aerobic biodegradation experiments, it is not clear what fraction of 4-AP was mineralized in the integrated anaerobic/aerobic EGSB bioreactor. The partial recovery of 4-AP in the anaerobic control bioreactor was due to the autoxidation of this compound in the sampler prior to the HPLC measurements. Other bioreactor parameters like COD and ethanol removal were high during all three periods indicating that there were no negative effects of the addition of aniline and 4-PAP.

CONCLUSIONS

The results clearly show the applicability of an integrated anaerobic/aerobic EGSB bioreactor for the degradation of the azo dye 4-PAP. Furthermore, this bioreactor setup is potentially promising for the complete degradation of other aromatic compounds with multiple electron-withdrawing substituents.

REFERENCES

Field, J. A., A. J. M. Stams, M. Kato, and G. Schraa. 1995. "Enhanced Biodegradation of Aromatic Pollutant in Coculture of Anaerobic and Aerobic Bacterial Consortia." *Antonie van Leeuwenhoek 67*: 47-77.

Jensen, J., C. Cornett, C. E. Olsen, J. Tjornelund, and S. H. Hansen. 1992. "Identification of Major Degradation Products 5-ASA Formed in Aqueous Solutions." *Int. J. Pharmacol. 88*: 177-187.

Kato, M. T., J. A. Field, and G. Lettinga. 1993. "Methanogenesis in Granular Sludge Exposed to Oxygen." *FEMS Microbiol. Lett. 114*(3): 317-323.

Kudlich, M., P. L. Bishop, H. J. Knackmuss, and A. Stolz. 1996. "Simultaneous Anaerobic and Aerobic Degradation of the Sulfonated Azo Dye Mordant Yellow 3 by Immobilized Cells from a Naphthalenesulfonate-degrading Mixed Culture." *Appl. Microbiol. Biotechnol. 46*(5-6): 597-603.

O'Reilly, A. M., and J. A. Scott. 1995. "Defined Coimmobilization of Mixed Microorganism Cultures." *Enzyme and Microbiol. Technol. 17*(7): 636-646.

Razo-Flores, E., B. A. Donlon, J. A. Field, and G. Lettinga. 1996. "Biodegradability of N-substituted Aromatics and Alkylphenols under Methanogenic Conditions Using Granular Sludge." *Water Sci. Technol. 33*(3): 47-57.

Shen, C. F., and S. R. Guiot. 1995. "Long Term Impact of Dissolved O_2 on the Activity of Anaerobic Granules." *Biotechnol. Bioeng. 49*: 611-620.

Tan, N. C. G., G. Lettinga, and J. A. Field. 1999a. "Reduction of the Azo Dye Mordant Orange 1 by Methanogenic Granular Sludge Exposed to Oxygen." *Biores. Technol. 67*(1): 35-42.

Tan, N. C. G., F. X. Prenafeta-Boldu, J. L. Opsteeg, G. Lettinga, and J. A. Field. 1999b. "Biodegradation of Azo Dyes in Cocultures of Anaerobic Granular Sludge with Aerobic Aromatic Amine Degrading Enrichment Cultures." *Appl. Microbiol. Biotechnol. (in press)*.

BIOAVAILABILITY AND REDUCTIVE DECHLORINATION OF HEXACHLOROBENZENE SORBED TO AN ESTUARINE SEDIMENT

Mark T. Prytula and *Spyros G. Pavlostathis*
Georgia Institute of Technology, Atlanta, GA, USA

ABSTRACT: The rate and extent of desorption of hexachlorobenzene (HCB) from a historically contaminated estuarine sediment as well as the effect of desorption on the bioavailability and sequential reductive dechlorination of sediment-sorbed HCB were investigated using laboratory microcosms. The rate of both desorption and reductive dechlorination of the sediment-bound HCB was enhanced by the physical manipulation of the sediment (air drying and grinding) and the addition of an exogenous carbon source, respectively. An enriched, methanogenic culture -- developed with estuarine contaminated sediment as inoculum -- reductively dechlorinated all chlorobenzene congeners with three or more chlorine substituents resulting in the accumulation of dichlorobenzene isomers. The sequential reductive dechlorination of HCB to dichlorobenzene was modeled using Michaelis-Menten kinetics. The sediment-bound HCB desorption kinetics were described using a modified, spherical coordinates diffusional model. Then, by combining this desorption model with the Michaelis-Menten reductive dechlorination model, the combined desorption and biotransformation of sediment-bound HCB was successfully simulated. The experimental data and the model simulation confirmed that the overall transformation rate of the sediment-bound HCB is limited by its desorption from the sediment matrix and the availability of degradable organic matter.

INTRODUCTION

Aquatic sediments have long acted as natural sinks for hydrophobic anthropogenic contaminants released into the environment. The fate of these contaminants is of environmental concern since biologically mediated reactions occurring in the sediments can transform strongly sorbed and relatively immobile compounds into less hydrophobic and more mobile forms. These new contaminants are then more available to affect benthic organisms and can also be released back into the water column. The objective of this study was to assess the rate and extent of desorption of hexachlorobenzene (HCB) from a historically contaminated estuarine sediment as well as to assess the effect of desorption on the bioavailability and sequential reductive dechlorination of sediment-sorbed HCB.

Estuarine sediment samples were collected from the Bayou d'Inde, a tributary of the Calcasieu River near Lake Charles, Louisiana. These sediments have been contaminated with a wide range of chlorinated aliphatic, chlorinated aromatic, and polyaromatic compounds. Major chlorinated contaminants in these sediments include HCB, other chlorinated benzene congeners, and hexachloro-1,3-butadiene. These contaminants are strongly bound into the organic matrix of the sediments as a result of their hydrophobicity and long residence time in this highly organic sediment (Pereira et al. 1988; Cunningham et al. 1990; Murray et al. 1992; Prytula and Pavlostathis 1996a).

MATERIALS AND METHODS

HCB Bioavailability Assay. The bioavailability of the strongly sorbed HCB in the study sediment was tested using sediment-slurry microcosms. Microcosms were prepared with intact (native) sediment from the study site and with site sediment that had been air dried and finely ground to pass through a 500 μm sieve. Slurry incubations were carried out in modified 2-L glass flasks fitted with serum bottle crimp openings and Chemglass sampling ports. The slurries were prepared in the flasks which were flushed with helium after adding the sediment materials, then capped with Teflon-lined stoppers and aluminum crimps and then diluted using anaerobic media (Prytula and Pavlostathis 1996b) to create slurries with final solids concentrations of 10% by weight. Three types of microcosms were prepared with each sediment type (native and dried/ground): a) unamended; b) amended with an acetate/lactate carbon source solution every 7 to 14 days; and c) amended with 2 g/L (final concentration) of sodium azide to inhibit any microbial activity. Samples were periodically withdrawn in triplicate and the liquid and solid phases separated by centrifugation in 50-mL Teflon tubes. Details on the analytical methods used have been presented elsewhere (Prytula and Pavlostathis 1996a; Prytula and Pavlostathis 1996b; Prytula 1998).

Enriched Culture Development. A HCB dechlorinating, enrichment culture was developed by inoculating anaerobic media with 100 mL of sediment in a helium-flushed 9-L glass bottle. The culture was kept at 22°C in the dark and fed with glucose and yeast extract, sodium bicarbonate for pH control, and HCB dissolved in methanol. Weekly additions of HCB and electron donor, and 1-L volume replacements every 14 days resulted in a solids retention time of 84 days.

Reductive Dechlorination Kinetics. In order to assess the reductive dechlorination kinetics of each individual chlorobenzene congener, aliquots from the enriched culture were flushed with a $He:CO_2$ (90:10) gas mixture to strip out all dechlorination products remaining from previous feeding cycles. Samples of the flushed culture were dispensed into He-flushed 500-mL serum bottles, fed with the electron donor solution, and spiked with a methanolic stock solution of one chlorinated benzene congener. The amended culture samples were then dispensed anaerobically in 20-mL aliquots into 28-mL serum tubes, incubated at 22°C and sacrificially sampled via extraction with isooctane.

Desorption and Biotransformation Assays of Sediment-Bound Contaminants. To assess the biotransformation potential of sediment sorbed HCB, two sediment slurry microcosms were prepared. One microcosm was prepared with native sediment from the study site and the other was prepared with site sediment that had been air dried and finely ground to pass through a 500 μm sieve. Incubations were carried out in modified 2-L glass flasks. The slurries were prepared in the flasks which were flushed with helium after adding the sediment materials, then sealed with Teflon-lined stoppers and aluminum crimps. Both flasks were inoculated with 100 mL of the enriched, methanogenic culture and then diluted with media to create slurries with final solids concentrations of 10% by weight. The cultures were fed weekly with glucose, yeast extract and sodium bicarbonate to maintain a near-neutral pH. The contaminant levels in each slurry microcosm were monitored over a 146-day incubation period at 22°C. The gas, liquid, and solid phases of the microcosms were periodically sampled and extracted. Gas samples were passed through a Tenax TA adsorbent trap to collect the volatile organic transformation products. These products were then extracted from the trap using hexane. Slurry samples were centrifuged and the supernatant was liquid/liquid extracted with isooctane. The sediment pellets were extracted using a hot solvent technique, in which the pellet was suspended in methanol and isooctane in a glass serum tube sealed with an aluminum-lined septum and an aluminum crimp and heated for 20 h at 95°C. The extracted sample was then centrifuged, and the isooctane layer was collected for analysis.

In the above described HCB bioavailability experiments, the aqueous phase concentrations of HCB were kept at nearly zero, indicating that the microbial consortium was able to dechlorinate HCB as fast as it was released from the sediment (Prytula and Pavlostathis 1996b). To confirm that the rate of HCB desorption was in fact limiting the overall rate of transformation, a means of measuring the release of HCB from the sediment under these conditions (i.e., desorption into a solution in which the desorbing material does not accumulate) was necessary. Two sediment slurries (one with natural and one with dried and ground sediment) were prepared as described above, but without the culture inoculum, and with sodium azide added to prevent any biological activity by the indigenous sediment microbial population. The sediment slurries were then dispensed in 20 mL aliquots into 28-mL helium-flushed serum tubes into which 300 mg of 30-40 mesh size Tenax TA polymeric resin material had been added. The Tenax resin facilitated the removal of any desorbing contaminant from the aqueous phase and maintained a high driving force for mass transfer as observed in the biologically active systems (Pignatello 1990). The tubes were placed on a rotator in a 22°C temperature controlled room, and sacrificially sampled over 146 days alongside the biologically active microcosms. At each sampling event, triplicate tubes of each slurry were removed and centrifuged to separate the Tenax, aqueous, and solid phases. The Tenax was removed and extracted with hexane. The supernatant was liquid/liquid extracted with isooctane, and the sediment pellets were hot solvent extracted as described above.

RESULTS AND DISCUSSION

HCB Bioavailability. Desorption in the azide-amended sediment slurries after 205 days of incubation was limited to only 6.5 and 6.9% of the sediment-bound HCB in the intact and dried/ground sediment microcosms, respectively. In contrast, microbially active microcosms frequently supplemented with electron donor achieved HCB removals of 32.4 and 93.3% in the intact and dried/ground sediment microcosms, respectively, indicating that the bioavailability of the sediment-sorbed HCB was enhanced by the physical manipulation of the sediment. In contrast to the carbon-amended microcosms, those microcosms incubated without the addition of exogenous carbon sources achieved a much lower extent of HCB desorption and transformation, which indicated the limiting effect of the sediment organic matter low biodegradability. Accumulation of HCB in the liquid phases of all azide-amended microcosms was observed, whereas sequential reductive dechlorination of the desorbed HCB and accumulation of dichlorobenzene isomers (mainly 1,3-dichlorobenzene) took place in both microbially active microcosms.

Reductive Dechlorination Kinetics. Dechlorination did not take place in the absence of biomass (media-only controls) or in sodium azide inactivated cultures. The predominant HCB dechlorination pathway was: HCB → Pentachlorobenzene → 1,2,3,5- and 1,2,4,5-Tetrachlorobenzene (TeCB) → 1,3,5- and 1,2,4-Trichlorobenzene (TCB) → 1,3- and 1,4-dichlorobenzene (DCB). Dechlorination of all three DCB isomers was not observed under the conditions of this study. A plot showing the sequential dechlorination of HCB by the enriched culture is shown in Figure 1A. Since the process is biologically linked, but not metabolically (growth) associated, the Michaelis-Menten enzyme kinetics model was used to describe the observed reaction rates. The dechlorination data for each chlorobenzene congener was analyzed using a nonlinear regression algorithm to estimate the values of the Michaelis-Menten kinetic parameters (maximum velocity V_{max} and half-velocity constant K_M) for each chlorobenzene congener (Prytula 1998). The kinetic parameters were then used in conjunction with the observed dechlorination pathway to simulate the sequential dechlorination pattern of HCB, as shown in Figure 1B.

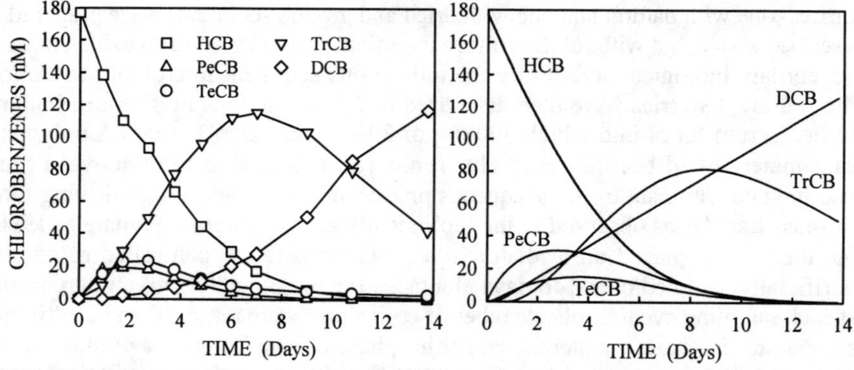

FIGURE 1. Sequential reductive dechlorination of aqueous-phase HCB by a methanogenic, enriched culture at 22°C (A, Experimental data; B, Model simulation).

Desorption and Biotransformation of Sediment-Bound Contaminants. The sediment-bound HCB concentration over the incubation time for both the native as well as the dried sediment and for both the bioactive and abiotic, Tenax-carrying microcosms are shown in Figure 2. The removal of HCB occurred much more rapidly in the dried sediment slurries where the contaminant was rendered more available through the drying and grinding process. Although a fast HCB desorption rate was initially observed in the bioactive dried sediment system, the observed relatively slow HCB dechlorination rate within the first 15 d of incubation (data not shown) led to a temporary increase of the liquid-phase HCB concentration and a cessation of HCB desorption. However, at about 30 d of incubation, the increased rate of reductive dechlorination of HCB allowed further desorption of the sediment-bound HCB to take place and beyond this incubation time the overall HCB transformation rate was desorption limited. The HCB in the wet, native sediment microcosms was much more recalcitrant and was slowly removed from the sediment matrix. The data scatter and replicate sample error are indicative of the non-homogeneous nature of the natural sediment as well as the strong affinity of the contaminant for this material.

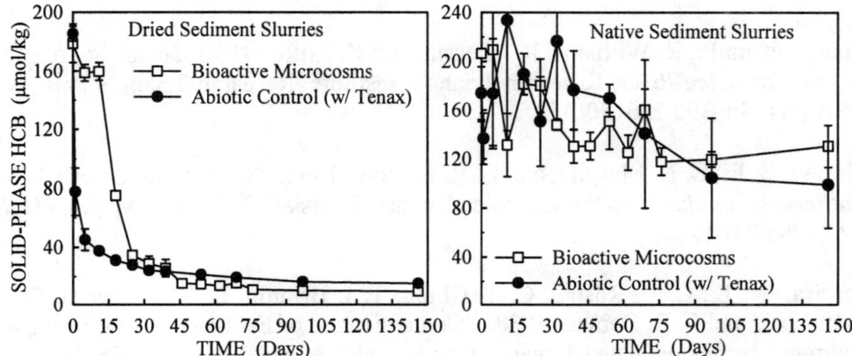

FIGURE 2. Comparison of HCB removal from sediment solids in bioactive microcosms and abiotic, Tenax-carrying controls (Error bars represent one standard deviation of the means).

Desorption/Biotransformation Modeling. The sediment-bound HCB desorption kinetics were described using a modified, spherical coordinates diffusional model (Yiacoumi and Tien 1994; 1995). Then, by combining this desorption model with the above-described Michaelis-Menten kinetic model (adjusted for the increase in enzyme concentration resulting from the increase in biomass during the long-term incubation of the periodically fed sediment-slurry microcosms), the combined desorption and biotransformation of sediment-bound HCB was successfully simulated. Figure 3 shows the results of the application of the combined desorption/biotransformation model in comparison to actual contaminant profiles in intact (natural) and dried/ground sediment slurry microcosms.

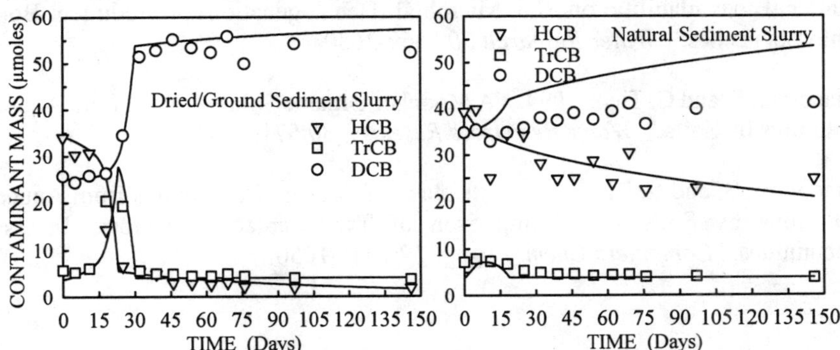

FIGURE 3. Model fit of experimental data by combining a modified diffusion model and Michaelis-Menten kinetics of the sequential reductive dechlorination of HCB.

The experimental data and the model simulation confirmed that the rate-limiting factor to the overall chlorobenzene transformation in the historically contaminated sediment is the rate of contaminant desorption when a biodegradable carbon source is available.

REFERENCES

Cunningham, P., R. Williams, R. Chessin, and K. Little. 1990. *Toxics Study of the Lower Calcasieu River*. Research Triangle Institute, Research Triangle Park, NC, USA (NTIS PB90 226150/AS).

Murray, H. E., C. N. Murphy, and G. R. Gaston. 1992. "Concentration of HCB in *Callinectes sapidus* from the Calcasieu Estuary, Louisiana." *J. Environ. Sci. Health* *A27*:1095-1101.

Pereira, W. E., C. E. Rostad, C. T. Chiou, T. I. Brinton, L. B. Barber II, D. K. Demcheck, and C. R. Demas. 1988. "Contamination of Estuarine Water, Biota, and Sediment by Halogenated Organic Compounds: A Field Study." *Environ. Sci. Technol. 22*:772-778.

Pignatello, J.J. 1990. "Slowly Reversible Sorption of Aliphatic Halocarbons in Soils. I. Formation of Residual Fractions." *Environ. Toxicol. Chem. 9*:1107-1115.

Prytula, M.T. 1998. *Bioavailability and Microbial Dehalogenation of Chlorinated Benzenes Sorbed to Estuarine Sediments*, Ph.D. Dissertation, Georgia Institute of Technology, UMI Dissertation Services Microform No. 9827356.

Prytula, M.T. and S. G. Pavlostathis. 1996a. "Extraction of Sediment-bound Chlorinated Organic Compounds: Implications on Fate and Hazard Assessment." *Wat. Sci. Technol. 33*(6):247-254.

Prytula, M.T. and S. G. Pavlostathis. 1996b. "Effect of Contaminant and Organic Matter Bioavailability on the Microbial Dehalogenation of Sediment-Bound Chlorobenzenes." *Water Research 30*:2669-2680.

Yiacoumi, S. and C. Tien. 1994. "A Model of Organic Solute Uptake from Aqueous Solutions by Soils." *Water Resources Research 30*:571-580.

Yiacoumi, S. and C. Tien. 1995. "Uptake of Organic Compounds from Aqueous Solutions by Soils - A Comparison of Two Laplace Transform Inversion Techniques." *Computers Chem. Engng. 19*:1041-1050.

AUTHOR INDEX

This index contains names, affiliations, and volume/page citations for all authors who contributed to the eight-volume proceedings of the Fifth International In Situ and On-Site Bioremediation Symposium (San Diego, California, April 19–22, 1999). Ordering information is provided on the back cover of this book. The citations reference the eight volumes as follows:

5(1): Alleman, B.C., and A. Leeson (Eds.), *Natural Attenuation of Chlorinated Solvents, Petroleum Hydrocarbons, and Other Organic Compounds*. Battelle Press, Columbus, OH, 1999. 402 pp.

5(2): Leeson, A., and B.C. Alleman (Eds.), *Engineered Approaches for In Situ Bioremediation of Chlorinated Solvent Contamination*. Battelle Press, Columbus, OH, 1999. 336 pp.

5(3): Alleman, B.C., and A. Leeson (Eds.), *In Situ Bioremediation of Petroleum Hydrocarbon and Other Organic Compounds*. Battelle Press, Columbus, OH, 1999. 588 pp.

5(4): Leeson, A., and B.C. Alleman (Eds.), *Bioremediation of Metals and Inorganic Compounds*. Battelle Press, Columbus, OH, 1999. 190 pp.

5(5): Alleman, B.C., and A. Leeson (Eds.), *Bioreactor and Ex Situ Biological Treatment Technologies*. Battelle Press, Columbus, OH, 1999. 256 pp.

5(6): Leeson, A., and B.C. Alleman (Eds.), *Phytoremediation and Innovative Strategies for Specialized Remedial Applications*. Battelle Press, Columbus, OH, 1999. 340 pp.

5(7): Alleman, B.C., and A. Leeson (Eds.), *Bioremediation of Nitroaromatic and Haloaromatic Compounds*. Battelle Press, Columbus, OH, 1999. 302 pp.

5(8): Leeson, A., and B.C. Alleman (Eds.), *Bioremediation Technologies for Polycyclic Aromatic Hydrocarbon Compounds*. Battelle Press, Columbus, OH, 1999. 358 pp.

Al-Abed, Souhail R. (U.S. EPA/USA) 5(2):263

Al-Awadhi, Nader (Kuwait Institute for Scientific Research/KUWAIT) 5(5):13

Al-Daher, Reyad (Kuwait Institute for Scientific Research/KUWAIT) 5(5):13

Al-Hakak, A. (Biotechnology Research Institute/CANADA) 5(3):439

Alleman, Bruce (Battelle/USA) 5(2):7; 5(3):201

Allen, Harry L. (U.S. EPA/USA) 5(6):89

Alsaleh, Esmaeil (King's College/UK) 5(8):111

Altman, Denis J. (Westinghouse Savannah River Company/USA) 5(2):107

Alvarez, Pedro J. (University of Iowa/ USA) 5(1):171; 5(4):79; 5(6):139

Ambert, Jack (Battelle Europe/ SWITZERLAND) 5(1):159

Ambus, Per (Risø National Laboratory/ DENMARK) 5(6):15

Amerson, Illa L. (Oregon Graduate Institute/USA) 5(3):123

Ampleman, Guy (National Defense/CANADA) 5(7):15

Anasawa, T.A. (Campinas State University-UNICAMP/BRAZIL) 5(8):105

Anderson, Bruce N. (RMIT University/ AUSTRALIA) 5(3):517

Anderson, David W. (Roy F. Weston, Inc./USA) 5(2):157

Andreotti, Giorgio (ENI/AGIP S.p.A./ ITALY) 5(1):159; 5(5):69, 183

Andretta, Massimo (Centro Ricerche Ambientali Montecatini/ITALY) 5(3):511

Andrews, Elizabeth J. (University of Minnesota/USA) 5(2):35

Andrews, Scott (SECOR International Inc./USA) 5(1):71

Angelmi, Barbara (ARBES-Umwelt GmbH/GERMANY) 5(3):487

Anthony, John W. (Parsons Engineering Science, Inc./USA) 5(1):121

Antia, Jimmy (University of Cincinnati/ USA) 5(8):283

Araneda, Edgardo (Jordforsk/ NORWAY) 5(8):99

Arienzo, Michele (Universita degli Studi di Napoli/ITALY) 5(7):197, 203

Arlotti, Daniele (Foster Wheeler Environmental Italia S.R.L./ITALY) 5(5):183

Arrocha, Alfonso (PDVSA Manufactura y Mercadeo/VENEZUELA) 5(3):215

Ashcom, Dave W. (AGI Technologies/USA) 5(3):189

Atalay, Ferhan Sami (Ege University/TURKEY) 5(5):149

Atwater, Jim (University of British Columbia/CANADA) 5(1):231

Auria, Richard (Institut de Recherche pour le Developpement/FRANCE) 5(3):31

Autenrieth, Robin L. (Texas A&M University/USA) 5(3):209; 5(6):57; 5(7):21

Aziz, Carol E. (Groundwater Services, Inc./USA) 5(1):83

Babel, Wolfgang (Umweltforschungszentrum/ GERMANY) 5(7):143, 235

Babin, J. (Golder Associates/HONG KONG) 5(8):309

Bächle, Arthur (Mannheimer Versorgungs und Verkehrsgesellschaft mbH (MVV)/GERMANY) 5(3):337

Baker, Ralph S. (ENSR Consulting & Engineering/USA) 5(3):149, 155

Balba, M. Talaat (Kuwait Institute for Scientific Research/KUWAIT) 5(5):13

Balcer, Denis (ARCADIS Geraghty & Miller, Inc./USA) 5(2):135

Bandala, Erick R. (Instituto Mexicano de Technologia del Agua/MEXICO) 5(7):137

Banwart, Steven A. (University of Sheffield/UK) 5(1):277, 307

Bär, Eckart (Environmental Expert Office/GERMANY) 5(3):363, 369, 517

Barbaro, A. (EniTecnologie/ITALY) 5(8):129

Barbonio, Miriam L. (Chevron Chemical Company/USA) 5(4):91

Barcelona, Michael J. (University of Michigan/USA) 5(1):171, 201; 5(3):345

Bare, Richard E. (Exxon Research & Engineering/USA) 5(3):227

Barker, Craig S. (IT Environmental Pty Ltd./SOUTH AUSTRALIA) 5(2):199

Barkovskii, Andrei (The University of Michigan/USA) 5(6):75

Barlaz, Morton A. (North Carolina State University/USA) 5(1):165

Barnes, Paul (Waste Management, Inc./ USA) 5(7):39, 45, 63

Barnett, J. Stephen (Exponent Environmental Group/USA) 5(7):83

Bartholomae, Philip G. (BP Amoco Company/USA) 5(1):97, 219

Bartlett, Craig L. (DuPont Company/ USA) 5(1):29

Baartmans, R. (TNO Institute of Environmental Sciences/ NETHERLANDS) 5(2):141

Basel, Michael D. (Montgomery Watson/ USA) 5(1):41

Beard, A.G. (Biotechnology Research Institute/CANADA) 5(3):439

Beaton, Yvonne (University of Aberdeen/ UK) 5(7):247

Beaudet, R. (Institut Armand-Frappier/ CANADA) 5(7):107

Beaulieu, Maude (INRS, Institut Armand-Frappier/CANADA) 5(7):89

Beaulieu, Sophie (École Polytechnique de Montréal/CANADA) 5(6):211

Bécaert, Valérie (École Polytechnique de Montréal/CANADA) 5(7):89

Beck, Frank P. (U.S. EPA/USA) 5(1):103

Beck, James W. (U.S. EPA/USA) 5(7):95

Becker, Bob (Colt Engineering Corporation/CANADA) 5(5):167

Becker, David J. (U.S. Army Corps of Engineers/USA) 5(3):155

Becker, James M. (Battelle PNNL/USA) 5(1):359

Becker, Jennifer G. (Lehigh University/ USA) 5(1):343

Becker, P. (Battelle Marine Sciences Laboratory/USA) 5(6):63

Bedient, Philip B. (Rice University/USA) 5(1):7

Bell, J.N.B. (Imperial College of Science, Technology, and Medicine/UK) 5(8):123

Belloso, Claudio (Facultad Católica de Química e Ingeniería/ARGENTINA) 5(3):331; 5(5):75, 177

Belote, J.G. (Campinas State University-UNICAMP/BRAZIL) 5(6):271

Belton, Darren (U.S. Navy/USA) 5(3):131, 383

Bender, Judith (Clark Atlanta University/ USA) 5(4):109

Bennetzen, Susan (VKI/DENMARK) 5(6):15

Benson, Sally M. (Lawrence Berkeley National Laboratory/USA) 5(4):141

Bergeron, Eric (Sodexen Group/ CANADA) 5(8):1

Bergersen, Ove (SINTEF Oslo/ NORWAY) 5(5):37; 5(6):187

Beveridge, Terry J. (University of Guelph/CANADA) 5(4):121

Beyer, Lothar (University of Kiel/GERMANY) 5(3):325, 487

Bhadra, Rajiv (Rice University/USA) 5(6):121

Bhatt, Manish (Academy of Sciences of the Czech Republic/CZECH REPUBLIC) 5(6):69

Bidgood, Jason B. (Parsons Engineering Science, Inc./USA) 5(1):121

Biehle, Alfred A. (DuPont Company/ USA) 5(1):35; 5(2):55

Bienkowski, Lisa A. (IT Group, Inc./ USA) 5(2):165

Biggar, Kevin (University of Alberta/ CANADA) 5(1):195

Bil, Jerzy (Military University of Technology/POLAND) 5(5):1

Bisaillon, Jean-Guy (Université du Quebec/CANADA) 5(8):1

Bittoni, A. (EniTecnologie/ITALY) 5(8):129

Bizzell, Cydney (Texas A&M University/ USA) 5(6):57

Gorder, Kyle (CELTIC Technologies Ltd./UK) 5(1):65

Gordon, E. Kinzie (Parsons Engineering Science, Inc./USA) 5(1):121

Gordon, Milton P. (University of Washington/USA) 5(6):133

Gosselin, Christian (Sodexen Group/ CANADA) 5(8):1

Gossett, James M. (Cornell University/ USA) 5(2):27

Gourdon, Remy P. (Institut National des Sciences Appliquees/FRANCE) 5(8):235

Govind, Rakesh (University of Cincinnati/USA) 5(4):37

Graham, David W. (University of Kansas/USA) 5(1):283

Graham, William H. (University of Kansas/USA) 5(1):283

Granade, Steve (U.S. Navy/USA) 5(2):165

Gravel, Marie-Julie (École Polytechnique de Montréal/CANADA) 5(5):135

Graves, Duane (IT Corporation/USA) 5(1):141

Gray, A.L. (Zeneca Corp./CANADA) 5(7):125

Gray, Julian (Integrated Science & Technology, Inc./USA) 5(1):97

Gray, Neil C.C. (Zeneca Corp./ CANADA) 5(7):125

Green, Roger B. (Waste Management, Inc./USA) 5(7):39, 45

Greene, Mark R. (Evirogen, Inc./USA) 5(7):7

Greenwald, Robert M. (HSI GeoTrans, Inc./USA) 5(6):169

Gregory, Betsy (The University of Tennessee/USA) 5(7):155, 161

Gregory, G.E. (URS Greiner Woodward-Clyde/USA) 5(1):35

Greis, Rolf (Institut fur Biologische Sanierung GmbH/GERMANY) 5(3):517

Griffin, Gerv C. (EA Engineering Science & Technology, Inc./USA) 5(6):251

Griffioen, Jasper (TNO Institute of Applied Geoscience/NETHERLANDS) 5(1):189; 5(3):463, 481

Griffiths, Evan (University of Idaho/ USA) 5(1):1

Grishchenkov, Vladimir G. (Russian Academy of Sciences/RUSSIA) 5(3):209

Groenwold, J. (AB-DLO/ NETHERLANDS) 5(3):529

Grøn, Christian (Risø National Laboratory/DENMARK) 5(6):15

Grossman, Matthew J. (Exxon Research & Engineering Co,/USA) 5(3):227

Grotenhuis, Tim C. (Wageningen Agricultural University/ NETHERLANDS) 5(3):195, 529; 5(8):69, 241, 265

Groudev, Stoyan N. (University of Mining & Geology/BULGARIA) 5(4):103; 5(5):7

Groudeva, Veneta I. (University of Sofia/ BULGARIA) 5(5):7

Grundl, Timothy J. (University of Wisconsin-Milwaukee/USA) 5(1):253

Guénette, Chantal C. (SINTEF Applied Chemistry/NORWAY) 5(3):227

Guérin, Valérie A.E. (Laboratoire de Géomécanique/FRANCE) 5(3):433

Guiler, H. (Zeneca Corp./CANADA) 5(7):125

Guiot, Serge R. (Biotechnology Research Institute/CANADA) 5(3):439; 5(7):15, 107

Gundrum, Jack (The Pennsylvania State University/USA) 5(2):237

Gusek, James (Knight Piesold, LLC/ USA) 5(6):217

Guy, Christophe (École Polytechnique de Montréal/CANADA) 5(5):135

Guy, John H. (Inland Pollution Services PR/USA) 5(3):319

Gwinn, Rosa E. (Dames & Moore, Inc./USA) 5(7):51

Haas, Patrick E. (U.S. Air Force/USA) 5(1):83, 313; 5(2):185

Haasnoot, C. (Logisticon Water Treatment/NETHERLANDS) 5(2):225

Haeseler, Frank (Institut Français du Pétrole/FRANCE) 5(8):117

Hagley, Michelle (Texaco Ltd./UK) 5(3):239

KEYWORD INDEX

This index contains keyword terms assigned to the articles in the eight-volume proceedings of the Fifth International In Situ and On-Site Bioremediation Symposium (San Diego, California, April 19-22, 1999). Ordering information is provided on the back cover of this book.

In assigning the terms that appear in this index, no attempt was made to reference all subjects addressed. Instead, terms were assigned to each article to reflect the primary topics covered by that article. Authors' suggestions were taken into consideration and expanded or revised as necessary. The citations reference the eight volumes as follows:

5(1): Alleman, B.C., and A. Leeson (Eds.), *Natural Attenuation of Chlorinated Solvents, Petroleum Hydrocarbons, and Other Organic Compounds*. Battelle Press, Columbus, OH, 1999. 402 pp.

5(2): Leeson, A., and B.C. Alleman (Eds.), *Engineered Approaches for In Situ Bioremediation of Chlorinated Solvent Contamination*. Battelle Press, Columbus, OH, 1999. 336 pp.

5(3): Alleman, B.C., and A. Leeson (Eds.), *In Situ Bioremediation of Petroleum Hydrocarbon and Other Organic Compounds*. Battelle Press, Columbus, OH, 1999. 588 pp.

5(4): Leeson, A., and B.C. Alleman (Eds.), *Bioremediation of Metals and Inorganic Compounds*. Battelle Press, Columbus, OH, 1999. 190 pp.

5(5): Alleman, B.C., and A. Leeson (Eds.), *Bioreactor and Ex Situ Biological Treatment Technologies*. Battelle Press, Columbus, OH, 1999. 256 pp.

5(6): Leeson, A., and B.C. Alleman (Eds.), *Phytoremediation and Innovative Strategies for Specialized Remedial Applications*. Battelle Press, Columbus, OH, 1999. 340 pp.

5(7): Alleman, B.C., and A. Leeson (Eds.), *Bioremediation of Nitroaromatic and Haloaromatic Compounds*. Battelle Press, Columbus, OH, 1999. 302 pp.

5(8): Leeson, A., and B.C. Alleman (Eds.), *Bioremediation Technologies for Polycyclic Aromatic Hydrocarbon Compounds*. Battelle Press, Columbus, OH, 1999. 358 pp.

A

A. bisporus, see Agaricus bisporus
accumulation *5(5):*117
acid mine drainage, *see* mine waste
acidophiles *5(6):*225
activation *5(7):*89
adsorption *5(8):*271
adsorption/desorption equilibria and kinetics *5(8):*283
aeration *5(3):*103

aerobic *5(2):*81; *5(7):*27; *5(8):*197
aerobic bioremediation *5(3):*295
aerobic degradation *5(6):*265
aerobic stabilization *5(6):*245
*Agaricus bisporus 5(8):*87
aging *5(8):*235
air channeling *5(3):*103
air sparging *5(3):*47, 59, 83, 89, 103, 109, 123, 169, 445, 541
air sparging/soil vapor extraction pilot study (AS/SVE pilot study) *5(3):*307

R

Ra, *see* radium
radiation *5(5):*63
radioactive isotopes *5(4):*103
radio-respirometry *5(3):*545
radiowave *5(3):*487
radium (Ra) *5(4):*103
radius of influence *5(3):*103
Raoult's law *5(3):*19
rates, *see* kinetics
RBCA, *see* Risk-Based Corrective Action
RDX *5(7):*1, 15, 21, 33, 45, 57, 217
reactive barrier, *see* reactive wall
reactive wall *5(2):*205; *5(3):*541; *5(4):*19
recirculation *5(6):*251
redox, *see* oxidation/reduction potential
reduction *5(4):*79, 121, 135
reductive dechlorination *5(1):*1, 53, 83,
    89, 253; *5(2):*35, 47, 55, 67, 73, 121,
    129, 147, 165, 171,181, 185, 225;
    *5(6):*101; *5(7):*259
reductive transformation *5(7):*95
reeds *5(6):*109
regulatory acceptance *5(1):*355
regulatory strategy *5(1):*295
remote monitoring *5(6):*259
renewable energy *5(5):*149
resistive heating *5(3):*403
resource damage *5(6):*289
respiration *5(1):*153; *5(3):*109, 183
respiration test *5(3):*109
respirometry *5(3):*115; *5(5):*183
reverse osmosis *5(4):*97
rhizodegradation *5(6):*239
rhizosphere *5(6):*27, 33
risk assessment *5(1):*359, *5(6):*231, 283
Risk-Based Corrective Action (RBCA)
    *5(1):*219, 349
risk-based design *5(1):*349
risks *5(8):*57
RT3D *5(2):*165
Ruelene™ *5(6):*151

S

salt tolerance *5(6):*21
sand/sawdust *5(2):*237
sanitary landfill *5(6):*181

sediment(s) *5(6):*39, 187, *5(7):*95, 167;
    5(8): 31, 223, 271, 283, 289, 309
selenium *5(4):*141
semiarid *5(1):*89
semifield scale *5(3):*529
semiquinone *5(8):*75
septic tank *5(4):*85
SEQUENCE, *see* visualization tool(s)
sequencing batch reactor, *see* batch reactor
sequestration *5(8):*283, 289
sewage sludge *5(6):*15
SF6, *see* sulfurhexafluoride
signature lipid biomarkers *5(4):*1
site management *5(6):*283
sludge *5(5):*37
sludge stabilization *5(6):*277
slurry *5(7):*33, 179; *5(8):*1
slurry reactors *5(6):*75
SOBs, *see* sulfur-oxidizing bacteria
soil heating *5(3):*487
soil moisture *5(3):*201
soil productivity *5(5):*189
soil resistance *5(6):*175
soil shearing *5(2):*249
soil vapor extraction (SVE) *5(3):*19
soil washing *5(5):*19; *5(7):*101, 161
solid waste *5(6):*181
solid-phase microextraction (SPME) *5(3):*71
solid-state bioreactor *5(5):*155
solvents *5(1):*89
sparging, *see* air sparging
species richness *5(6):*33
SPME, *see* solid-phase microextraction
16SrDNA *5(5):*143
stability *5(4):*115
stable isotope(s) *5(1):*207; *5(5):*183
starvation *5(4):*13
starved cells *5(2):*211
statistical study *5(1):*337
styrene *5(5):*161
sulfate *5(4):*127
sulfate reduction *5(2):*165; *5(4):*19
sulfate-reducing bacteria *5(6):*193, 205,
    211, 225; *5(7):*235
sulfide *5(8):*309
sulfurhexafluoride *5(3):*123
sulfur/limestone *5(4):*85
sulfur-oxidizing bacteria (SOBs) *5(4):*127
sulphate, *see* sulfate

sulphide, *see* sulfide
surface hydrophobicity *5(6):*163
surfactants *5(5):*167, 177; *5(7):*155, 161;
   *5(8):*81
sustainable remediation *5(6):*283
SVE, *see* soil vapor extraction

T

T. versicolor, see Trametes versicolor
tannery wastewater *5(6):*157
TBA, *see tert*-butyl alcohol
TCA, *see* trichloroethane
1,1,1-TCA, *see* 1,1,1-trichloroethane
TCE, *see* trichloroethene and
   trichloroethylene
temperature, *see* cold temperature
tert-butanol *5(3):*25
tert-butyl alcohol (TBA) *5(1):*103
tetrachloroethene (PCE) *5(1):*7, 29, 253;
   *5(2):*21, 27, 47, 55, 121, 129, 141,
   181, 205, 225
Th, *see* thorium
thermodynamic *5(1):*201
thermodynamic model *5(6):*163
thermophilic *5(6):*277
Thiessen Polygon Method *5(1):*141
thorium (Th) *5(4):*103
tides *5(1):*147
TNB, *see* trinitrobenzene
TNT, *see* 2,4,6-trinitrotoluene
toluene *5(1):*189, 207; *5(3):*71, 469;
   *5(5):*117, 123
toluene ortho-monooxygenase *5(2):*293
total petroleum hydrocarbons (TPH,
   TPHC) *5(1):*153; *5(3):*277; *5(5):* 7, 13,
   37, 43, 87, 167; *5(6):*51, 63; *5(8):*309
town gas *5(8):*7, 123
toxaphene *5(6):*89; *5(7):*125
toxicity *5(3):*351, 369, 389, 395; *5(4):*7,
   *5(5):*37, *5(7):*229, *5(8):*185, 217
TPH, *see* total petroleum hydrocarbons
TPHC, *see* total petroleum hydrocarbons
Trametes versicolor *5(8):*93
transformer oil *5(7):*149
translocation *5(6):*127
transport *5(3):*433
transport model *5(2):*27
trichloroethane (TCA) *5(1):*259
1,1,1-trichloroethane (1,1,1-TCA) *5(1):*71
trichloroethene and trichloroethylene
   (TCE) *5(1):*1, 41, 47, 65, 77, 241; *5(2):*

89, 95, 101, 107, 121, 141, 147, 157, 181,
   191, 199, 217, 225, 263, 275, 293,
   *5(5):*161; *5(6):*101, 127
trinitrobenzene (TNB) *5(7):*33
2,4,6-trinitrotoluene (TNT) *5(6):*9; *5(7):*
   27, 33, 39, 45, 57, 63, 197, 203, 211
two-phase extraction *5(3):*155

U

uranium *5(4):*103, 115; *5(7):*235

V

vacuum *5(3):*143
vacuum-enhanced *5(3):*163
vacuum-enhanced free-product recovery
   *5(3):*149
vadose zone *5(1):*89, 213, 219, 225, 231;
   *5(7):*241
vapor phase *5(5):*123, 135
vapor-phase bioreactor *5(5):*129, 135
vapor-phase transport *5(1):*219
vegetable oil *5(4):*47, 53
vinyl chloride *5(1):*77; *5(2):*15, 81
viral sorption *5(6):*163
visualization tool(s) *5(1):*325
vitamin B$_{12}$ *5(2):*205
VOC(s), *see* volatile organic
   compound(s)
volatile organic compound(s) [VOC(s)]
   *5(5):*135
volatilization *5(4):*141
volatilization-to-biodegradation ratio
   *5(3):*195

W

waste gas *5(5):*161
wastewater *5(5):*203
weathered crude oil *5(8):*51
wetland *5(3):*325; *5(6):*57
white-rot fungi *5(6):*271; *5(7):*33, 69;
   *5(8):*69, 81, 87, 93, 99, 185
wood preservative(s) *5(5):*97; *5(7):*89;
   *5(8):*167

Z

zero-valent iron *5(2):*205
zinc (Zn) *5(4):*25; *5(6):*217
Zn, *see* zinc